D1579367

Springer

Berlin
Heidelberg
New York
Hong Kong
London
Milan
Paris
Tokyo

Peter E. Kloeden   Eckhard Platen
Henri Schurz

Numerical Solution
of SDE Through
Computer Experiments

With 55 Figures

Springer

Peter E. Kloeden

Fachbereich Mathematik
Johann Wolfgang Goethe-Universität
60054 Frankfurt am Main, Germany
e-mail: kloeden@math.uni-frankfurt.de

Eckhard Platen

Department of Mathematical Sciences
University of Technology
Broadway, NSW 2007, Australia
e-mail: Eckhard.Platen@uts.edu.au

Henri Schurz

Department of Mathematics
Southern Illinois University
Carbondale, IL 62901-4408, USA
e-mail: hschurz@math.siu.edu

Corrected Third Printing 2003

Cataloging-in-Publication Data applied for

Bibliographic information published by Die Deutsche Bibliothek

Die Deutsche Bibliothek lists this publication in the Deutsche Nationalbibliografie;
detailed bibliographic data is available in the Internet at <http://dnb.ddb.de>.

Mathematics Subject Classification (2000): 60H10, 93E30, 93E11, 90A09, 93D05

ISBN 3-540-57074-8 Springer-Verlag Berlin Heidelberg New York

Springer-Verlag Berlin Heidelberg New York
a member of BertelsmannSpringer Science+Business Media GmbH

http://www.springer.de

© Springer-Verlag Berlin Heidelberg 1994
Printed in Germany

Cover design: *design & production* GmbH, Heidelberg
Typeset by the authors

Printed on acid-free paper   SPIN 10897492   41/3142LK - 5 4 3 2 1 0

Contents

Preface

The numerical solution of stochastic differential equations is becoming an indispensible worktool in a multitude of disciplines, bridging a long-standing gap between the well advanced theory of stochastic differential equations and its application to specific examples. This has been made possible by the much greater accessibility to high-powered computers at low-cost combined with the availability of new, effective higher order numerical schemes for stochastic differential equations. Many hitherto intractable problems can now be tackled successfully and more realistic modelling with stochastic differential equations undertaken.

The aim of this book is to provide a computationally oriented introduction to the numerical solution of stochastic differential equations, using computer experiments to develop in the readers an ability to undertake numerical studies of stochastic differential equations that arise in their own disciplines and an understanding, intuitive at least, of the necessary theoretical background. It is related to, but can also be used independently of the monograph

P.E. Kloeden and E. Platen, *Numerical Solution of Stochastic Differential Equations,* Applications of Mathematics Series Vol. 23, Springer-Verlag, Heidelberg, 1992,

which is more theoretical, presenting a systematic treatment of time-discretized numerical schemes for stochastic differential equations along with background material on probability and stochastic calculus. To facilitate the parallel use of both books, the presentation of material in this book follows that in the monograph closely. The results of over one hundred personal computer exercises were used in the monograph to illustrate many of the concepts and theoretical matters under discussion. An immediate objective of the present book is to provide the solutions for these PC-Exercises in the form of Borland TURBO PASCAL programs on a floppy diskette together with commentaries on the structure of the programs and hints on their use.

The book is directed at beginners to stochastic differential equations and their numerical solution. Readers will typically be upper level undergraduates or graduates in diverse disciplines with several semesters background in deterministic calculus and basic mathematical methods together with an exposure to probability and statistics. A familiarity with elementary programming and the ability to interact with IBM compatible PCs using MS-DOS and Borland TURBO PASCAL are also assumed (see the end of the preface for other languages and other computing environments). The emphasis is quite practical and mathematical proofs are avoided, but, as with all meaningful numerical work, an understanding of the underlying theoretical issues is as important for the numerical solution of stochastic differential equations as are hands-on computational skills. While non-mathematicians amongst the readers may not aspire to a mastery of the stochastic calculus on which the numerical schemes for stochastic differential equations are based, they should be aware that it has some peculiar features which are quite different from those of the familiar deterministic calculus and may lead to serious errors if disregarded. Readers are urged to delve more deeply into these important theoretical matters if they have plans for specific, nontrivial numerical investigations.

The organisation and contents of the book are based on courses that the co-authors have taught at the Australian National University, the Humboldt University in Berlin, the University of Hamburg and the University of Technology in Sydney. The PC-Exercises from the above monograph, which are called Problems here, form the backbone of the book. Background material on probability, random number generation, stochastic processes and statistics are reviewed in Chapter 1 by means of a large number of these Problems. The programs, in particular the key computational and graphics subroutines that reappear repeatedly in the sequel, are explained in some detail in this chapter, which should be read if only for this reason. Chapter 2 provides a descriptive introduction to stochastic calculus and stochastic differential equations, including stochastic Taylor expansions and their constituent multiple stochastic integrals from which the higher order numerical schemes for stochastic differential equations are derived. The numerical solution of initial value problems for differential equations via time-discretized numerical schemes is surveyed in Chapter 3, first for deterministic ordinary differential equations and then for stochastic differential equations. Computing exercises for differential equations with known explicit solutions are used to illustrate theoretical convergence rates and the effects of round-off error and numerical instability. For stochastic differential equations a fundamental distinction is made between strong, pathwise approximations and weak, distributional approximations. Different requirements for each are discussed in the context of the Euler scheme applied to a simple linear stochastic differential equation. Chapters 4 and 5 are devoted to higher order schemes for strong and weak approximations, respectively. Comparative numerical studies are presented there, along with many practical hints based on the experience of the co-authors and their co-workers over the past 15 years. The final chapter, Chapter 6, introduces the reader to the vast range of situations in which the numerical simulation of stochastic differential equations can play a useful role, including stability and bifurcation in stochastic dynamical systems, filtering, testing parametric estimators, calculating invariant measures, and finance modelling. In each case several applied simulation projects are formulated as Exercises for the reader.

The diskette contains Borland TURBO PASCAL programs for all of the Problems in Chapters 1 to 5. They are based on programs and subroutines prepared by the co-authors and their co-workers, who are not professional programmers, over the past five years and are made available here solely for the convenience of the readers, to spare them the tedium of preparing their own programs and to enable them to adapt or modify the programs to their own problems. The PASCAL language was chosen because of its universality and to highlight the transparency of the structure and the adaptability of the programs, which usually took priority over computational efficiency in their preparation. Preliminary instructions on their use are given in the 'README' file on the diskette. The programs typically require at most a few minutes running time on a 386 PC with numerics processor, though some of those in the later chapters may need an hour or two to give acceptable accuracy in terms of small confidence intervals. In most cases the equations and parameters in the programs can be easily changed. Readers are encouraged to do so in a copy of the original program and to experiment as much as possible.

The authors would like to thank Katrin Platen who word-processed the manuscript and Steven Dutch, David Heath, John van der Hoek, Norbert Hofmann, Ivan Huerta, Igor Shkurko, Robert Stine and Raymond Webster who developed some of the programs, tested many of the others, and carefully read the manuscript. In addition they would like to thank Ludwig Arnold, Daryl Daley and Chris Heyde and their respective institutions for support and encouragement.

Canberra, June 1993
              Peter E. Kloeden
              Eckhard Platen
              Henri Schurz

Authors' Comments on the Reprinting

We are delighted by the overwhelming response to this book from readers in all areas of applications, in particular in finance, that has led to its being sold out within a few years of its first appearance at the end of 1994. This reprinting has provided us with the opportunity to correct some minor misprints and to update the status of some papers previously listed in the References as to appear. Unfortunately other commitments have prevented us from developing versions of the software in other programming languages as we had hoped to do and had mentioned in a now deleted paragraph in the original version of the above Preface. We thank all readers for their helpful feedback and welcome further comments from readers.

May 1997

Authors' Comments on the Third Printing

Earlier printings of this book included a diskette containing the TURBO PAS-
CAL programs that are used. The rapid developments of the Internet in recent
years now make it both cheaper and more convenient to make these programs
available on each of our homepages

http://www.math.uni-frankfurt.de/~numerik/kloeden/

http://www.business.uts.edu.au/finance/staff/eckhard.html

http://www.math.siu.edu/schurz/SOFTWARE/

Readers can also find other useful software and information about stochastic
differential equations, their applications and the methods for their numerical
solution on these homepages.

Note that wherever "diskette" is used in the book, it should be interpreted
as "downloadable software". There are too many occurrences of this, naturally,
too numerous to make a correction of each and every one possible in practical
term. We hope that this will not be too bothersome for the reader.

August 2002

Legal Matters

We make no warranties, expressed or implied, that the programs contained in this book and the downloadable software are free of error, or are consistent with any particular standard of merchantability, or that they will meet your requirements for any particular application. They should not be relied on for solving a problem that could result in injury to a person or loss of property. If you do use the programs in such a manner, it is at your own risk. The authors and publisher disclaim all liability for direct or consequential damages resulting from your use of the programs.

Several registered trademarks appear in the book: IBM is a trademark of International Business Machines Corporation, MS-DOS and Microsoft C are trademarks of Microsoft Corporation, THINK PASCAL is a trademark of Symantec Corporation, TURBO PASCAL and TURBO C are trademarks of Borland International, Inc., UNIX is a trademark of AT&T Bell Laboratories, Macintosh is a trademark of Apple Computing, Inc. and MATHEMATICA is a trademark of Wolfram Research, Inc.

Computer programs can be protected by copyright like artistic and literary compositions. Generally it is an infringement for you to copy into your computer a program from a copyrighted source. Although this book and its programs and their future versions in other programming languages are copyrighted, we specifically authorize you, a reader of the book, to make one machine-readable copy of the programs for your own use.

We have been careful and have tried to avoid mistakes, but if you find any errors or bugs, please document them and tell us.

The downloadable software requires the user to have installed Borland TURBO PASCAL 5.0 or later versions. Comments on how to run the programs can be found in the 'README' file.

Introduction

The theory of stochastic differential equations (SDEs) has been in existence for nearly fifty years now. Complicated finite dimensional stochastic dynamics can now be modelled and understood theoretically through the Ito stochastic calculus and the more general theory of semimartingales. SDEs now find application as a matter of course in diverse disciplines such as in civil and mechanical engineering, economics and finance, environmental science, signal processing and filtering, chemistry and physics, population dynamics and psychology, pharmacology and medicine, to mention just a few. Their role in modelling continuous time stochastic dynamics is comparable to that of deterministic ordinary differential equations (ODEs) in nonrandom differentiable dynamics. In some applications, such as to stock market prices, the deterministic component in the dynamics is often only of secondary interest. Other fields, such as random vibrations in mechanical systems, require stochastic models more sophisticated than earlier, purely deterministic ones for an adequate description of the noise effects.

Very few specific SDEs have explicitly known solutions, these being mainly reducible by appropriate transformation to the solution of linear SDEs. The computation of important characteristics such as moments or sample paths for a given SDE is thus crucial for the effective practical application of SDEs. This requires methods that are specific not only to SDEs but also for the desired task.

The visualization of sample paths of SDEs, as in stochastic flows or phase portraits of noisy oscillators can be achieved by use of quite natural discrete time strong approximation methods. These provide algorithms in which the SDE is discretized in time and approximations to its solution at the discretization instants are computed recursively, with values at intermediate times being obtained through interpolation. An approximate sample path of the SDE is thus obtained for a given sample path of the driving Wiener process, which is usually simulated with an appropriately chosen pseudo random number generator. In filtering the observation process is playing the role of the driving Wiener process in the Zakai equation. Higher order methods can be derived from stochastic Taylor expansions which involves multiple stochastic integrals of the driving Wiener process in the SDEs. Such schemes are, however, only useful when they are stable numerically, an issue that attracts much attention.

Many applications require the computation of a functional of a solution of an SDE such as a moment or the expectation of a terminal pay-off. Numerically this is easier because only a probability measure of the solution has to be approximated rather than its highly irregular sample paths, but there are nevertheless still many practical complications. Higher order schemes for this weak kind of convergence can also be constructed systematically from stochastic Taylor expansions, but now the driving noise integrals can be approximated by simpler, more easily generated random variables depending on the desired order of the scheme.

Other numerical methods are also available for the computation of such functionals, which are often the solution of a partial differential equation (PDE) related to the Kolmogorov backward or Fokker-Planck equation. In principle, standard finite-difference or finite-element methods could be used, but these require both spatial and temporal discretization and their complexity increases exponentially in the spatial dimension. Such an approach remains intractable even with present day supercomputers for most higher dimensional SDEs. A more probabilistic method based on Markov chains acting at discrete time instants on a discretized state space has been developed, but this is also limited in higher dimensions and it is difficult to derive higher order schemes and to undertake a stability analysis. In principle, the binomial and trinomial tree methods in finance modelling have much in common with this Markov chain approach. The classical Monte-Carlo methods may seem more attractive in comparsion since their complexity increases only polynomially with the spatial dimension, but accuracy of statistical estimates of the simulated functionals is still only inversely proportional to the square root of the number of samples used and a huge number of sample paths may be required when large variances are involved. Higher order methods and a stability analysis also do not fit well within the framework of such Monte-Carlo methods even though variance reduction techniques are available.

The emphasis in this book will focus on strong and weak numerical schemes that are based on discrete time approximations of SDEs. Unlike the PDE and Markov chain approaches, they involve the whole state space since only the time variable is discretized. While the complexity of weak schemes also increases polynomially in the state space dimension like that of the Monte-Carlo methods, the direct use of the specific structure of the given SDE greatly facilitates the derivation of more efficient higher order schemes and numerically stable algorithms. This approach allows to use much from the internal structur of the problem at hand by exploiting specific properties and tools based on stochastic analytic ideas as martingales and corresponding measure transformations. Thus modern stochastic analysis allows maximal use of the information in a given SDE to be incorporated in such weak time discretized methods, in particular with control over variances considerably superior to those of Monte-Carlo methods.

Chapter 1

Background on Probability and Statistics

Some basic facts from probability and statistics, including methods for the generation and testing of random variables, are reviewed in this chapter. Even if the reader is familiar with probability theory, this chapter should be perused, at least, because it provides an explanation of how the book is organized, in particular the computer programs, as well as hints and useful background remarks.

1.1 Probability and Distributions

A. Events and Probabilities

In everyday life we meet many situations where randomness plays a crucial role. Obvious examples are games of chance. If we toss a die, we always observe one of six basic outcomes: it lands with its upper face indicating one of the numbers 1, 2, 3, 4, 5 or 6. We shall denote these outcomes by ω_1, ω_2, ω_3, ω_4, ω_5 and ω_6, respectively, and call the set of outcomes $\Omega = \{\omega_1, \omega_2, \omega_3, \omega_4, \omega_5, \omega_6\}$ the *sample space*. If we toss the die N times and count the number of times N_i that outcome ω_i occurs we obtain a relative frequency $f_i(N) = N_i/N$. This number usually varies considerably with N, but experience tells us that as N becomes larger it approaches a limit

$$\lim_{N \to \infty} f_i(N) = p_i,$$

which we call the *probability* of the outcome ω_i. Clearly $0 \le p_i \le 1$ for each $i = 1, 2, \ldots, 6$ and $\sum_{i=1}^{6} p_i = 1$. For a fair die we have each $p_i = 1/6$, giving a uniform distribution of probabilities over the outcomes.

Often we are interested in combinations of outcomes, that is subsets of the sample space Ω such as the subset $\{\omega_1, \omega_3, \omega_5\}$ of odd indexed outcomes. If we can distinguish such a combination by either its occurence or its nonoccurence we call it an *event*.

We can often determine the probability $P(A)$ of each event A from frequency records of either the event itself or of its constituent outcomes. Such events and their probabilities must satisfy certain properties. Theoretical difficulties are encountered, however, if we continue with this intuitive approach, particularly when the sample space Ω is uncountable. Instead we shall follow Kolmogorov's axiomatic approach to probability theory which assumes that the events and their probabilities are given.

We call a triple $(\Omega, \mathcal{A}, P)$ a *probability space*. Here the sample space Ω is a nonempty set, the set of events $\mathcal{A}$ is a σ-algebra of subsets of Ω and P is a probability measure on $\mathcal{A}$, where we say that a set $\mathcal{A}$ of subsets of Ω is a *σ-algebra* if

$$(1.1) \qquad \Omega \; \in \; \mathcal{A}$$

$$(1.2) \qquad A^c \; = \; \Omega \setminus A \in \mathcal{A} \quad \text{if } A \in \mathcal{A}$$

$$(1.3) \qquad \bigcup_{n=1}^{\infty} A_n \; \in \; \mathcal{A} \quad \text{if } A_1, A_2, \ldots, A_n, \ldots \in \mathcal{A}.$$

The set $\Omega \setminus A$ contains those elements of Ω which do not belong to A, whereas the set $A_1 \cup A_2$ represents all elements which are contained in A_1 or A_2 or in both A_1 and A_2.

We call a function $P : \mathcal{A} \to [0, 1]$ a *probability measure* if

$$(1.4) \qquad P(\Omega) \; = \; 1$$

$$(1.5) \qquad P(A^c) \; = \; 1 - P(A)$$

$$(1.6) \qquad P(\bigcup_{n=1}^{\infty} A_n) \; = \; \sum_{n=1}^{\infty} P(A_n)$$

for mutually exclusive $A_1, A_2, \ldots, A_n, \ldots \in \mathcal{A}$, that is with $A_i \cap A_j = \emptyset$ if $i \neq j$. Here $A_1 \cap A_2$ is just the set of elements which belong to both A_1 and A_2. We note that $P(\emptyset) = 0$ for the empty set $\emptyset \in \mathcal{A}$. When $P(A) = 1$ for some $A \in \mathcal{A}$ we say that the event A occurs *almost surely* (a.s.) or *with probability one* (w.p.1).

Regardless of how we actually evaluate it, the probability $P(A)$ of an event A is an indicator of the likelihood that A will occur. Our estimate of this likelihood may change if we possess some additional information, such as that another event has occured. For example, if we toss a fair die the probability of obtaining a 6 is $P(\{\omega_6\}) = p_6 = 1/6$ and the probability of obtaining an even number, that is the probability of the event $E = \{\omega_2, \omega_4, \omega_6\}$, is $P(E) = p_2 + p_4 + p_6 = 1/2$. If we know that an even number has been thrown, then, since this occurs in one of three equally likely ways, we might now expect that the probability of its being the outcome ω_6 is $1/3$. We call this the conditional probability of the event $\{\omega_6\}$ given that the event E has occured and denote it by $P(\{\omega_6\}|E)$, noting that

$$P(\{\omega_6\}|E) = \frac{P(\{\omega_6\} \cap E)}{P(E)}$$

where $P(E) > 0$. In general, we define the *conditional probability* $P(A|B)$ of A given that an event B has occured by

$$(1.7) \qquad P(A|B) = \frac{P(A \cap B)}{P(B)}$$

provided $P(B) > 0$. This definition is readily suggested from the relative frequencies

$$\frac{N_{A \cap B}}{N_B} = \frac{N_{A \cap B}}{N} \Big/ \frac{N_B}{N},$$

where $N_{A \cap B}$ and N_B are the numbers of times that the events $A \cap B$ and B, respectively, occur out of N repetitions of what we usually call an *experiment*.

It is possible that the occurence or not of an event A is unaffected by whether or not another event B has occured. Then its conditional probability $P(A|B)$ should be the same as $P(A)$, which from (1.7) implies that

(1.8) $$P(A \cap B) \stackrel{c}{=} P(A)P(B)$$

In this case we say that the events A and B are *independent*. For example, if $P(A) = P(B) = 1/2$ and $P(A \cap B) = 1/4$, then the events A and B are independent. This particular situation occurs if we toss a fair coin twice, with A the event that we obtain a head on the first toss and B a head on the second toss, provided that the way in which we toss the coin the second time is not biased by the outcome of the first toss.

B. Random Variables and Distributions

Usually we are interested in numerical quantities associated with the outcome of a probabilistic experiment, such as our winnings in a gambling game based on tossing a die or the revenue made by a telephone company based on the number of calls made. These numbers, $X(\omega)$ say, provide us with information about the experiment, which, of course, can never exceed that already summarized in its probability space $(\Omega, \mathcal{A}, P)$. They correspond to the values taken by some function $X : \Omega \to \Re$, which we call a *random variable* if its information content is appropriately restricted, that is if $\{\omega \in \Omega : X(\omega) \le a\}$ is an event for each $a \in \Re$. Then we call the function $F_X : \Re \to \Re$ defined for each $x \in \Re$ by

(1.9) $$F_X(x) = P_X((-\infty, x)) = P(\{\omega \in \Omega : X(\omega) < x\}),$$

the *distribution function* of a random variable X.

The simplest nontrivial random variable X takes just two distinct real values x_1 and x_2, where $x_1 < x_2$, with probabilities p_1 and $p_2 = 1 - p_1$, respectively. It is often called a *two-point random variable* and its distribution function is given by

(1.10) $$F_X(x) = \begin{cases} 0 & : \ x < x_1 \\ p_1 & : \ x_1 \le x < x_2 \\ 1 & : \ x_2 \le x \end{cases}$$

For instance, the indicator function I_A of an event A, which is defined by

(1.11) $$I_A(\omega) = \begin{cases} 0 & : \ \omega \notin A \\ 1 & : \ \omega \in A, \end{cases}$$

is such a random variable with $x_1 = 0$ and $x_2 = 1$ and $P(A) = p_1$.

When the random variable X takes values $0, 1, 2, \ldots$ without any upper bound, the probabilities $p_n = P(X = n)$ sometimes satisfy those of the *Poisson distribution* with

$$(1.12) \qquad\qquad p_n = \frac{\lambda^n}{n!} \exp(-\lambda)$$

for $n = 0, 1, 2, \ldots$, for a given parameter $\lambda > 0$.

The two preceding examples are typical of a discrete random variable X taking a finite or countably infinite number of distinct values.

We call a random variable X taking all possible values in $\Re$ a *continuous random variable* if the probability $P(\{\omega \in \Omega : X(\omega) = x\})$ is zero for each $x \in \Re$. In this case the distribution function F_X is often differentiable and there exists a nonnegative function p, called the *density function*, such that $F_X'(x) = p(x)$ for each $x \in \Re$; when F_X is only piecewise differentiable this holds everywhere except at certain isolated points. Then we have

$$(1.13) \qquad\qquad F_X(x) = \int_{-\infty}^{x} p(s)\, ds$$

for all $x \in \Re$.

A random variable X which only takes values in a finite interval $a \leq x \leq b$, such that the probability of its being in a given subinterval is proportional to the length of the subinterval is said to be *uniformly distributed* on $[a, b]$, denoted by $X \sim U(a, b)$. Its distribution function given by

$$(1.14) \qquad\qquad F_X(x) = \begin{cases} 0 & : \quad x < a \\ \frac{x-a}{b-a} & : \quad a \leq x \leq b \\ 1 & : \quad b < x \end{cases}$$

is differentiable everywhere except at $x = a$ and $x = b$ and the corresponding density function is

$$(1.15) \qquad\qquad p(x) = \begin{cases} 0 & : \quad x \notin (a, b) \\ \frac{1}{b-a} & : \quad x \in (a, b) \end{cases}$$

Alternatively, we say that X has a *rectangular* density function.

A random variable X is said to have an *exponential distribution* if

$$(1.16) \qquad\qquad F_X(x) = \begin{cases} 0 & : \quad x < 0 \\ 1 - \exp(-\lambda x) & : \quad x \geq 0 \end{cases}$$

for some intensity parameter $\lambda > 0$. F_X is differentiable everywhere except for $x = 0$ and the density function is

$$(1.17) \qquad\qquad p(x) = \begin{cases} 0 & : \quad x < 0 \\ \lambda \exp(-\lambda x) & : \quad x \geq 0. \end{cases}$$

Finally, the *Gaussian density* function

(1.18) $$p(x) = \frac{1}{\sqrt{2\pi}} \exp\left(-\frac{1}{2}x^2\right)$$

has a bell-shaped graph which is symmetric about $x = 0$. The corresponding distribution function $F_X(x)$ is differentiable everywhere and has a sigmoidal-shaped graph, but must be evaluated numerically or taken from tables since no anti-derivative is known in analytical form for (1.18). A random variable with this density function is called a *standard Gaussian random variable*.

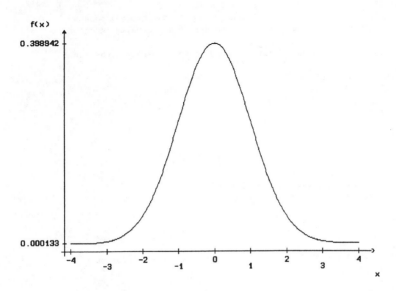

Figure 1.1.1   Gaussian density function.

Gaussian random variables occur so commonly in applications, for example as measurement errors in laboratory experiments, that they are often said to be *normally distributed*.

To familiarize the reader with the graphics routines that will be used repeatedly in the TURBO PASCAL programs in this book we shall now generate the standard Gaussian density function (1.18) on the screen of our PC. As will be typical, we shall first state the problem, then list the program, in full or just its new, essential parts, and finally comment on the structure and implementation of the program, as well as on points of mathematical interest, where appropriate. Full listings of all programs can be found on the accompanying diskette.

Problem 1.1.1   *Generate the Gaussian density function (1.18) on the interval* $[-4, 4]$.

We obtain Figure 1.1.1 on our PC screen with the following program.

Program 1.1.1   Plotting a Gaussian density function

```
USES CRT,DOS,GRAPH,INIT,SETSCR,SERVICE,AAGRAPHS;

CONST
 X1=-4.0;  { left end point  }
 X2=+4.0;  { right end point }

VAR
 CR:STRING;        { help string                                }
 K:INTEGER;        { counter                                    }
 DELTA:REAL;       { step size                                  }
 X:REAL;           { actual interval point                      }
 ABSCISSA:VECTOR;  { vector of the values of the interval points }
 FX:VECTOR;        { vector of the values of the density function }

{ Generates one value of the standard Gaussian density function }

FUNCTION F(X:REAL):REAL;
BEGIN
 F:=EXP(-X*X/2)/SQRT(PI+PI);
END;{ F }

{ Main program : }

BEGIN

 INITIALIZE;  { standard initialization }
 MAINWINDOW(''); { sets a view port      }

{ Generation of data : }

 DELTA:=NUMINV;
 DELTA:=(X2-X1)/DELTA; { highest possible screen resolution }
 X:=X1-DELTA;
 K:=-1;
 REPEAT
  K:=K+1;
  X:=X+DELTA;
  ABSCISSA[K]:=X; { values of the X-axis }
  FX[K]:=F(X);    { values of the Y-axis }
 UNTIL K=NUMINV;

{ Plots the graph of the density function using AAGRAPHS routine }
{ ! Scaling of the x-axes is done in powers of 2                 }
{ ! Scaling of the y-axes corresponds to the extrema of data     }

 GRAPH111(FX,ABSCISSA,'f(x)','x','');

 CR:='The Gaussian density function f(x) = exp(-x*x/2)/sqrt(2*pi)';
 STATUSLINE(CR);  { draws the bottom line   }
 WAITTOGO; { waits for <ENTER> to be pressed }
         { ! <ESC> terminates the program  }

 MYEXITPROC; { closes graphics mode and sets the old procedure address }
END.
```

Notes 1.1.1

(i) The graphics routine GRAPH111 from the unit AAGRAPHS produces Figure 1.1.1 for the given function F. Its input parameters are the vector ABSCISSA of x-axis values and the vector FX of corresponding values of the function, with the option of an additional caption on the screen provided by the empty string ". We do not describe the internal structure of GRAPH111 here, the listing of which can be found on the diskette. The function F and x-axis interval can be easily changed to produce graphs of other density functions.

(ii) The constant PI $= \pi$ is provided automatically by TURBO PASCAL. The procedure INITIALIZE from the unit INIT initializes the graphics card and fixes graphics parameters, while the procedure MAINWINDOW(") provides a viewing window on the screen, with the frame being omitted when the input string is empty. A caption from the string CR is provided beneath the figure produced by the routine STATUSLINE. Finally, WAIT-TOGO and MYEXITPROC allow the user to maintain the picture on the screen and to exit when desired.

(iii) The constant NUMINV from the unit INIT can be set equal to $512 = 2^9$ for EGA and VGA graphics cards, their highest possible resolution to a power of 2 along the x-axis. A corresponding power of 2 should be used for other cards. (In later programs we shall usually use step sizes which are powers of 2).

(iv) Units such as CRT, DOS and GRAPH are standard units in TURBO PAS-CAL developed by BORLAND INTERNATIONAL, INC. Our own units INIT, SETSCR, SERVICE and AAGRAPHS are described briefly in the 'README' file on the diskette.

1.2 Random Number Generators

Our objective in this book is to investigate the basic properties of stochastic differential equations quantitatively. Often only numerical simulations of such a complicated probabilistic system provide sufficient information about the behaviour of the underlying model, and hopefully of the original system itself, which cannot be obtained directly or easily by other means. Numerical values of each of the random variables involved must be provided for a test run of the model, which requires the generation of many random numbers with specified statistical properties.

Originally such numbers were taken directly from actual random variables, generated, for example, mechanically by tossing a die or electronically by the noisy output of a valve. This proved impractical for large scale simulations as the numbers were not always statistically reliable and a particular sequence of random numbers could not always be reproduced as is required for comparative studies.

Electronic digital computers allow the implementation of simple deterministic algorithms to generate sequences of random variables, quickly and reproducably. Such numbers are consequently not truly random, but with sufficient care they can be made to resemble random numbers in most properties, in which case they are called *pseudo-random numbers*.

A. Congruential Random Number Generators

Most modern digital computers and related software include a *linear congruential pseudo-random number generator* of the recursive form

$$(2.1) \qquad X_{n+1} = aX_n + b \quad (\bmod\ c)$$

where a and c are positive integers and b is a nonnegative integer. For an integer initial value or *seed* X_0, the algorithm (2.1) generates a sequence taking integer values from 0 to $c-1$, the remainders when the $aX_n + b$ are divided by c. When the coefficients a, b and c are chosen appropriately the numbers

$$(2.2) \qquad U_n = X_n/c$$

often seem to be uniformly distributed on the unit interval $[0, 1]$.

A much used example was the *RANDU* generator of the older *IBM* Scientific Subroutine Package with multiplier $a = 65,539 = 2^{16} + 3$, $b = 0$ and modulus $c = 2^{31}$. The more recent *IBM* System 360 uniform random number generator uses the multiplier $a = 16,807 = 7^5$, $b = 0$ and modulus $c = 2^{31} - 1$, which is a prime number.

Problem 1.2.1 *Use the random number generator in your computer to generate a sequence of 10 independent U(0,1) uniformly distributed pseudo-random numbers.*

Program 1.2.1 Uniform random number generation

```
USES CRT,DOS,GRAPH,INIT,SERVICE;

CONST
 N=10; { number of random numbers to be generated }

VAR
 CR:STRING; { help string                          }
 K:INTEGER; { counter                              }
 U:REAL;    { current value of the random variable }

{ Generates a uniformly distributed random number }

PROCEDURE GENEROO(VAR XU:REAL);
BEGIN
 XU:=RANDOM;
END;{ GENEROO }

{ Main program : }

BEGIN
```

```
INITIALIZE;CLEARDEVICE; { standard initialization not necessary here }
RESTORECRTMODE;          { restores the CRT mode                     }

CLRSCR;
CR:='Uniformly distributed random numbers';
WRITELN(CR);
K:=0;
REPEAT
 K:=K+1;
 GENEROO(U);
  WRITELN(K:4,'. value  U = ',CHCR(U)):
  IF ((K MOD 20 = 0) AND (K<N)) THEN
   BEGIN
     WAITE;CLRSCR;
     WRITELN(CR);
   END;
UNTIL K=N;
WAITE; { waits for <ENTER> to be pressed }

SETGRAPHMODE(GETGRAPHMODE); { returns to graphics mode }

MYEXITPROC; { closes graphics mode and sets the old procedure address }
END.
```

Notes 1.2.1

(i)   The structure of Program 1.2.1 is more general than is required for the task
      at hand. For instance, it would suffice to remain in graphics mode, but we
      include the switch to the standard CRT mode to demonstrate how easily
      this can be achieved.

(ii) The command RANDOM is standard to PASCAL and provides a $U(0,1)$
      uniformly distributed pseudo-random number. The parameter $N = 10$
      can be changed to any other integer from 1 to MAXINT, the maximum
      value available. Later we shall save generated random numbers as a vector.
      We can do this by including the command $F[K] := U$; having of course
      declared the vector first. Alternatively we could call GENER00($F[K]$) and
      use $F[K]$ directly without mentioning it.

(iii) If nothing appears on the screen, then press either the <ENTER> or
      <RETURN> key to continue. In graphics mode the <ESC> key can be
      used to terminate the program whenever desired.

Exercise 1.2.1   *Generate a sequence of 10 pseudo-random numbers by the
linear congruential generator (2.1) with $a = 1229$, $b = 1$ and $c = 2048$.*

B. Generators for Discrete Random Variables

For the remainder of this section we shall assume that we have a subroutine RANDOM which provides us with $U(0,1)$ uniformly distributed pseudo-
random numbers by means of (2.1) and (2.2). We shall see how we can then
use this subroutine to generate pseudo-random numbers with other commonly
encountered distributions, in particular those described in Section 1.1.

A two-point random variable X (1.10) taking values $x_1 < x_2$ with probabilities p_1 and $p_2 = 1 - p_1$ can be generated easily from a $U(0,1)$ random variable U with

$$(2.3) \qquad X = \begin{cases} x_1 & : \quad 0 \le U \le p_1 \\ x_2 & : \quad p_1 < U \le 1. \end{cases}$$

This idea extends readily to an N-state random variable X taking values $x_1 < x_2 < \cdots < x_N$ with nonzero probabilities $p_1, p_2, \ldots, p_N$ where $\sum_{i=1}^{N} p_i = 1$. With $s_0 = 0$ and $s_j = \sum_{i=1}^{j} p_i$ for $j = 1, 2, \ldots, N$ we set $X = x_{j+1}$ if $s_j < U \le s_{j+1}$ for $j = 0, 1, 2, \ldots, N-1$.

Problem 1.2.2   *Generate 50 independent two-point distributed pseudo-random numbers, where the outcomes $x_1 = -1$ and $x_2 = 1$ occur with probabilities $p_1 = p_2 = 0.5$.*

The program required here is essentially the same as Program 1.2.1 so we list only the new procedure.

Program 1.2.2   Generation of two-point random numbers

```
   ...
{ Generates a two-point distributed random number }

PROCEDURE GENER01(P,X1,X2:REAL;VAR XU:REAL);
BEGIN
 XU:=RANDOM;
 IF XU <= P THEN XU:=X1 ELSE XU:=X2;
END;{ GENER01 }
   ...
```

Notes 1.2.2   Here we simply replace the random number generating procedure GENER00 in Program 1.2.1 by GENER01, which is taken from the unit RANDNUMB, later to be installed automatically through the USES line at the beginning of each PASCAL program. We need to change the caption string and to declare the parameters P, $X1$, $X2$ as constants, where $X1$ and $X2$ are the two values taken by the random variable X and P is the probability that $X = X1$.

Exercise 1.2.2   *Generate 50 independent three-point distributed pseudo-random numbers, where the outcomes $x_1 = -\sqrt{3}$ and $x_2 = \sqrt{3}$ both occur with probability $\frac{1}{6}$ and the outcome $x_3 = 0$ with probability $\frac{2}{3}$.*

C. Generators for Exponential Random Variables

For a continuous random variable X the corresponding method requires the probability distribution function F_X to be inverted when this is possible. For a number $0 < U < 1$ we define $x(U)$ by $U = F_X(x(U))$, so $x(U) = F_X^{-1}(U)$ if F_X^{-1} exists, or in general

$$(2.4) \qquad x(U) = \inf\{x : U \le F_X(x)\}.$$

This is called the *inverse transform method* and is best used when (2.4) is easy to evaluate. For example, the exponential random variable with parameter $\lambda > 0$ and distribution given in (1.16) has an invertible distribution function with

$$x(U) = F_X^{-1}(U) = -\ln(1-U)/\lambda \quad \text{for} \quad 0 < U < 1.$$

Problem 1.2.3 *Generate a sequence of 10 independent exponentially distributed pseudo-random numbers with parameter $\lambda = 2.0$.*

Program 1.2.3   Generation of exponential random numbers

```
...
{ Generates an exponentially distributed random number }
{ with parameter Lambda > 0.0                          }

PROCEDURE GENERO2(LAMBDA:REAL;VAR XU:REAL);
BEGIN
 XU:=RANDOM;
 IF XU > 0. THEN XU:=-LN(XU)/LAMBDA ELSE XU:=0.;
END;{ GENERO2 }
...
```

Notes 1.2.3
(i) As in the previous program we just replace the generator, this time by GENER02, which is also from the unit RANDNUMB. We also have to change the caption string and declare the parameter LAMBDA as a constant.
(ii) Since $1 - U$ is also $U(0,1)$ distributed when U is, we save a little computational effort by generating $1 - U$ directly in the GENER02 procedure.
(iii) The GENER02 procedure produces a very slight bias caused by computer truncation error.

D. Box-Muller Method

In principle the inverse transform method can be used for any continuous random variable, but may require too much computational effort to evaluate (2.4). This occurs with the standard Gaussian random variable since the integrals for its distribution function must be evaluated numerically. The *Box-Muller method* for generating standard Gaussian pseudo-random numbers avoids this problem. It is based on the observation that if U_1 and U_2 are two independent $U(0,1)$ uniformly distributed random variables, then G_1 and G_2 defined by

$$(2.5) \qquad G_1 = \sqrt{-2\ln(U_1)}\ \cos(2\pi U_2)$$
$$G_2 = \sqrt{-2\ln(U_1)}\ \sin(2\pi U_2)$$

are two independent standard Gaussian random variables (pairs of independent random variables will be defined in Subsection 1.3.C as having their joint distribution function equal to just the product of their marginal distribution functions).

Problem 1.2.4 *Generate 10 pairs of independent standard Gaussian distributed random variables by the Box-Muller method.*

Program 1.2.4 Box-Muller method

```
   ...
PROCEDURE GENERO3(VAR G1,G2:REAL);
VAR
 U1,U2:REAL;
BEGIN
 REPEAT U1:=RANDOM UNTIL U1>0.;U2:=RANDOM;
 U1:=LN(U1);U2:=PI*U2;
 U1:=SQRT(-U1-U1);U2:=U2+U2;
 G1:=U1*COS(U2);G2:=U1*SIN(U2);
END;{ GENERO3 }
   ...
```

Notes 1.2.4

(i) We now use the generator GENER03 from the unit RANDNUMB with the obvious caption string and declaration changes.

(ii) The program differs from the previous three in that a pair rather than a single random number is generated by each application of the generator. The output is presented in pairs using CRT mode, which requires changes to the output text string.

E. Polar Marsaglia Method

A variation of the Box-Muller method which avoids the time-consuming calculation of trigonometric functions is the *Polar Marsaglia method*. It is based on the facts that $V = 2U - 1$ is $U(-1,1)$ uniformly distributed if U is $U(0,1)$ distributed and that for two such random variables V_1 and V_2 with

$$W = V_1^2 + V_2^2 \leq 1,$$

W is $U(0,1)$ distributed and $\theta = \arctan(V_1/V_2)$ is $U(0,2\pi)$ distributed. Since the inscribed unit circle has $\pi/4$ of the area of the square $[-1,1]^2$, the point (V_1, V_2) will take values inside this circle with probability $\pi/4 = 0.7864816\cdots$. We only consider these points, discarding the others. Using

$$\cos\theta = \frac{V_1}{\sqrt{W}}, \quad \sin\theta = \frac{V_2}{\sqrt{W}}$$

when $W = V_1^2 + V_2^2 \leq 1$ we can rewrite (2.5) as

$$(2.6) \qquad \begin{aligned} G_1 &= V_1\sqrt{-2\ln(W)/W} \\ G_2 &= V_2\sqrt{-2\ln(W)/W} \end{aligned}$$

Although a proportion of the generated uniformly distributed numbers are discarded, this method is often computationally more efficient than the Box-Muller method when a large quantity of numbers is to be generated.

Problem 1.2.5   *Generate a sequence of 10 pairs of independent standard Gaussian distributed pseudo-random numbers by the Polar Marsaglia method.*

Program 1.2.5   Polar Marsaglia method

```
    . . .
PROCEDURE GENERATE(VAR G1,G2:REAL);
VAR
 V1,V2,W,LW:REAL;
BEGIN
 REPEAT
  V1:=2.*RANDOM-1.;V2:=2.*RANDOM-1.;W:=V1*V1+V2*V2;
 UNTIL ((W<=1.0) AND (W>0.0)); { obviously LN(0) does not exist. }
 LW:=LN(W)/W;LW:=SQRT(-LW-LW);G1:=V1*LW;G2:=V2*LW;
END;{ GENERATE }
    . . .
```

Notes 1.2.5   The program is identical to Program 1.2.4 for the Box-Muller method with the generator GENER03 there replaced by the Polar Marsaglia generator GENERATE. We call it this way instead of GENER04 since it is the main generator that we shall use in most of the simulations in the sequel.

F. A Combined Program

We shall use the above methods in the following parts of the book under the assumption that they do indeed generate random numbers with the asserted properties. In Section 5 of this chapter we shall examine the validity of this assumption using several tests. Simulations based on pseudo-random number generators are experimental in nature, but as we shall see, this approach turns out to be quite successful. One should, however, always be aware of the fact that the underlying pseudo-random numbers do not reflect completely all properties of their theoretical counterparts. Thus it might be sometimes necessary to gain more confidence in an experimental simulation result by repeating the simulation with other random number generators.

Now let us combine the above generators in one program.

Problem 1.2.6 (PC-Exercise 1.3.1)   *Write a single program for the generation of two-point, exponential and Gaussian pseudo-random numbers based on the above methods. Generate a list of 50 pseudo-random numbers from each type.*

The combined program, Program 1.2.6, which can be found on the diskette, is just one possibility for combining the four preceding programs. Additional user input control procedures could easily be included, but would considerably lengthen the program.

G. Comparing Computer Times

It is useful and often necessary to compare the efficiency of different algorithms. This can be done by measuring the necessary computer time. Here we compare the efficiency of the Polar Marsaglia and the Box-Muller methods for the generation of Gaussian pseudo-random numbers.

Problem 1.2.7   *Compare the computer times required to generate 10^4 standard Gaussian random numbers by the Box-Muller and by the Polar Marsaglia methods.*

Program 1.2.7   Comparing computer times

```
   ...
{ Takes the elapsed time in seconds for run times not exceeding one day }
{ Assumes GETTIME(HOUR,MINUTE,SECOND,SEC100) has been called before      }

PROCEDURE TIMEINSEC(VAR TIME:STRING);
VAR
 ABSSEC,ABSSEC100,OLDHOUR,OLDMINUTE,OLDSEC100,OLDSECOND:WORD;
 CR:STRING;
BEGIN
 OLDHOUR:=HOUR;OLDMINUTE:=MINUTE;OLDSECOND:=SECOND;OLDSEC100:=SEC100;
 GETTIME(HOUR,MINUTE,SECOND,SEC100);
 IF SEC100<OLDSEC100 THEN
   BEGIN OLDSECOND:=OLDSECOND+1;ABSSEC100:=100-OLDSEC100+SEC100; END
   ELSE ABSSEC100:=SEC100-OLDSEC100;
 IF SECOND<OLDSECOND THEN
   BEGIN OLDMINUTE:=OLDMINUTE+1;ABSSEC:=60-OLDSECOND+SECOND; END
   ELSE ABSSEC:=SECOND-OLDSECOND;
 IF MINUTE<OLDMINUTE THEN
   BEGIN OLDHOUR:=OLDHOUR+1;ABSSEC:=ABSSEC+60*(60-OLDMINUTE+MINUTE); END
   ELSE ABSSEC:=ABSSEC+60*(MINUTE-OLDMINUTE);
 IF HOUR<OLDHOUR THEN ABSSEC:=ABSSEC+3600*(24-OLDHOUR+HOUR)
   ELSE ABSSEC:=ABSSEC+3600*(HOUR-OLDHOUR);
 STR(ABSSEC,CR);TIME:=CR+'.';STR(ABSSEC100,CR);
 IF ((LENGTH(CR)=1) AND (CR<>'0')) THEN CR:='0'+CR;
 TIME:=TIME+CR;
END;{ TIMEINSEC }

{ Main program : }
   ...

{ Random number generation by the Box-Muller method : }

 GETTIME(HOUR,MINUTE,SECOND,SEC100);
 K:=0;
 REPEAT
  K:=K+1;
  GENERO3(U1,U2);
 UNTIL K=N;
 TIMEINSEC(TIME);CR:='Box-Muller CPU-run-time in seconds : '+TIME;
 OUTTEXTXY(TRUNC(MAXX/2),TRUNC(MAXY/2)-2*TEXTHEIGHT('M'),CR);

{ Random number generation by the Polar-Maraglia method : }
```

```
GETTIME(HOUR,MINUTE,SECOND,SEC100);
K:=0;
REPEAT
 K:=K+1;
 GENERATE(U1,U2);
UNTIL K=N;
TIMEINSEC(TIME);CR:='Polar Marsaglia CPU-run-time in seconds : '+TIME;
OUTTEXTXY(TRUNC(MAXX/2),TRUNC(MAXY/2)-2*TEXTHEIGHT('M'),CR);
...
```

Notes 1.2.7

(i)   The generation of random numbers is usually faster by the Polar Marsaglia method than by the Box-Muller method, requiring approximately 80 per cent of the time taken by the latter.

(ii) Program 1.2.7 first generates the random numbers by the Box-Muller method and then by the Polar-Marsaglia method using the random number generators in the RANDNUMB unit. The procedure TIMEINSEC measures elapsed time in seconds assuming that GETTIME has already been called. Of course, the PC needs to have a built-in clock.

H. TURBO PASCAL Random Number Generator

The random number generator used with version 5.0 of Borland TURBO PASCAL is based on a linear congruential method of the form (2.1). The actual values for the constants a, b and c can be obtained by setting the seed of the random number generator to an appropriate set of values and then solving a simple set of linear equations. Thus using an initial value of $X_0 = 0$ we obtain $X_1 = b$ and using an initial value of $X_0 = 1$ we obtain $X_1 = a + b \pmod{c}$. From these and other similar equations we can compute the values for the multiplier a, additive constant b and modulus c. Because TURBO PASCAL uses a signed 32-bit number for the generator's integer sequence a widely used value is $c = 2^{31}$. In fact, it may be considered a likely candidate for the modulus. In fact, because TURBO PASCAL allows negative integer values to be generated, the integer sequence obtained from the generator is not a true linear congruential recurrence. The actual algorithm used for the TURBO PASCAL generator can be expressed as

$$(2.7) \qquad X_{n+1} = \left((aX_n + b + c) \pmod{2c}\right) - c$$

where $a = 134,775,813$, $b = 1$ and $c = 2^{31}$. The corresponding normalized sequence produced by calls to the TURBO PASCAL random number generator RANDOM (without the RANGE parameter) consists of the numbers (see (2.2))

$$(2.8) \qquad U_n = |X_n/c|;$$

$n = 0, 1, 2, \ldots$ which are within the interval $[0, 1]$.

Exercise 1.2.3   *Using an initial value of $X_0 = 1$ compute a sequence of the first 20 integer and floating point numbers obtained from the recurrences (2.7)*

and (2.8) and compare these with the values obtained using the system vari-
ables RANDSEED (which stores the integer sequence for the TURBO PASCAL
random number generator) and the TURBO PASCAL *generator RANDOM.*

The generator parameters a, b and c associated with the linear congruential
generators for other languages or implementations can usually also be found
by a similar analysis.

I. Lagged - Fibonacci Generators

One problem with linear congruential generators is that its *period*, which can
be defined as the smallest positive integer p which satisfies $X_{i+p} = X_i$ for all
$i > k$ for some $k \geq 0$, cannot exceed the value for the modulus c. This can be a
major weakness if supercomputers are being used. For example, a 32-bit linear
congruential generator will have a period less than 2^{32}. For a supercomputer
this sequence would typically be exhausted within a few minutes and for a fast
workstation within a few hours.

In recent years research into random number generators has focussed on
what are generally referred to as lagged–Fibonacci generators. These are defined using the recurrence

$$(2.9) \qquad\qquad X_n = X_{n-r} \, \text{op} \, X_{n-s}$$

where $0 < s < r$ are the lags with $n \geq r$ and op is a binary operator, for example   addition $(\text{mod } c)$ or   substraction $(\text{mod } c)$. Extremely long periods
are possible with these generators and several have been shown to exhibit good
global properties if the parameters are chosen carefully.

Exercise 1.2.4   *Using the values $r = 17$, $s = 5$, op $=$ addition $(\text{mod } 2^{31})$ com-*
pute the first 100 integer values of the corresponding lagged–Fibonacci genera-
tor. Use the first 17 nonnegative integers produced from the standard TURBO
PASCAL *generator, with an initial value or seed of $X_0 = 1$, to initialize the*
lagged–Fibonacci generator.

It should be noted that all simulations described in this book could, in principle,
be performed by replacing the standard TURBO PASCAL generator with this
lagged–Fibonacci generator. This has a period of $(2^{17} - 1) \, 2^{31}$ if at least one
member of the initialization sequence $X_0, X_1, \ldots, X_{16}$ is odd.

1.3   Moments and Conditional Expectations

The random number generators just described allow us to perform numerical
experiments to illustrate important notions from probability theory and mathematical statistics.

A. Moments

The values taken by a random variable X can vary considerably, so it is useful if we can isolate some salient features of this variability such as the average value and the way in which the values spread out about this average value.

The first of these is an average weighted by the likelihood of occurence and is usually called the *mean value* or *expected value* and denoted by $E(X)$. For a discretely distributed random variable with $P(X_i = x_i) = p_i$ it is defined as

$$(3.1) \qquad E(X) = \sum_{i \in I} x_i \, p_i,$$

where $I = \{0, \pm 1, \pm 2, \dots\}$, which is readily suggested by the relative frequency interpretation of the probabilities. When the random variable has an absolutely continuous distribution the corresponding definition for its mean value is

$$(3.2) \qquad E(X) = \int_{-\infty}^{\infty} x \, p(x) \, dx,$$

since, roughly speaking, $p(x) \, dx$ is the probability that X takes its value in the interval $(x, x + dx)$.

For a particular random variable X we often use the notation $\mu = E(X)$ for the mean value. A measure of the spread about μ of the values taken by X is given by its *variance* which is defined as

$$(3.3) \qquad \mathrm{Var}(X) = E\left((X - \mu)^2\right),$$

or equivalently as
$$(3.4) \qquad \mathrm{Var}(X) = E\left(X^2\right) - \mu^2.$$

The variance is consequently always nonnegative and is often denoted by $\sigma^2 = \mathrm{Var}(X)$, where σ is called the *standard deviation* of X.

Problem 1.3.1 (PC-Exercise 1.4.4) *Use the generator on your PC to generate $N = 10^3$ uniformly $U(0,1)$ distributed pseudo-random numbers. Partition the interval $[0,1]$ into subintervals of equal length 10^{-2} and count the number of generated numbers falling into each subinterval. Plot a histogram, that is the proportion of generated numbers falling into each subinterval normalized by the length of the subinterval. Does this histogram resemble the graph of the density function of a $U(0,1)$ random variable? In addition, evaluate the sample average $\hat{\mu}_N$ and sample variance $\hat{\sigma}_N^2$ for the generated numbers $x_1, x_2, \dots, x_N$, where*

$$(3.5) \qquad \hat{\mu}_N = \frac{1}{N} \sum_{i=1}^{N} x_i \quad and \quad \hat{\sigma}_N^2 = \frac{1}{N-1} \sum_{i=1}^{N} (x_i - \hat{\mu}_N)^2.$$

How do these compare with the mean $1/2$ and variance $1/12$ of a truly $U(0,1)$ distributed random variable?

Program 1.3.1   Histogram for uniformly distributed random numbers

```
USES CRT,DOS,GRAPH,INIT,SETSCR,SERVICE;

CONST
  N=1000;              { number of random numbers  }
  INVLENGTH=0.01;  { length of subinterval        }
  ABSCMIN=0.0;      { left end point               }
  ABSCMAX=1.0;      { right end point              }
  ORDMIN=0.0;        { minimum of the ordinate     }
  ORDMAX=20.0;       { maximum of the ordinate      }
  ORDPOINT=10.0;    { significant ordinate point  }

TYPE
  VECTOR=ARRAY[1..N] OF REAL;

VAR
  CR:STRING;                    { help string                  }
  I:INTEGER;                    { counter                      }
  AXEX,AXEY:INTEGER;            { location of the axes         }
  DISTX,DISTY:INTEGER;          { scale parameters             }
  INVEND:REAL;                  { actual subinterval end       }
  AVERAGE,VARIANCE:REAL;   { statistical parameters       }
  F:VECTOR;                     { vector of the random numbers }


PROCEDURE COORDSYS;
PROCEDURE STATDATATOSCR(AVERAGE,VARIANCE,EXACTAVERAGE,EXACTVARIANCE:REAL);
PROCEDURE QSORT(DOWN,UP:INTEGER;VAR F:VECTOR);


PROCEDURE HISTOGRAM(NN:INTEGER;F:VECTOR); { assumes a sorted vector F }
VAR
  CR:STRING;
  ENDS:BOOLEAN;
  I,IH,IL,IR,K,KLOWER,KUPPER:INTEGER;
  FACTORX,FACTORY:REAL;
BEGIN
  FACTORX:=DISTX/(ABSCMAX-ABSCMIN);
  FACTORY:=DISTY/(ORDMAX-ORDMIN);
  INVEND:=ABSCMIN;

{ Control over data used and the printout of the histogram : }

  ENDS:=FALSE; { permitted data }
  KLOWER:=0;KUPPER:=NN+1;
  I:=NN+1;
  IF F[NN]<ABSCMIN THEN ENDS:=TRUE ELSE
   IF F[1]>ABSCMAX THEN ENDS:=TRUE ELSE
    BEGIN
      WHILE ((I>2) AND (F[I-1]>ABSCMAX)) DO I:=I-1;
      IF I>1 THEN KUPPER:=I-1
        ELSE ENDS:=TRUE; { out of range }
      I:=0;
      WHILE ((I<NN-1) AND (F[I+1]<ABSCMIN)) DO I:=I+1;
      IF I<NN THEN KLOWER:=I
        ELSE ENDS:=TRUE; { out of range }
    END;
```

```
IF ENDS=FALSE THEN
   BEGIN
     I:=KLOWER;
     REPEAT
      INVEND:=INVEND+INVLENGTH;I:=I+1;K:=0;
      WHILE ((I+K<NN) AND (F[I+K]<=INVEND)) DO K:=K+1;
      IF ((I+K=NN) AND (F[I+K]<=INVEND)) THEN K:=K+1;
      IF K>0 THEN
        BEGIN
          I:=I+K-1;
          IL:=AXEY+10+TRUNC((INVEND-INVLENGTH-ABSCMIN)*FACTORX);
          IR:=AXEY+10+TRUNC((INVEND-ABSCMIN)*FACTORX);
          IH:=AXEX-10-TRUNC((((K/NN)/INVLENGTH)-ORDMIN)*FACTORY);
          LINE(IL,AXEX-10,IL,IH);LINE(IL,IH,IR,IH);LINE(IR,IH,IR,AXEX-10);
          LINE(IL,AXEX-10,IR,AXEX-10);
        END
      ELSE I:=I-1;
     UNTIL ((INVEND>=ABSCMAX) OR (I>=KUPPER));
   END
  ELSE { terminates the program because of nonallowable data configuration }
    BEGIN
      CR:='Please, use another interval!';
      OUTTEXTXY(TRUNC(MAXX/2),TRUNC(MAXY/2),CR);
      CR:='Data out of range! Press <ESC> and check data';
      STATUSLINE(CR);
      WAITTOGO;
    END;
END;{ HISTOGRAM }

PROCEDURE COMPSAMPLEPARA(NN:INTEGER;X:VECTOR;VAR SAVERAGE,SVARIANCE:REAL);
VAR
 J:INTEGER; { data index    }
 SQ:REAL;   { help variable }
BEGIN
 SAVERAGE:=0.0;SVARIANCE:=0.0;SQ:=0.0; { initialization }
 FOR J:=1 TO NN DO
  BEGIN
    SAVERAGE:=SAVERAGE+X[J];SQ:=SQ+X[J]*X[J];
  END;
 SVARIANCE:=(SQ-SAVERAGE*SAVERAGE/NN)/(NN-1);
 SAVERAGE:=SAVERAGE/NN;
END;{ COMPSAMPLEPARA }

{ Main program : }
  ...

{ Generation of the random numbers                }
{ Calculation of the sample average and variance : }

 FOR I:=1 TO N DO F[I]:=RANDOM;
 COMPSAMPLEPARA(N,F,AVERAGE,VARIANCE); { computes the sample parameters }

{ Printout : }

 QSORT(1,N,F); { sorts the vector F }
 CLEARDEVICE;
 COORDSYS; { draws the coordinate system }
 HISTOGRAM(N,F); { plots the histogram    }
```

```
STATDATATOSCR(AVERAGE,VARIANCE,(ABSCMAX+ABSCMIN)/2,SQR(ABSCMAX-ABSCMIN)/12);
CR:='Histogram of relative frequencies divided by subinterval length';
STATUSLINE(CR);
    ...
```

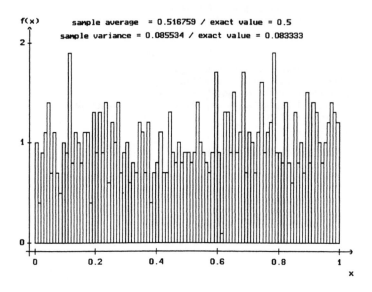

Figure 1.3.1   Histogram of U(0,1) pseudo-random numbers.

Notes 1.3.1

(i) Listings for the procedures COORDSYS, STATDATATOSCR and QSORT can be found on the diskette. COORDSYS plots the coordinate axes on the screen in preparation for the histogram, whereas STATDATATOSCR displays the sample average and variance, calculated by the procedure COMP-SAMPLEPARA, on the screen along with the true values. QSORT implements a quick-sorting algorithm to convert the generated data vector into a sorted input vector for the procedure HISTOGRAM which calculates the relative frequencies and prepares the histogram, aborting and displaying a failure message if the inputed data falls outside the designated range.

(ii) The scaling parameters for the histogram must be chosen with care. Peaks may occur in the histogram, depending on the inputed random numbers, but will generally be smaller than the reciprocal of INVLENGTH, the length of the histogram subintervals. Random numbers are not counted when they fall outside of the assigned base interval of the histogram, as in the exponential and Gaussian histograms which follow.

Looking at Figure 1.3.1 one realizes that the noise still obscures much of the uniform shape which one should expect.

Exercise 1.3.1   *Repeat Problem 1.3.1 for $N = 10^4$ and $N = 10^5$.*

We now repeat the above experiment for exponentially distributed random numbers.

Problem 1.3.2 (PC-Exercise 1.4.5) *Repeat Problem 1.3.1 for exponentially distributed random numbers with the parameter value* $\lambda = 2.0$ *and subinterval length* 0.05.

Since the density function $p(x) > 0$ for all $x \geq 0$, we can partition the finite interval $[0, 5]$, say, into subintervals of equal length, with all of the values larger than 5 being discarded in the count. It is recommended to repeat the above computation for larger N.

Notes 1.3.2 We omit Program 1.3.2 as its structure is similar to that of Program 1.3.1. Apart from changing the generator and the text strings, the distribution parameters are now calculated for the declared value of the parameter LAMBDA.

Finally we check the Gaussian pseudo-random number generators.

Problem 1.3.3 (PC-Exercise 1.4.6) *Repeat Problem 1.3.1 for standard Gaussian random numbers generated by the Box-Muller and Polar Marsaglia methods, using* $[-2.5, 2.5]$ *as the histogram base interval and subintervals of length* 0.05.

Notes 1.3.3 The required program is obtained by obvious modifications to Program 1.3.1 and is listed as Program 1.3.3 on the diskette. It is easy to change it for other larger N.

The standard Gaussian distribution with density considered in (1.18) is a special case of the general Gaussian distribution with mean μ and variance σ^2. Its density function

$$(3.6) \qquad p(x) = \frac{1}{\sqrt{2\pi}\,\sigma} \exp\left(\frac{-(x-\mu)^2}{2\sigma^2}\right)$$

still has a bell-shaped graph, but is now centered on the mean value $x = \mu$ and is stretched or compressed according to the magnitude of σ^2 with the maximum value being $1/(\sqrt{2\pi}\,\sigma)$ and the points of inflection at $\mu \pm \sigma$. We shall write $X \sim N(\mu; \sigma^2)$ for a random variable with this distribution. It is not hard to see that the transformed random variable

$$Z = (X - \mu)/\sigma$$

satisfies the standard Gaussian distribution. The transformation

$$X = \sigma Z + \mu$$

converts a standard Gaussian random variable Z into an $N(\mu; \sigma^2)$ distributed random variable X. We can use it to obtain an $N(\mu; \sigma^2)$ distributed pseudo-random variable X from the output of the Box-Muller or Polar Marsaglia methods.

Typical functions of random variables are the polynomials $g(x) = x^p$ or $g(x) = (x - \mu)^p$ for integers $p \geq 1$. The resulting expected values of $Y = g(X)$ are called the *pth-moment* or the *pth-central moment* respectively and the variance $\text{Var}(X)$ is the *2nd-order* or *squared central moment* of X. Such moments convey information about the random variable, but the higher order moments need not always provide additional information. For example, the Gaussian distribution is completely characterized by its first two moments, its mean μ and variance σ^2, and the Poisson distribution by its mean λ.

For Gaussian $X \sim N(\mu; \sigma^2)$ and $k = 0, 1, 2, \ldots$ we note that

$$(3.7) \qquad\qquad E(X - \mu)^{2k+1} = 0$$

and
$$(3.8) \qquad\qquad E(X - \mu)^{2k} = 1 \cdot 3 \cdot 5 \cdots (2k - 1)\sigma^{2k}.$$

B. Conditional Expectations

The mean value or expectation $E(X)$ is one of the coarsest characteristics that we have for a random variable X. If we know that some event has occured we may be able to improve on this. For instance, if the event $A = \{\omega \in \Omega : a \leq X(\omega) \leq b\}$ has occured, then in expecting the value of X we need only consider these values of X and weight them according to their likelihood of occurence, which is the conditional probability given this event A. The resulting characteristics is called the *conditional expectation* of X given the event A and is denoted by $E(X|A)$. For a discretely distributed random variable the conditional probabilities for the event $A = \{\omega \in \Omega : a \leq X(\omega) \leq b\}$ satisfy

$$P(X = x_i|A) = \begin{cases} 0 & : \quad x_i < a \text{ or } b < x_i \\ p_i \Big/ \sum_{a \leq x_j \leq b} p_j & : \quad a \leq x_i \leq b \end{cases}$$

with $P(X = x_i) = p_i$, so the conditional expectation is

$$(3.9) \qquad E(X|A) = \sum_{i \in I} x_i P(X = x_i|A) = \sum_{a \leq x_i \leq b} x_i p_i \Big/ \sum_{a \leq x_j \leq b} p_j .$$

For a continuously distributed random variable with a density function p the corresponding *conditional density* is

$$p(x|A) = \begin{cases} 0 & : \quad x < a \text{ or } b < x \\ p(x) \Big/ \int_a^b p(s)\, ds & : \quad a \leq x \leq b \end{cases}$$

and the conditional expectation has the form

$$(3.10) \qquad E(X|A) = \int_{-\infty}^{\infty} x\, p(x|A)\, dx = \int_{a}^{b} x\, p(x)\, dx \left/ \int_{a}^{b} p(x)\, dx \right. .$$

As an illustration we note that a random variable X distributed exponentially with parameter λ has conditional expectation

$$(3.11) \qquad E(X|X \geq a) = a + E(X) = a + \lambda^{-1}$$

for any $a > 0$, where the event $A = \{\omega \in \Omega : X(\omega) \geq a\}$ is abbreviated to $\{X \geq a\}$. This identity characterizes the exponential distribution, saying that the expected remaining life of a light bulb, for example, is independent of how old the bulb is when the life is exponentially distributed.

Problem 1.3.4 (PC-Exercise 1.4.10) *Generate 10^3 exponentially distributed pseudo-random numbers with parameter $\lambda = 0.5$. Calculate the average of these numbers and the averages of those numbers $\geq a$ where $a = 1, 2, 3$ and 4. Compare these with the identity (3.11).*

Program 1.3.4   Conditional expectation

```
...
FOR I:=1 TO 4 DO { initialization }
BEGIN
  L[I]:=0;    { index counter          }
  E[I]:=0.0; { conditional expectation }
END;
I:=0;
E[0]:=0.0;
REPEAT
 I:=I+1;
 GENERO2(LAMBDA,X);
 E[0]:=E[0]+X;
 IF X>=1.0 THEN
   BEGIN
    L[1]:=L[1]+1;
    E[1]:=E[1]+X;
    IF X>=2.0 THEN
      BEGIN
       L[2]:=L[2]+1;
       E[2]:=E[2]+X;
       IF X>=3.0 THEN
         BEGIN
          L[3]:=L[3]+1;
          E[3]:=E[3]+X;
          IF X>=4.0 THEN
            BEGIN
             L[4]:=L[4]+1;
             E[4]:=E[4]+X;
            END;
         END;
      END;
   END;
```

```
    END;
    UNTIL I=NUMBER;
    E[0]:=E[0]/NUMBER;
    FOR I:=1 TO 4 DO IF L[I]>0 THEN E[I]:=E[I]/L[I];
    ...
```

Notes 1.3.4   We list only the essential counting section of Program 1.3.4 here, the remaining details being on the diskette. The procedure SETTABLETOSCR is used there and prints out a table on the screen comparing the calculated sample parameters with the corresponding exact values.

C. Correlated Random Variables

We mentioned that the random variables G_1 and G_2 generated by the Box-Muller (2.5) and Polar Marsaglia (2.6) methods are independent Gaussian with zero mean vector when U_1 and U_2 are independent $U(0,1)$ distributed random variables. It is also often useful to have pairs (X_1, X_2) of correlated Gaussian random variables, that is with given covariance

$$(3.12) \qquad E(X_1 X_2) = \mathrm{cov}(X_1 X_2).$$

These can be obtained from a pair (G_1, G_2) of independent standard Gaussian random variables by an appropriate linear transformation.

Problem 1.3.5 (PC-Exercise 1.4.12)   *Write a program to generate pairs of Gaussian pseudo-random numbers (X_1, X_2) with means $\mu_1 = E(X_1) = 0$, $\mu_2 = E(X_2) = 0$ and covariances $E(X_1^2) = h$, $E(X_2^2) = h^3/3$ and $E(X_1 X_2) = h^2/2$ for any $h > 0$. Generate 10^3 pairs of such numbers for $h = 0.1$, 1.0 and 10.0. Evaluate their sample averages and sample covariances, comparing these with the desired theoretical values.*

In the following program we use the fact that if G_1 and G_2 are two independent standard Gaussian random variables, then

$$(3.13) \qquad X_1 = \sqrt{h}\, G_1 \quad \text{and} \quad X_2 = \frac{1}{2} h^{\frac{3}{2}} \left(G_1 + \frac{1}{\sqrt{3}} G_2 \right)$$

form a pair of correlated Gaussian random variables with the desired properties.

Program 1.3.5   Correlated Gaussian random numbers

```
USES CRT,DOS,GRAPH,INIT,SETSCR,SERVICE,RANDNUMB;

CONST
  NUMBER=1000; { number of pairs of random numbers }
  H1=0.1;      { constants for covariances         }
  H2=1.0;
  H3=10.0;

TYPE
  MATRIX=ARRAY[1..2,1..2] OF REAL;
```

```
VAR
  I,K:INTEGER;              { counters                          }
  V1,V2,X1,X2:REAL;         { random numbers                    }
  COV:MATRIX;               { covariances                       }
  H:ARRAY[1..3] OF REAL;    { constants of the covariances      }
  EX1,EX2:REAL;             { estimates of the first moments    }
  E2X1,E2X2:REAL;           { estimates of the second moments   }
  EX1X2:REAL;               { estimate of the mixed moment      }

PROCEDURE SETTABLETOSCR;
PROCEDURE SETESTIMATESTOSCR;

PROCEDURE TRANSFOR(V1,V2:REAL;VAR X1,X2:REAL;COV:MATRIX);
VAR A,B,C,D:REAL;
BEGIN
  A:=0;B:=SQRT(COV[1,1]);
  D:=COV[1,2]/B;C:=SQRT(COV[2,2]-D*D);
  X1:=A*V1+B*V2;
  X2:=C*V1+D*V2;
END;{ TRANSFOR }

{ Main program : }

BEGIN
  ...

  K:=0;
  REPEAT
   K:=K+1;
   CASE K OF { initialization of some parameters }
     1 : H[K]:=H1;
     2 : H[K]:=H2;
     3 : H[K]:=H3;
    END;
   COV[1,1]:=H[K];COV[1,2]:=H[K]*H[K]/2; { initialization of the covariances }
   COV[2,1]:=COV[1,2];COV[2,2]:=H[K]*H[K]*H[K]/3;
   EX1:=0;EX2:=0;E2X1:=0;E2X2:=0;EX1X2:=0; { initialization of the moments }
   I:=0;
   REPEAT
    I:=I+1;
    GENERATE(V1,V2); { generates the Gaussian random numbers }
    TRANSFOR(V1,V2,X1,X2,COV); { transforms to dependent random numbers }
    EX1:=EX1+X1;EX2:=EX2+X2;E2X1:=E2X1+X1*X1;E2X2:=E2X2+X2*X2;
    EX1X2:=EX1X2+X1*X2;
   UNTIL I=NUMBER;
   EX1:=EX1/NUMBER;EX2:=EX2/NUMBER;
   E2X1:=E2X1/(NUMBER-1);E2X2:=E2X2/(NUMBER-1);
   EX1X2:=EX1X2/NUMBER;
   ...
  UNTIL K=3;
  ...
```

Notes 1.3.5

(i) Only the key procedure TRANSFOR in Program 1.3.5 is listed above. It transforms the two independent standard Gaussian random numbers G_1 and G_2 into two correlated Gaussian random numbers X_1 and X_2 with the covariances given in the matrix COV, the components of which can be changed if desired.

(ii) A table for comparing the calculated sample values and the desired exact values is displayed on the screen by the procedure SETTABLETOSCR. The current estimates are inserted in the table by the procedure SETES-TIMATETOSCR.

We say that two random variables X_1 and X_2 are *independent* if their joint and marginal distribution functions satisfy

$$(3.14) \qquad F_{X_1,X_2}(x_1, x_2) = F_{X_1}(x_1) F_{X_2}(x_2)$$

for all $x_1, x_2 \in \Re$. If both F_{X_1} and F_{X_2} have density functions p_1 and p_2, respectively, and if X_1 and X_2 are independent, then their joint distribution function F_{X_1,X_2} has the density function

$$(3.15) \qquad p(x_1, x_2) = p_1(x_1) p_2(x_2).$$

Moreover, the product $X_1 X_2$ of two independent random variables X_1 and X_2 has expectation

$$(3.16) \qquad E(X_1 X_2) = E(X_1) E(X_2)$$

and the sum $X_1 + X_2$ has variance

$$(3.17) \qquad \mathrm{Var}(X_1 + X_2) = \mathrm{Var}(X_1) + \mathrm{Var}(X_2).$$

Hence for two independent random variables the means and variances are additive, and the means are also multiplicative.

Problem 1.3.6 (PC-Exercise 1.4.13) *Check numerically whether or not the jointly Gaussian random variables X_1, X_2 with means and covariances given in Problem 1.3.5 are independent.*

Notes 1.3.6   To do this in the Gaussian case we simply compare estimates for the left and right hand sides of (3.16) and of (3.17). Program 1.3.6, which is given on the diskette, uses the procedure SETTABLETOSCR to display the sample variances of the sum $X_1 + X_2$ with the sum of sample variances for X_1 and X_2 for different values of the parameter h as a table on the screen.

1.4   Random Sequences

Here we mention various convergence criteria for random sequences and state two fundamental limit theorems from probability theory.

A. Convergence of Random Sequences

Often we have an infinite sequence $X_1, X_2, \ldots, X_n, \ldots$ of random variables and are interested in its asymptotic behaviour, that is in the existence of a random variable X which is the limit of X_n for $n \to \infty$ in some sense. There are several different ways in which such convergence can be defined. Broadly speaking these fall into two classes, a stronger one in which the realizations of X_n are required to be close in some way to those of X and a weaker one in which only their probability distributions need be close.

The following convergences from the strong class are commonly used.

I. *Convergence with probability one* (w.p.1):

$$(4.1) \qquad P\left(\{\omega \in \Omega : \lim_{n \to \infty} |X_n(\omega) - X(\omega)| = 0\}\right) = 1.$$

This is also known as *almost sure convergence*.

II. *Mean-square convergence:* $E(X_n^2) < \infty$ for $n = 1, 2, \ldots$, $E(X^2) < \infty$ and

$$(4.2) \qquad \lim_{n \to \infty} E\left(|X_n - X|^2\right) = 0.$$

III. *Strong convergence:* $E(|X_n|) < \infty$ for $n = 1, 2, \ldots$, $E(|X|) < \infty$ and

$$(4.3) \qquad \lim_{n \to \infty} E(|X_n - X|) = 0.$$

The last criterion is a straight-forward generalization of the deterministic convergence, and we shall use this criterion in discussing the convergence of numerical schemes. By the Lyapunov inequality

$$(4.4) \qquad E(|Y|) \leq \sqrt{E(|Y|^2)}$$

we see that mean-square convergence (4.2) implies strong convergence (4.3).

For the class of weaker convergences we do not need to know the actual random variables, just their distribution functions. We mention here the following representatives from this class.

IV. *Convergence in distribution:*

$$(4.5) \qquad \lim_{n \to \infty} F_{X_n}(x) = F_X(x) \quad \text{at all continuity points of } F_X.$$

This is also known as *convergence in law* and refers to the distributions of the random variables X_n and X.

V. *Weak convergence:*

$$(4.6) \qquad \lim_{n \to \infty} \int_{-\infty}^{\infty} f(x)\, dF_{X_n}(x) = \int_{-\infty}^{\infty} f(x)\, dF(x)$$

for all *test functions* $f : \Re \to \Re$, usually continuous functions vanishing outside of a bounded interval which may depend on the particular function. Sometimes it is useful to consider weak convergence with respect to special classes of test functions such as the class of all polynomials which ensure that all moments converge.

Problem 1.4.1 (PC-Exercise 1.5.3)   *Let $Y_n = X + Z_n$ where X is a $U(0,1)$ uniformly distributed random variable and Z_n an independent $N(0; 1/n)$ Gaussian random variable for $n = 1, 2, 3, \ldots$. For fixed but large n. say $n = 10^3$, generate n realizations of Y_n. Calculate relative frequencies and plot a histogram with subinterval length 0.05 to obtain an indication of the graph of the density function of Y_n. Then estimate the mean-square error $E(|Y_n - X|^2)$ for a larger n, say $n = 10^4$. Does this suggest that Y_n converges to X in the mean-square sense?*

Program 1.4.1   Mean-square convergence

```
  ...
CONST
  N=1000;          { parameter n }
  NUMBER=1000;     { sample size }
  ...
VAR
  ...
  ROOTN:REAL;              { square root of 1/n         }
  U1,U2:REAL;              { random numbers             }
  AVERAGE,VARIANCE:REAL;   { sample average and variance }
  E2YNX,E2YN:REAL;         { sample 2nd moments         }
  F:VECTOR;                { vector of the random numbers }
  ...
PROCEDURE COORDSYS;
PROCEDURE QSORT(DOWN,UP:INTEGER;VAR F:VECTOR);
PROCEDURE HISTOGRAM(F:VECTOR);
PROCEDURE STATDATATOSCR(AVERAGE,VARIANCE,MEANSQRERROR:REAL);

{ Main program : }
  ...
ROOTN:=SQRT(1/N);
AVERAGE:=0.0;VARIANCE:=0.0;E2YNX:=0.0;E2YN:=0.0;
FOR I:=1 TO NUMBER DO
  BEGIN
    IF I MOD 2 = 1 THEN
      BEGIN
        GENERATE(U1,U2);U1:=U1*ROOTN;U2:=U2*ROOTN;
        F[I]:=RANDOM+U1;
        E2YNX:=E2YNX+U1*U1;
      END
    ELSE
      BEGIN
        F[I]:=RANDOM+U2;
        E2YNX:=E2YNX+U2*U2;
      END;
    AVERAGE:=AVERAGE+F[I];
    E2YN:=E2YN+F[I]*F[I];
  END;
VARIANCE:=(E2YN-AVERAGE*AVERAGE,'NUMBER)/(NUMBER-1);
AVERAGE:=AVERAGE/NUMBER;
E2YNX:=E2YNX/(NUMBER-1);
```

```
QSORT(1,NUMBER,F); { sorts the vector F }
CLEARDEVICE;
COORDSYS; { draws the coordinate system }
HISTOGRAM(F); { plots the histogram    }
  ...
```

Notes 1.4.1

(i) The Gaussian random numbers are provided by the Polar Marsaglia generator GENERATE.

(ii) For information about the histogram preparation procedures see the notes for Problem 1.3.1 - 1.3.3.

(iii) The procedure SETTABLETOSCR is used in a slightly different way here, now providing the estimated mean-square error.

(iv) The comparison of the run with $n = 10^3$ with that of the larger $n = 10^4$ should suggest the mean square convergence.

B. Law of Large Numbers

The intuitive idea of defining probabilities as the limits of relative frequencies determined from many repetitions of a given probabilistic experiment can be justified theoretically in the case of sequences of *independent identically distributed* (i.i.d.) random variables X_1, X_2, X_3,.... Suppose that $\mu = E(X_n)$ and $\sigma^2 = \text{Var}(X_n)$. From independence the averaged random variables

$$(4.7) \qquad A_n = \frac{1}{n} S_n = \frac{1}{n}(X_1 + X_2 + \cdots + X_n)$$

also have mean $E(A_n) = \mu$ and variance $\text{Var}(A_n) = \sigma^2/n$. The *Law of Large Numbers* says that

$$(4.8) \qquad A_n \to \mu \quad \text{as} \quad n \to \infty$$

with the convergence taken in mean-square sense (4.2).

Problem 1.4.2 (PC-Exercise 1.5.4) *Use the random number generators in Section 1.2 for uniform, two-point, exponential and Gaussian pseudo-random numbers to verify the limit (4.8) in the mean-square sense by averaging $(A_n - \mu)^2$ for 10^2 different runs with $n = 10$, 10^2 and 10^3.*

Notes 1.4.2

(i) Much of the required program is already contained in Program 1.4.1, so the listing is omitted here.

(ii) The procedure SETTABLETOSCR is now used just to prepare a table on the screen, with the error estimates being added separately for the different n values as they are calculated.

(iii) More realizations and larger sequence indices n could be used. If the index n is very large, TURBO PASCAL requires it to be declared as a LONGINT.

C. Central Limit Theorem

Another fundamental result, the *Central Limit Theorem*, says that the normalized i.i.d. random variables

(4.9)
$$Z_n = \frac{S_n - n\mu}{\sigma\sqrt{n}},$$

for which $E(Z_n) = 0$ and $\mathrm{Var}(Z_n) = 1$, converge in distribution to a standard Gaussian random variable Z. This is also true under weaker assumptions than the i.i.d. assumption on the original random variables X_1, X_2, X_3, ... and provides an explanation for the dominant role of the Gaussian distribution in probability and statistics.

Problem 1.4.3 (PC-Exercise 1.5.6)   *Generate sequences of random numbers as in Problem 1.4.2 and calculate their Z_n values, where Z_n is defined by (4.9). Use 10^3 realizations and compute histograms with subinterval length 0.05 for the relative frequencies to obtain simulated approximations of the graphs of the density functions for the Z_n and show that these approach the graph of the density function of the standard Gaussian distribution as n is taken larger and larger.*

Program 1.4.3   Central Limit Theorem

```
    ...
PROCEDURE COORDSYS;
PROCEDURE QSORT(DOWN,UP:INTEGER;VAR F:VECTOR);
PROCEDURE HISTOGRAM(NN:INTEGER;F:VECTOR);
PROCEDURE SETPARATOSCR(AVERAGE,VARIANCE,CURRENTNUMBER:REAL);

{ Main program : }
    ...
CASE M OF
  1 : BEGIN MU:=0.5;SIGMA:=SQRT(1/12);END;
  2 : BEGIN MU:=X1*P+X2*(1-P);SIGMA:=SQRT(X1*X1*P+X2*X2*(1-P)-MU*MU);END;
  3 : BEGIN MU:=1/LAMBDA;SIGMA:=MU;END;
  4 : BEGIN MU:=0.0;SIGMA:=1.0;END;
END;
FOR I:=1 TO NUMBER DO SN[I]:=0.0;
N1:=0;
REPEAT
 N1:=N1+INDEX;
 ROOTN1:=SQRT(N1);
 FOR I:=1 TO NUMBER DO
   BEGIN
    L:=N1-INDEX;
    REPEAT
     L:=L+1;
     CASE M OF
       1 : X:=RANDOM;
       2 : GENER01(P,X1,X2,X);
       3 : GENER02(LAMBDA,X);
       4 : IF I MOD 2 = 1 THEN GENERATE(X,Y) ELSE X:=Y;
     END;
```

```
    SN[I]:=SN[I]+X;
    UNTIL L=N1;
    ZN[I]:=(SN[I]-N1*MU)/(SIGMA*ROOTN1);
    END;
...
UNTIL N1=N;
...
```

Notes 1.4.3

(i) The full listing of Program 1.4.3 is on the diskette.

(ii) The desired distribution is provided by the appropriate choice of the parameter M $(= 1, 2, 3, 4)$ in the program, with the Gaussian numbers being generated by the Polar Marsaglia method where $M = 4$. Successive histograms appear automatically on the screen for n increasing in steps of integer INDEX until n reaches 10^3. For low values of n peaks in the histograms may sometimes exceed the size of the screen.

(iii) As in Problems 1.3.1 - 1.3.3 the histograms are plotted on a prescribed base interval, with values of the random variables falling outside this interval not being counted. Remarks on the histogram preparation procedures can be found in the notes to Problems 1.3.1 - 1.3.3.

1.5   Testing Random Numbers

So far we have glossed over some very basic issues concerning the use of pseudo-random number generators, assuming that they actually do generate independent random numbers with the desired distribution. In addition, we have never really specified just how many terms of a random sequence are required to provide a good approximation or estimation of their limit. These issues are closely interconnected and an extensive theory and array of tests have been proposed for their resolution. Unfortunately none of the tests is completely definitive and an element of subjectivity is often involved in their use. Moreover, where answers can be provided they are in the form of confidence intervals and levels of significance rather than certainties.

It should be understood that a sequence of numbers can be interpreted as a sequence of random numbers only with respect to an array of specific tests which it passes. In what follows we shall briefly describe several useful tests.

A. Confidence Intervals

We can calculate an estimate of an unknown mean value and determine the number of terms of an approximating sequence needed for a good estimate with the aid of the Law of Large Numbers (4.8) or the Central Limit Theorem (4.9).

Consider the Bernoulli trials with a sequence of i.i.d. random variables $X_1, X_2, \ldots, X_n, \ldots$ taking the value 1 with probability p and the value 0 with probability $1-p$, where p is unknown. Then $E(X_n) = p$ and $\text{Var}(X_n) = p(1-p)$ and so, by independence, $E(S_n) = np$ and $\text{Var}(S_n) = np(1-p)$ for the sum $S_n = X_1 + X_2 + \ldots + X_n$. Alternatively, $E(A_n) = p$ and $\text{Var}(A_n) = p(1-p)/n$

for the sample averages $A_n = S_n/n$. The Law of Large Numbers tells us that A_n converges to p in mean-square sense, but gives no information about some rate of convergence. To obtain an indication of this we can apply the well-known Chebyshev inequality to $A_n - p$ and the approximation $p(1-p) \le 1/4$ for $0 \le p \le 1$ to get

$$(5.1) \qquad P\left(\{\omega \in \Omega : |A_n - p| \ge a\}\right) \le \frac{\text{Var}(A_n)}{a^2} = \frac{p(1-p)}{na^2} \le \frac{1}{4na^2}$$

for all $a > 0$ and so (omitting the ω)

$$P\left(|A_n - p| < a\right) = 1 - P\left(|A_n - p| \ge a\right) \ge 1 - \frac{1}{4na^2}.$$

Thus for any $0 < \alpha < 1$ and $a > 0$ we can conclude that the unknown mean p lies in the interval $(A_n - a, A_n + a)$ with at least probability $1 - \alpha$ when $n \ge n(a, \alpha) = 1/(4\alpha a^2)$. In statistical terminology we say that the hypothesis that p lies in the interval $(A_n - a, A_n + a)$ for $n \ge n(a, \alpha)$ is acceptable at a $100\alpha\%$ *level of significance*, and call $(A_n - a, A_n + a)$ a $100(1 - \alpha)\%$ *confidence interval*. For example, $(A_n - 0.1, A_n + 0.1)$ is a 95% confidence interval when $n \ge n(0.1, 0.05) = 500$. We note that the confidence interval is random.

The number $n(a, \alpha)$ above is usually larger than necessary because of the coarseness of the inequalities in (5.1). We can obtain a shorter interval by using the Central Limit Theorem instead of the Law of Large Numbers. Since the inequality $|A_n - p| \ge a$ is equivalent to $|Z_n| \ge b$ where $Z_n = (S_n - np)/\sqrt{np(1-p)}$ and $b = a\sqrt{n/p(1-p)}$ and since, by the Central Limit Theorem, Z_n is approximately standard Gaussian for large n we have

$$(5.2) \qquad\qquad P\left(|A_n - p| < a\right) = P\left(|Z_n| < b\right) \approx 2\Phi(b)$$

for sufficiently large n, where

$$\Phi(b) = \frac{1}{\sqrt{2\pi}} \int_0^b \exp\left(-\frac{1}{2}x^2\right) dx.$$

For a given $100\alpha\%$ significance level we read the standard Gaussian statistical tables backwards to determine a value $b = b(\alpha) > 0$ satisfying $\Phi(b) = (1-\alpha)/2$. Then if, say, $p \in (1/4, 3/4)$ for a given a we solve the inequality

$$b(\alpha) = a\sqrt{n/p(1-p)} \le 4a\sqrt{n}$$

for

$$n \ge \bar{n}(a, \alpha) = b^2(\alpha)/16a^2.$$

This will give us a $100(1 - \alpha)\%$ confidence interval $(A_n - a, A_n + a)$ when $n \ge \bar{n}(a, \alpha)$. For example, when $\alpha = 0.05$ we solve $\Phi(b) = 0.475$ for $b \approx 1.96$, so for $a = 0.1$ we calculate $\bar{n}(0.1, 0.05) \approx 25$.

If we do not know the variance σ^2 or do not have an estimate for it, we can sometimes use the sample variance $\hat{\sigma}^2$ instead. Let $X_1, X_2, \ldots, X_n$ be

n independent Gaussian random variables with known mean μ and unknown variance σ^2. As before the *sample mean* is $\hat{\mu}_n = A_n = \sum_{j=1}^{n} X_j/n$ and the *sample variance* has the form

$$(5.3) \qquad \hat{\sigma}_n^2 = \frac{1}{n-1} \sum_{j=1}^{n} (X_j - \hat{\mu}_n)^2 .$$

Henceforth we shall use a hat "^" on sample statistics to distinguish them from the unknown parameter. Then for $n > 3$ the random variable

$$T_n = \frac{\hat{\mu}_n - \mu}{\sqrt{\hat{\sigma}_n^2/n}}$$

satisfies the *Student t-distribution* with $n - 1$ degrees of freedom. Similarly to (5.2) we have

$$P\left(|\hat{\mu}_n - \mu| < a\right) = P\left(|T_n| < t\right)$$

where $t = a\sqrt{n/\hat{\sigma}_n^2}$. For a given $100\alpha\%$ significance level, we check whether or not the test variable

$$T_n^0 = \frac{\hat{\mu}_n - \mu_0}{\sqrt{\hat{\sigma}_n^2/n}}$$

with hypothesized mean μ_0 satisfies the inequality

$$|T_n^0| < t_{1-\alpha, n-1}$$

where $t_{1-\alpha, n-1}$ can be found in statistical tables; some typical values of $t_{1-\alpha, n-1}$ are given in Table 1.5.1. If this is not the case, then we reject the *null hypothesis* H_0 that $\mu = \mu_0$. Otherwise, we accept it on the basis of this test. In addition, we form the corresponding $100(1 - \alpha)\%$ confidence interval $(\hat{\mu}_n - a, \hat{\mu}_n + a)$ with

$$a = t_{1-\alpha, n-1}\sqrt{\hat{\sigma}_n^2/n}.$$

This contains all of the values of μ_0 for which the null hypothesis would be accepted in this test.

n	10	20	30	40	60	100	200
$t_{0.9, n-1}$	1.83	1.73	1.70	1.68	1.67	1.66	1.65
$t_{0.99, n-1}$	3.25	2.86	2.76	2.70	2.66	2.62	2.58

Table 1.5.1 Values of $t_{1-\alpha, n-1}$ for the Student t-distribution with $n - 1$ degrees of freedom for $\alpha = 0.1$ and 0.01.

The t-test requires the original random variables to be Gaussian. When they are not, we can resort to the Central Limit Theorem and use the test asymptotically. We take n batches of m random variables $X_1^{(j)}$, $X_2^{(j)}$, ..., $X_m^{(j)}$ for $j = 1, 2, \ldots, n$, which are independent and identically distributed (i.i.d.) with mean μ and variance σ^2. Then we form the sample means $\hat{A}_m^{(j)}$ and the sample variances $(\hat{\sigma}_m^{(j)})^2$ for each batch $j = 1, 2, \ldots, n$ and use the Central Limit Theorem to conclude that the $\hat{A}_m^{(j)}$ are approximately Gaussian. The preceding t-test is then approximately valid for these batch averages (rather than for the original $X_i^{(j)}$). In practice it has been found that each batch should have at least $m \geq 15$ terms. For pseudo-random number generators each batch could be determined from a different seed or starting value, thus allowing different sequences to be tested. If a parallel computer or a networked pool of PCs is available, the batches and their sample statistics can be calculated simultaneously.

Problem 1.5.1 (PC-Exercise 1.9.2) *Simulate $M = 20$ batches of length $N = 100$ of $U(0, 1)$ distributed pseudo-random numbers and evaluate the 90% confidence interval for their mean value.*

Program 1.5.1 Confidence intervals

```
   ...
CONST
   M=20;              { number of batches                                   }
   N=100;             { batch length                                        }
   QUANTILE=1.73;  { percentage point of the t distribution for alpha = 0.1 }
   ...
I:=0; { initialization of the batch index }
MU:=0.0;SIGMA:=0.0;SQ:=0.0;
REPEAT
 I:=I+1;
 K:=0; { initialization of the sample index }
 SI:=0.0;
 REPEAT
  K:=K+1;
  X:=RANDOM; { replace this statement for other random number generators }
  SI:=SI+X;
 UNTIL K=N;
 SI:=SI/N;MU:=MU+SI;SQ:=SQ+SI*SI;
UNTIL I=M;
SIGMA:=(SQ-MU*MU/M)/(M-1);MU:=MU/M;
SIGMA:=SQRT(SIGMA);A:=QUANTILE*SIGMA/SQRT(M);
   ...
```

Notes 1.5.1

(i)   Only the arithmetic core of the program is listed above, with a variation of formula (5.3) being used so the variance and mean can be calculated simultaneously.

(ii) The results are printed on the screen both as an interval with stated endpoints and as the midpoint plus/minus half of the interval length.

(iii) The sample size and confidence level can be easily changed at the beginning of the program.

(iv) Other types of random numbers can be considered by replacing the RANDOM generator by the appropriate generator, such as one of those listed in RANDNUMB.

B. χ^2 - Test

Problems 1.3.1, 1.3.2 and 1.3.3 required frequency histograms for the outputs of various pseudo-random number generators to be plotted and compared visually with the graphs of the density functions that they were supposed to simulate. There are various statistical tests which allow less subjective comparisons, the χ^2-*goodness-of-fit test* being one of the most commonly used. To apply it we need a large number N of values of i.i.d. random variables. From these we form a cumulative frequency histogram $F_N(x)$ which we wish to compare with the supposed distribution $\bar{F}(x)$. We subdivide our data values into $k + 1$ mutually exclusive categories and count the numbers $N_1, N_2, \ldots, N_{k+1}$ terms falling into these categories; obviously $N_1 + N_2 + \cdots + N_{k+1} = N$. We compare these with the expected numbers $N\bar{p}_1, N\bar{p}_2, \ldots, N\bar{p}_{k+1}$ for each category for the distribution $\bar{F}$. To do this we calculate the Pearson statistic

$$(5.4) \qquad \chi^2 = \sum_{j=1}^{k+1} \frac{(N_j - N\bar{p}_j)^2}{N\bar{p}_j}$$

which should be small if our null hypothesis H_0 that the data generating mechanism has $\bar{F}(x)$ as its distribution function is to be acceptable at a reasonable significance level. It is known that the Pearson statistic is distributed asymptotically in N according to the χ^2-*distribution* with k-degrees of freedom with $E(\chi^2) = k$ and $\text{Var}(\chi^2) = 2k$. To complete the test we pick a $100\alpha\%$ significance level and determine from statistical tables a value $\chi^2_{1-\alpha,k}$ such that $P(\chi^2 < \chi^2_{1-\alpha,k}) = 1 - \alpha$. If our χ^2 value in (5.4) satisfies $\chi^2 < \chi^2_{1-\alpha,k}$, then we accept our null hypothesis at the significance level $100\alpha\%$; otherwise we reject this hypothesis. Table 1.5.2 contains some values $\chi^2_{1-\alpha,k}$ with $k = 30$ degrees of freedom.

α	0.01	0.02	0.05	0.10
$\chi^2_{1-\alpha,30}$	50.892	47.962	43.773	40.256

Table 1.5.2 Values of $\chi^2_{1-\alpha,30}$ for the χ^2 - distribution.

Problem 1.5.2 (PC-Exercise 1.9.3) *Use the χ^2-goodness-of-fit test with $k = 30$ degrees of freedom, $N = 10^3$ generated numbers and significance levels 1% and 5% to test the goodness-of-fit of the $U(0,1)$ uniformly distributed pseudo-random number generator on your PC. Repeat these tests for the exponentially distributed random number generator with parameter $\lambda = 1.0$ and for the Box-Muller and Polar Marsaglia $N(0;1)$ generators.*

Program 1.5.2 Chi-squared test

```
...
CONST
  N=1000;          { sample size              }
  ALPHA1=0.01;     { significance level       }
  ALPHA2=0.05;     { significance level       }
```

```
    QUANTIL1=43.8; { percentage point of the CHI-square distribution }
    QUANTIL2=50.9; { percentage point of the CHI-square distribution }
    K=30;          { with K degrees of freedom                       }
    LAMBDA=1.0;    { parameter of the exponential distribution       }
    ...
VAR
    ...
    X:VECTOR;    { sample vector                                     }
    CHI:VECTOR1; { test values for different random generators }

PROCEDURE ROMBERG(A,B:REAL;VAR INTVALUE:REAL);
PROCEDURE QSORT(DOWN,UP:INTEGER;VAR F:VECTOR);
PROCEDURE SETTABLETOSCR;

PROCEDURE CHITEST(DEGREE,INDEX,SIZE:INTEGER;X:VECTOR;VAR TESTVALUE:REAL);
VAR
    NOEND:BOOLEAN; { control if the current data belongs to the category }
    KK:INTEGER;    { counter                                             }
    J:INTEGER;     { category index                                      }
    NJ:INTEGER;    { number of data falling into a category              }
    NECLEFT,NECRIGHT:INTEGER; { numbers of extra categories              }
    PJ:REAL;       { probability of a category                           }
    SUMPJ:REAL;    { sum of probabilities of the categories              }
    TJ:REAL;       { left end point of the j'th category                 }
    INVLENGTH:REAL; { length of subinterval                              }
BEGIN
    TESTVALUE:=0.0;
    CASE INDEX OF { choice of finite categories and control over its lengths }
      1 : BEGIN TJ:=0.0;INVLENGTH:=1./(DEGREE+1); END;
      2 : BEGIN TJ:=0.0;INVLENGTH:=10./(DEGREE*LAMBDA); END;
      3 : BEGIN TJ:=-3.;INVLENGTH:=6./DEGREE; END;
      4 : BEGIN TJ:=-3.;INVLENGTH:=6./DEGREE; END;
    END;
    KK:=1; { pointer to the current data index }
    NOEND:=TRUE;
    IF ((INDEX=3) OR (INDEX=4)) THEN { correction of the data index }
     BEGIN
      KK:=0;
      WHILE NOEND DO
       IF X[KK+1]<=TJ THEN
         BEGIN
           KK:=KK+1;IF KK=SIZE THEN NOEND:=FALSE;
         END
       ELSE NOEND:=FALSE;
      NECLEFT:=KK;KK:=KK+1;
     END;
    J:=0; { index of the current category }
    SUMPJ:=0.0;
    REPEAT
     J:=J+1;
     TJ:=TJ+INVLENGTH; { right end point of the current category }
     NJ:=0;NOEND:=TRUE;
     IF KK=SIZE+1 THEN NOEND:=FALSE;
     WHILE NOEND DO { provides the number of data falling into the category }
      IF X[KK]<=TJ THEN
        BEGIN
          KK:=KK+1;NJ:=NJ+1;IF KK=SIZE+1 THEN NOEND:=FALSE;
        END
```

```
    ELSE NOEND:=FALSE;
  CASE INDEX OF { generation of the probability of the current category }
      1 : PJ:=INVLENGTH;
      2 : PJ:=EXP(-LAMBDA*(TJ-INVLENGTH))-EXP(-LAMBDA*TJ);
      3,4 : ROMBERG(TJ-INVLENGTH,TJ,PJ);
    END;
  SUMPJ:=SUMPJ+PJ; { summation of the single probability }
  TESTVALUE:=TESTVALUE+SQR(NJ-SIZE*PJ)/(SIZE*PJ);
  UNTIL J=DEGREE;
  NECRIGHT:=SIZE+1-KK;
  IF ((INDEX=3) OR (INDEX=4)) THEN NJ:=NECLEFT+NECRIGHT ELSE NJ:=NECRIGHT;
  PJ:=1.-SUMPJ; { probability of the extra categories }
  TESTVALUE:=TESTVALUE+SQR(NJ-SIZE*PJ)/(SIZE*PJ);
END;{ CHITEST }

{ Main program : }
  ...
  L:=0; { index of the random number generator }
  REPEAT
   L:=L+1;
   FOR I:=1 TO N DO { initializes data vector X }
     CASE L OF
       1 : X[I]:=RANDOM;
       2 : GENERO2(LAMBDA,X[I]);
       3 : IF I MOD 2 = 1 THEN GENERO3(X[I],U) ELSE X[I]:=U;
       4 : IF I MOD 2 = 1 THEN GENERATE(X[I],U) ELSE X[I]:=U;
     END;
   QSORT(1,N,X); { sorts the vector X }
   CHITEST(K,L,N,X,CHI[L]); { calculates the current test value }
   UNTIL L=4;
  ...
```

Notes 1.5.2

(i) The procedure ROMBERG integrates the Gaussian density function over the interval $[A, B]$, terminating when two successive iterations differ by less than a prescribed margin EPS. The four different random number generators are installed by the appropriate value of the parameter INDEX, with the corresponding category subinterval length to be used accordingly. For the exponential and Gaussian random variables attention is restricted to values falling into the interval $[0, 10]$ and $[-3.0, 3.0]$, respectively.

(ii) The CHITEST procedure calculates the Pearson statistic (5.4) for the given DEGREEs of freedom, generator INDEX choice and sample SIZE. It assumes that the sample data vector X is a sorted vector, obtained from the generated random number sample here by the QSORT procedure. The results are tabulated on the screen by the SETTABLETOSCR procedure.

(iii) Care must be taken when changing the values of declared constants to ensure that matching values are used. In particular, the QUANTIL value $\chi^2_{1-\alpha,k}$ must correspond to the ALPHA and DEGREEs of freedom k values.

C. Kolmogorov-Smirnov Test

For continuously distributed random variables the discrete categories of the χ^2-goodness-of-fit test are artificial and subjective and do not take fully into account the variability in the data. These disadvantages are avoided in the *Kolmogorov-Smirnov test* which is based on the Glivenko-Cantelli theorem saying that

$$D_N = \sup_{-\infty < x < \infty} |F_N(x) - \bar{F}(x)| \to 0 \quad \text{as} \quad N \to \infty$$

almost surely. Note that the sample frequency histograms F_N here are random variables. To apply the test at a $100\alpha\%$ significance level we compare the value of $\sqrt{N}D_N$ calculated from our data with the value $x_{1-\alpha}$ satisfying $H(x_{1-\alpha}) = 1 - \alpha$, where H is the Kolmogorov distribution. If $\sqrt{N}D_N \leq x_{1-\alpha}$ we accept at the $100\alpha\%$ significance level the null hypothesis H_0 that the data generating mechanism has $\bar{F}(x)$ as its distribution function; otherwise we reject it. In general $N > 35$ suffices for this test, but for pseudo-random number generators a larger value can be easily taken and provides a more representative sample of generated numbers. Table 1.5.3 gives some values of $x_{1-\alpha}$.

α	0.01	0.02	0.05	0.10
$x_{1-\alpha}$	1.63	1.51	1.36	1.23

Table 1.5.3   Values of $x_{1-\alpha}$ for the Kolmogorov distribution.

Problem 1.5.3 (PC-Exercise 1.9.4)   *Repeat Problem 1.5.2 using the Kolmogorov-Smirnov test at 1% and 5% significance levels and $N = 10^3$ $U(0,1)$ random numbers. Compare the results with those for the χ^2-test.*

Program 1.5.3   Kolmogorov-Smirnov test

```
   ...
PROCEDURE KOLTEST(INDEX,SIZE:INTEGER;X:VECTOR;VAR TESTVALUE:REAL);
VAR
 J:INTEGER;       { current data index                                }
 DISTANCE:REAL;   { distance between the empirical and the test distribution }
 DN:REAL;         { maximum of these distances                        }
 F:REAL;          { value of the distribution to be tested            }
 PJ:REAL;         { probability of the increment of two neighbouring data }
 FNLEFT:REAL;     { left limit value of the empirical distribution    }
 FNRIGHT:REAL;    { value(=right limit value) of the empirical distribution }
 XMIN:REAL;       { help variable for the initialization if INDEX=3 or 4  }
BEGIN
 TESTVALUE:=0.0;F:=0.0;DN:=0.0;FNRIGHT:=0.0; { initialization }
 IF ((INDEX=3) OR (INDEX=4)) THEN
  BEGIN
   XMIN:=-10.0;IF X[1]<XMIN THEN XMIN:=X[1]-10.0
  END;
 DN:=0.0;
 J:=0; { indicates the current data }
 REPEAT
  J:=J+1;
  FNLEFT:=FNRIGHT;
```

```
FNRIGHT:=J/SIZE; { value of the empirical distribution }
CASE INDEX OF
    1 : F:=X[J];
    2 : F:=1-EXP(-LAMBDA*X[J]);
  3,4 : BEGIN
          IF J>1 THEN
            ROMBERG(X[J-1],X[J],PJ)
          ELSE
            ROMBERG(XMIN,X[J],PJ);
          F:=F+PJ;
        END;
   END;
  DISTANCE:=ABS(FNRIGHT-F); { current distance }
  IF DISTANCE<ABS(FNLEFT-F) THEN DISTANCE:=ABS(FNLEFT-F);
  IF DN<DISTANCE THEN DN:=DISTANCE; { maximum distance correction }
 UNTIL J=SIZE;
 TESTVALUE:=SQRT(SIZE)*DN;
END;{ KOLTEST }
 ...
```

Notes 1.5.3

(i) Only the core procedure KOLTEST is listed here since the program is essentially the same as Program 1.5.2 with it replacing the procedure CHITEST. The variables INDEX and SIZE play the same role here.

(ii) The ROMBERG procedure integrates the Gaussian density function as in Program 1.5.2.

(iii) See Figure 1.5.1 for the interpretation of the variables DISTANCE, FN-LEFT and FNRIGHT.

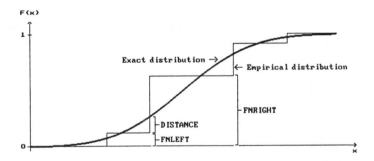

Figure 1.5.1   Exact and empirical distributions in the Kolmogorov-Smirnov test procedure KOLTEST.

D. A Test for Independence

Most commonly used pseudo-random number generators have been found to fit their supposed distributions reasonably well, but the generated numbers often seem not to be independent as they are supposed to be. This is not surprising

since, for congruential generators at least, each number is determined exactly
by its predecessor. In practice statistical independence is an elusive property to
confirm definitively for pseudo-random numbers generated by digital computers
and tests for it are nowhere near as satisfactory as those above for the goodness-
of-fit of distributions.

We shall restrict our remarks here to $U(0,1)$ uniformly distributed linear
congruential generators as described by (2.1) - (2.2) which have the form $U_n = X_n/c$ where

(5.5) $$X_{n+1} = aX_n + b \pmod{c}.$$

A simple test for independence involves plotting the successive pairs

$$(U_{2n-1}, U_{2n})$$

for $n = 1, 2, \ldots$ as points on the unit square with the U_{2n-1} as the x-coordinate
and the U_{2n} as the y-coordinate. These points lie on one of c different straight
lines of slope a/c and a large number of them should fairly evenly fill the unit
square. The presence of patches without any of these points is an indication of
bias in the generator.

One way to avoid such bias is to introduce a *shuffling procedure*. For this
we generate a string of 20 or more numbers and choose one number with equal
probability from the string. We take this number as the output of our shuffling
procedure and then generate a new number to replace it in the string. Repeating this step as often as required, we obtain a shuffled sequence of pseudo-
random numbers. We note that this requires more numbers to be generated
than for an unshuffled sequence of the same length since a random number
must also be generated at each step in order to choose the number to be taken
from the string. Shuffling procedures have been found to be quite effective in
reducing patchiness in poor generators as will be seen in the following problem.
They provide a possibility to lengthen the periods when using congruential generators.

Problem 1.5.4 (PC-Exercise 1.9.5)   *Plot 10^3 points (U_{2n-1}, U_{2n}) using the
linear congruential generator (5.5) with parameters $a = 1229$, $b = 1$ and $c = 2048$, using the seed $x_0 = 0$. Add a shuffling procedure to the generator and
repeat the above plots.*

Program 1.5.4   Shuffling random numbers

```
USES CRT,DOS,GRAPH,INIT,SETSCR,SERVICE;

CONST
  N=1000;        { number of random numbers                    }
  FREQUENCY=20;  { length of the shuffle vector                }
  X0=0;          { initial value of the congruential generator }
  A=1229;        { parameter of the congruential generator     }
  B=1;           { parameter of the congruential generator     }
  C=2048;        { parameter of the congruential generator     }
    . . .
```

```
VAR
  ...
  UHELP:VECTOR1;      { help vector for the shuffle subroutine    }
  U:VECTOR2;          { unshuffled random numbers                 }
  X:VECTOR3;          { Xn                                        }

PROCEDURE SETBOXESAPARATOSCR;

FUNCTION MODUS(XN:LONGINT):LONGINT;
VAR
 XX:LONGINT;
BEGIN
 XX:=A*XN+B;
 WHILE XX>=C DO XX:=XX-C;
 MODUS:=XX;
END;{ MODUS }

{ Main program : }
  ...
  SETBOXESAPARATOSCR; { draws boxes, sets scaling parameters }

{ Generation of the random numbers and output of the pixels on the screen : }

  QQ:=C; { converts into the real type }
  X[0]:=X0;
  I:=0;
  REPEAT
   I:=I+1;  { index of the random pair }
   X[2*I-1]:=MODUS(X[2*I-2]);X[2*I]:=MODUS(X[2*I-1]);
   Q1:=X[2*I-1];Q2:=X[2*I]; { converts into the real type          }
   U[2*I-1]:=Q1/QQ;U[2*I]:=Q2/QQ; { random numbers using the generator }
   PUTPIXEL(TRUNC(IX+U[2*I-1]*LX),TRUNC(IY+LY*(1.0-U[2*I])),MAXCOLOR);
  UNTIL I=N;

{ Shuffle subroutine and output of the pixels on the screen : }

  XX:=X[2*N]; { uses the last number generated                    }
  FOR I:=1 TO FREQUENCY DO { initializes the shuffle vector }
   BEGIN
    XX:=MODUS(XX);Q1:=XX;UHELP[I]:=Q1/QQ;
   END;
  I:=0;
  REPEAT
   I:=I+1; { index of the random pair }
   XX:=MODUS(XX); { picking out one random number of the shuffle vector    }
   U0:=XX/C;I1:=TRUNC(U0*FREQUENCY+1.0); { index of the first number picked }
   US1:=UHELP[I1];    { picks the random number    }
   UHELP[I1]:=U[2*I-1]; { prepares the shuffle vector }
   XX:=MODUS(XX); { picking out one random number of the shuffle vector    }
   U0:=XX/C;I2:=TRUNC(U0*FREQUENCY+1.0); { index of the second number picked }
   UHELP[I2]:=U[2*I]; { prepares the shuffle vector }
   US2:=UHELP[I2];    { picks the random number    }
   PUTPIXEL(TRUNC(5*IX+US1*LX),TRUNC(IY+LY*(1.0-US2)),MAXCOLOR);
  UNTIL I=N;
  ...
```

Notes 1.5.4

(i) The required random numbers are all generated first and stored as a vector. The unshuffled figure is then plotted and then the shuffling routine is applied with this vector as its input.

(ii) Experiments with different choices of the length FREQUENCY of the shuffling string underline its crucial role in determining the effectiveness and efficiency of the shuffling procedure.

(iii) Easy modifications to the program allow other generators, such as that in the PC, to be tested and coloured pictures for enhanced contrast if a colour monitor is available.

The generator in Problem 1.5.4 was chosen to highlight the effects of shuffling. We now test the random number generator installed in the PC.

Exercise 1.5.1 (PC-Exercise 1.9.6) *Plot 10^3 points in the unit square as in Problem 1.5.4 using the $U(0,1)$ random number generator on your PC. Add a shuffling procedure to the generator and repeat the above plots.*

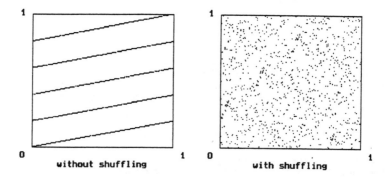

Figure 1.5.2   Linear congruential generator $a = 1229$, $b = 1$, $c = 2048$
and $x_0 = 0$ without and with shuffling.

If the pseudo random generator on your PC is not too bad you should obtain a fairly evenly filled unit square both without and with shuffling.

The preceding test is useful in eliminating glaringly biased generators, but is no guarantee of an unbiased generator. For example, the *RANDU* generator $X_{n+1} = 65,539 X_n \pmod{2^{31}}$ appears relatively unbiased, but successive triples (X_n, X_{n+1}, X_{n+2}) satisfy $X_{n+2} = (6X_n - 9X_{n+1}) \pmod{2^{31}}$. This relationship was however discovered long after the *RANDU* generator was introduced and lead to its demise. There are many other tests for independence, including the *runs test* in which the number and lengths of runs of successively increasing numbers are analysed statistically, but their validity is still unclear.

While independence is not guaranteed by pseudo-random generators, they have the significant advantage over generators based on physical noise sources in that they are reproducible. In general they are fairly successful at mimicking the salient properties of truly independent random sequences on digital computers and in applications the negative properties of such sequences often play no role. Thus, while far from perfect, pseudo-random number generators are often adequate for the task at hand. We shall assume in our simulations that they do have the asserted independence and distributions.

1.6 Markov Chains as Basic Stochastic Processes

A sequence of random variables $X_1, X_2, \ldots, X_n, \ldots$ often describes the evolution of a probabilistic system over discrete instants of time $t_1 < t_2 < \cdots < t_n < \cdots$. We then say that it is a discrete time stochastic process. Stochastic processes can also be defined for all time instants in a bounded interval such as $[0, 1]$ or in an unbounded interval such as $[0, \infty)$, in which case we call them *continuous time* stochastic processes. They form a family of random variables.

We shall denote the time set under consideration by T and assume that there is a common underlying probability space $(\Omega, \mathcal{A}, P)$. A stochastic process $X = \{X(t), t \in T\}$ is thus a function $X : T \times \Omega \to \Re$ of two variables, where $X(t) = X(t, \cdot)$ is a random variable for each $t \in T$ and $X(\cdot, \omega) : T \to \Re$ is a *realization*, *sample path* or *trajectory* of the stochastic process for each $\omega \in \Omega$.

For both continuous and discrete time sets T it is useful to distinguish various classes of stochastic processes. Markov chains are a very useful example of stochastic processes with specific temporal relationships.

A. Discrete Time Markov Chains

In a deterministic system governed by a first-order difference equation only the present value x_{i_n} of X_n is needed to determine the future value of X_{n+1}; the past values of $X_1, X_2, \ldots, X_{n-1}$ are involved only
indirectly in that they determine the value of X_n. This is just the common law of causality and there is a stochastic analogue called the *Markov property*, expressed by conditional probabilities

$$(6.1) \qquad P(X_{n+1} = x_j | X_n = x_{i_n})$$
$$= \; P(X_{n+1} = x_j | X_1 = x_{i_1}, X_2 = x_{i_2}, \ldots, X_n = x_{i_n})$$

for all possible $x_j, x_{i_1}, x_{i_2}, \ldots, x_{i_n}$ in a given state space $\mathcal{X}$ and all $n = 1, 2, 3, \ldots$. A sequence of discrete valued random variables with this property is an example of a *Markov chain*, in particular, a *discrete time Markov chain*.

For a discrete time Markov chain with a finite number of states $\mathcal{X} = \{x_1, x_2, \ldots, x_N\}$, we can define an $N \times N$ *transition matrix* $P(n) = [p^{i,j}(n)]$ componentwise by

$$(6.2) \qquad\qquad p^{i,j}(n) = P(X_{n+1} = x_j | X_n = x_i)$$

for $i, j = 1, 2, \ldots, N$ and $n = 1, 2, 3, \ldots$. Obviously $0 \leq p^{i,j}(n) \leq 1$ and, since X_{n+1} can only attain states in $\mathcal{X}$,

(6.3)
$$\sum_{j=1}^{n} p^{i,j}(n) = 1$$

for each $i = 1, 2, \ldots, N$ and $n = 1, 2, 3, \ldots$. So we call the components (6.2) the *transition probabilities* of the Markov chain at time n.

A *probability vector* on $\mathcal{X}$ is a row vector $p = (p_1, p_2, \ldots, p_N) \in \Re^N$ with $0 \leq p_i \leq 1$ for $i = 1, 2, \ldots, N$ and $\sum_{i=1}^{N} p_i = 1$. Thus if $p(n)$ is the probability vector corresponding to the random variable X_n, that is if $p_i(n) = P(X_n = x_i)$ for $i = 1, 2, \ldots, N$, then the probability vector $p(n+1)$ corresponding to X_{n+1} is given by

(6.4)
$$p(n+1) = p(n)P(n),$$

which involves a multiplication of a vector and a matrix. If we know the initial probability vector $p(1)$, then by applying (6.4) recursively we have

(6.5)
$$p(n) = p(1)P(1)P(2) \cdots P(n-1)$$

for $n = 2, 3, \ldots$. When the transition matrices are all the same, that is $P(n) \equiv P$ for $n = 1, 2, 3, \ldots$, we say that the Markov chain is *homogeneous*, in which case (6.5) can be written as

(6.6)
$$p(n + k - 1) = p(k)P^{n-1}$$

for any $k = 1, 2, 3, \ldots$ and $n = 2, 3, 4 \ldots$.

Consider an isolated city with constant population living in two districts E and W and suppose that in any given year the inhabitants of E move independently of each other with probability a to W and those of W tend to move with probability b to E. Taking E as the first state and W as the second, the matrix

(6.7)
$$\begin{bmatrix} 1-a & a \\ b & 1-b \end{bmatrix}$$

is the transition matrix of a homogeneous Markov chain for the probability distribution $p = (p_E, p_W)$ that a person lives in one of the two districts. If the probability vector of the person is initially $p(1)$, then by (6.6) after n years it is $p(n+1) = p(1)P^n$.

Problem 1.6.1 (PC-Exercise 1.6.5) *Let $a = 0.1$ and $b = 0.01$ in the Markov chain described above and consider a person originally living in district E of the city. Use a two-point random number generator to simulate step by step this person's yearly district of residence. Assuming extraordinary longevity, say 10^3 years, and using 10^2 sample paths, estimate the probabilities of this person's residing in districts E and W. Repeat the calculations for a person originally living in district W.*

Program 1.6.1   Discrete time Markov chain

```
   ...
CONST
 N=1000; { number of transitions                                    }
 M=100;  { sample size                                              }
 A=0.1;  { parameter a in the transition matrix                     }
 B=0.01; { parameter b in the transition matrix                     }
 P1=1.0; { initial probability to reside in E for the first simulation }
 P2=1.0; { initial probability to reside in W for the second simulation }
   ...
VAR
   ...
 PE,PW:REAL;    { probabilities to reside in E and W          }
 P:MATRIX;      { transition matrix                           }
 XT:REAL;       { value of the trajectory of the Markov chain }

PROCEDURE SETTABLETOSCR;
PROCEDURE SETDATATOSCR(COLUMN:INTEGER;DATA:REAL);

{ Main program : }
   ...
 P[1,1]:=1.0-A;P[1,2]:=A; { transition probabilities }
 P[2,1]:=B;P[2,2]:=1.0-B;
 K:=0;
 REPEAT
  K:=K+1; { simulation number }
  PE:=0.0;PW:=0.0;
  L:=0;
  REPEAT
   L:=L+1; { sample index }
   CASE K OF
     1 : U:=P1;
     2 : U:=1.0-P2;
    END;
   GENERO1(U,0,1,XT); { generates the initial state }
   I:=0;
   REPEAT
    I:=I+1; { number of transitions }
    IF TRUNC(XT)=0 THEN GENERO1(P[1,1],0,1,XT)
     ELSE GENERO1(P[2,1],0,1,XT); { generates the current state }
   UNTIL I=N;
   PE:=PE+1-XT;PW:=PW+XT;
  UNTIL L=M;
  PE:=PE/M;PW:=PW/M;
  IF K=1 THEN BEGIN CLEARDEVICE;SETTABLETOSCR; END;
  SETDATATOSCR(2,PE);SETDATATOSCR(3,PW);
  U:=PE;
  PE:=P[1,1]*U+P[2,1]*PW; { one further transition }
  PW:=P[1,2]*U+P[2,2]*PW;
  SETDATATOSCR(4,PE);SETDATATOSCR(5,PW);
 UNTIL K=2;
   ...
```

Notes 1.6.1

(i) Only the core of the program is listed here. When the probabilities after $N = 10^3$ transitions have been estimated from $M = 100$ samples, the probability vector is post-multiplied by the transition matrix P to verify, approximately at least, if it satisfies the identity (6.8) below.

(ii) Of course, better results will be possible for larger values of M and N, but will require considerably more computational time.

(iii) The procedure SETTABLETOSCR prepares a table on the screen, into which the results are inserted by the procedure SETDATATOSCR as they are calculated.

We call a probability vector $\bar{p}$ which satisfies the vector equation

$$(6.8) \qquad\qquad \bar{p} = \bar{p}\,P$$

a *stationary probability vector* for the homogeneous Markov chain. For example

$$\bar{p} = \left(\frac{b}{a+b}, \frac{a}{a+b} \right)$$

is a stationary probability vector for the Markov chain described by (6.7). Markov chains with a unique stationary probability vector $\bar{p}$ often have an important property called *ergodicity* which relates the long-term time averages of its realizations to the spatial averaging with respect to the stationary distribution. For any bounded function $f : \mathcal{X} \to \Re$ the time average of the values $f(X_t)$ taken by a sequence of random variables $X_1, X_2, \ldots, X_t, \ldots$ generated by the Markov chain is given by

$$(6.9) \qquad\qquad \frac{1}{T} \sum_{t=1}^{T} f(X_t)$$

We say that the Markov chain is *ergodic* if for every initial X_1 the limits of the time averages (6.9) exist and are equal to the average of f over $\mathcal{X}$ with respect to the stationary probability vector $\bar{p}$, that is if

$$(6.10) \qquad\qquad \lim_{T \to \infty} \frac{1}{T} \sum_{t=1}^{T} f(X_t) = \sum_{i=1}^{N} f(x_i)\bar{p}_i,$$

e.g. with the convergence in distribution.

Problem 1.6.2 (PC-Exercise 1.6.7) *Verify (6.10) numerically with $T = 10^3$ for the Markov chain used in Problem 1.6.1. Consider functions $f : \{E, W\} \to \Re$ with: (i) $f(E) = 0$, $f(W) = 1$; (ii) $f(E) = 1$, $f(W) = 0$; (iii) $f(E) = 1$, $f(W) = -1$; (iv) $f(E) = -1$, $f(W) = 1$. In each case try a variety of initial distributions $p(1) = (p_E(1), p_W(1)) = (p, 1 - p)$ for X_1, say, with $p = 0, 0.1, 0.2, \ldots, 0.9, 1$. Also try $p = 1/11 = 0.09090909\ldots$.*

Program 1.6.2   Ergodic Markov chain

```
   ...
CONST
 T=1000; { number of years = number of transitions }
 A=0.1;  { parameter a of the transition matrix   }
 B=0.01; { parameter b of the transition matrix   }
   ...
VAR
   ...
 LIMFXT:REAL;      { estimate of the long-time average          }
 PE,PW:REAL;       { probabilities of person's residing in E and W }
 XT:REAL;          { current realization of the Markov chain    }
 PO:VECTOR;        { initial probabilities                      }
 P:MATRIX;         { transition matrix                          }
 FE,FW:VECTOR1;    { function values of f used for the averaging }

PROCEDURE INITDATAAPRINT;
PROCEDURE SETDATATOSCR;

{ Main program : }
   ...
 P[1,1]:=1-A;P[1,2]:=A; { initialization of transition probabilities }
 P[2,1]:=B;P[2,2]:=1-B;
 L:=0;
 REPEAT
  L:=L+1; { index of the function values of f(X(t)) used }
  INITDATAAPRINT; { generates the values of f(X(t)) and prints out a table }
  K:=0;
  REPEAT
   K:=K+1; { index of the initial probabilities used }
   PE:=PO[K];PW:=1-PE; { sets the initial probabilities }
   GENERO1(PE,0,1,XT); { generates the realization of the Markov chain }
   I:=0;
   LIMFXT:=0.0; { initializes the estimate of the long-term time average }
   REPEAT
    I:=I+1; { counter of the transitions }
    LIMFXT:=LIMFXT+(1-XT)*FE[L]+XT*FW[L]; { corrects the long-term average }
    IF XT=0.0 THEN GENERO1(P[1,1],0,1,XT) { generates the next transition }
     ELSE GENERO1(P[2,1],0,1,XT);
   UNTIL I=T;
   LIMFXT:=LIMFXT/T;
   SETDATATOSCR; { prints out the current estimate of the long-term average }
  UNTIL K=12;
   ...
 UNTIL L=4;
   ...
```

Notes 1.6.2

(i)   The procedure INITDATAAPRINT selects the E and W values of the Lth function from the Lth components of the two 4-dimensional vectors FE and FW, respectively, and prepares a table on the screen.

(ii)   The calculated spatial and time averages in (6.10), with $T = 10^3$, for each of the twelve initial values are displayed in the table by the procedure SETDATATOSCR.

(iii) When the <ENTER> key is pressed after a table has been displayed the program will proceed to the next function. Pressing the <ESC> key terminates the program after the calculations for the twelve initial values for the current function.

B. Continuous Time Markov Chains

We can handle *continuous time Markov chains* in a similar way. Let $X(t)$ be distributed over a finite state space $\mathcal{X} = \{x_1, x_2, \ldots, x_N\}$ according to an N-dimensional probability vector $p(t)$ for each $t \geq 0$. In this context the Markov property (6.1) takes the form

$$(6.11) \qquad P\left(X(t_1) = x_j | X(s_1) = x_{i_1}, \ldots, X(s_n) = x_{i_n}, X(t_0) = x_i\right)$$
$$= P\left(X(t_1) = x_j | X(t_0) = x_i\right)$$

for all $0 \leq s_1 \leq s_2 \leq \cdots \leq s_n < t_0 \leq t_1$ and all $x_i, x_j, x_{i_1}, x_{i_2}, \ldots x_{i_n} \in \mathcal{X}$ where $n = 1, 2, 3, \ldots$. For each $0 \leq t_0 \leq t_1$ we can define an $N \times N$ transition matrix $P(t_0; t_1) = [p^{i,j}(t_0; t_1)]$ componentwise by

$$p^{i,j}(t_0; t_1) = P\left(X(t_1) = x_j | X(t_0) = x_i\right)$$

for $i, j = 1, 2, \ldots, N$. Clearly $P(t_0; t_0) = I$ and the probability vectors $p(t_0)$ and $p(t_1)$ are related by

$$p(t_1) = p(t_0)P(t_0; t_1).$$

For $t_0 \leq t_1 \leq t_2$ we also have $p(t_2) = p(t_0)P(t_0; t_2)$ and hence

$$p(t_2) = p(t_1)P(t_1; t_2) = p(t_0)P(t_0; t_1)P(t_1; t_2)$$

for any probability vector $p(t_0)$ from which we can conclude that

$$(6.12) \qquad P(t_0; t_2) = P(t_0; t_1)P(t_1; t_2).$$

When the transition matrices $P(t_0; t_1)$ depend only on the time difference $t_1 - t_0$, that is $P(t_0; t_1) = P(0; t_1 - t_0)$ for all $0 \leq t_0 \leq t_1$, we say that the continuous time Markov chain is *homogeneous* and write $P(t)$ for $P(0; t)$. Then (6.12) reduces to
$$(6.13) \qquad P(s + t) = P(s)P(t) = P(t)P(s)$$
for all $s, t \geq 0$.

There exists an $N \times N$ *intensity matrix* $A = (a^{i,j})$ with components

$$a^{i,j} = \begin{cases} \lim_{t \to 0} \dfrac{p^{i,j}(t)}{t} & : \quad i \neq j \\[2mm] \lim_{t \to 0} \dfrac{p^{i,i}(t) - 1}{t} & : \quad i = j \end{cases}$$

which, together with the initial probability vector $p(0)$. characterizes completely the homogeneous continuous time Markov chain. Moreover, the *waiting time* of a homogeneous continuous time Markov chain, that is the time between transitions from a state x_i to any other state, is exponentially distributed with intensity parameter $\lambda_i = \sum_{j \neq i} a^{i,j}$.

For example, the continuous time Markov chain X taking values $+1$ and -1 with probabilities $(p^+(t), p^-(t)) = p(t)$ and switching according to the homogeneous transition matrix

$$P(t) = \begin{bmatrix} (1 + e^{-t})/2 & (1 - e^{-t})/2 \\ (1 - e^{-t})/2 & (1 + e^{-t})/2 \end{bmatrix}$$

for $t \geq 0$ has intensity matrix

$$A = \begin{bmatrix} -0.5 & 0.5 \\ 0.5 & -0.5 \end{bmatrix}.$$

For the initial probability vector $p(0) = \bar{p} = (0.5, 0.5)$ we find that $\bar{p}P(t) = p(t) \equiv \bar{p}$ for all $t \geq 0$, so $\bar{p}$ is a stationary probability vector for this Markov chain. The stochastic process corresponding to this stationary probability vector is known as random telegraphic noise.

Problem 1.6.3 (PC-Exercise 1.6.9) *Use exponentially distributed waiting times to simulate the telegraphic noise in the above example on the time interval $[0, T]$ with $T = 10$. From a sample of 100 simulations calculate the relative frequency of being in the state $+1$ at time T.*

Program 1.6.3 Telegraphic noise

```
...
CONST
  N=100;      { number of simulations                        }
  T=10;       { endpoint of the time interval starting at zero }
  A11=-0.5;   { intensity parameter                           }
  A12=+0.5;   { intensity parameter                           }
  A21=+0.5;   { intensity parameter                           }
  A22=-0.5;   { intensity parameter                           }
  PROB=0.5;   { initial probability of starting in the state +1 }
  ...
VAR
  ...
  FREQUENCY:REAL; { frequency of being in the state +1 at time T }
  DT:REAL;        { exponentially distributed waiting time      }
  TI:REAL;        { time of the transition to the other state   }
  XT:REAL;        { current state                               }
  PO:VECTOR;      { initial probabilities                       }

PROCEDURE SETTEXTTOSCR;

{ Main program : }
  ...
  PO[1]:=PROB;PO[2]:=1.0-PROB; { sets the initial probabilities }
  I:=0;
  FREQUENCY:=0.0;
```

```
REPEAT
 I:=I+1; { counts the number of simulations }
 GENERO1(PO[1],1,-1,XT); { generates the starting state }
 TI:=0.0; { current time }
 REPEAT
  IF XT=1.0 THEN GENERO2(A12,DT) ELSE GENERO2(A21,DT); { exp. waiting time }
  TI:=TI+DT; { adds the current waiting time of the corresponding state    }
  IF (T-TI)>=0 THEN XT:=-XT; { transition to the new state                 }
 UNTIL TI>=T;
 IF XT=1.0 THEN FREQUENCY:=FREQUENCY+1.0; { corrects the frequency }
 UNTIL I=N;
 FREQUENCY:=FREQUENCY/N; { relative frequency }
 ...
```

Notes 1.6.3   The procedure SETTEXTTOSCR clears the screen and prints the calculated relative frequency as a string with the value FREQUENCY on the screen.

We say that a continuous time Markov chain is called *ergodic* if for each bounded f

$$(6.14) \qquad \lim_{T \to \infty} \frac{1}{T} \int_0^T f(X(t))\, dt = \sum_{i=1}^{N} f(x_i)\bar{p}_i,$$

that is with the time average (6.9) in (6.10) now written in integral form, where the convergence could be taken in distribution. We note that the random telegraphic noise process is an ergodic continuous time Markov chain.

1.7   Wiener Processes

The Wiener process was proposed by Norbert Wiener as a mathematical description of Brownian motion, the erratic motion of a grain of pollen on a water surface due to its being continually bombarded by water molecules. The Wiener process is sometimes called *Brownian motion*, but we will use separate terminology to distinguish between the mathematical and physical processes.

A. Approximate Wiener Paths

We define a *standard Wiener process* $W = \{W(t), t \geq 0\}$ to be a continuous Gaussian process with independent increments such that

$$(7.1) \quad W(0) = 0, \text{ w.p.1}, \quad E(W(t)) = 0, \quad \text{Var}(W(t) - W(s)) = t - s$$

for all $0 \leq s \leq t$. According to this definition, $W(t) - W(s)$ is $N(0; t-s)$ Gaussian distributed for $0 \leq s < t$ and the increments $W(t_2) - W(t_1)$ and $W(t_4) - W(t_3)$ are independent for all $0 \leq t_1 < t_2 \leq t_3 < t_4$.

Problem 1.7.1 (PC-Exercise 2.4.4)   *Generate and plot the linearly interpolated trajectory of a Wiener process on $[0,1]$ at the time instants $t_k = k2^{-9}$ for*

$k = 0, 1, \ldots, 2^9$ *using independent Gaussian increments* $W(t_{k+1}) - W(t_k) \sim N(0; 2^{-9})$.

Program 1.7.1   Linearly interpolated Wiener process

```
USES CRT,DOS,GRAPH,INIT,SETSCR,SERVICE,RANDNUMB,AAGRAPHS;

CONST
  TO=0.0;                 { left end point                        }
  T=1.0;                  { right end point                       }
  DELTA=(T-TO)/NUMINV; { minimum step size w.r.t. screen resolution }
  X0=0.0;                 { initial value of the trajectory       }
  ...
{ Main program : }
  ...
 SQDELTA:=SQRT(DELTA);
 U2:=0.0;WT:=0.0;
 TK:=TO-DELTA;
 K:=-1;
 REPEAT
  K:=K+1;                { counter                }
  TK:=TK+DELTA;          { time                   }
  ABSCISSA[K]:=TK; { values of the x-axis }
  IF K MOD 2 =1 THEN GENERATE(U1,U2)
   ELSE U1:=U2; { uses Polar Marsaglia }
  WT:=WT+U1*SQDELTA; { current value of the Wiener trajectory }
  XT[K]:=WT;           { values of the y-axis                  }
 UNTIL K=NUMINV;
  ...
```

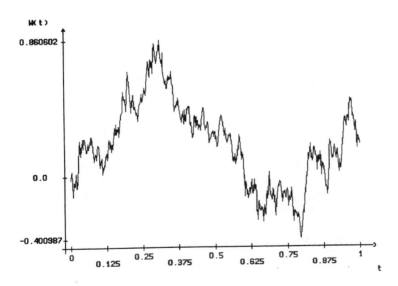

Figure 1.7.1   A linearly interpolated Wiener sample path.

Notes 1.7.1

(i) Only one trajectory can be plotted on the screen at a time with the graphics routine GRAPH111 from AAGRAPHS used in this program. It automatically adapts the vertical axis scale to the data available, thus allowing better visualization of the fine structure of the trajectory.

(ii) The choice of step size here corresponds to the maximum resolution on the screen for a VGA or EGA graphics card, with the data vector having dimension NUMINV+1 = 512 +1; it should be changed correspondingly if a CGA card is used.

Exercise 1.7.1   *Repeat Problem 1.7.1 using time instants* $t_k = k \, 2^{-7}$ *for* $k = 0, 1, \dots, 2^7$.

B. Random Walks

We can approximate a standard Wiener process in distribution on any finite time interval by means of a scaled random walk. For example, we can subdivide the unit interval $[0, 1]$ into N subintervals

$$0 = t_0^{(N)} < t_1^{(N)} < \cdots < t_k^{(N)} < \cdots < t_N^{(N)} = 1$$

of equal length $\Delta t = 1/N$ and construct a stepwise continuous random walk $S_N(t)$ by taking independent, equally probable steps of length $\pm\sqrt{\Delta t}$ at the end of each subinterval. We start with independent two-point random variables X_k taking values ± 1 with equal probability and define

$$(7.2) \qquad\qquad S_N(t_k^{(N)}) = (X_1 + X_2 + \cdots + X_k)\sqrt{\Delta t},$$

where we interpolate linearly by

$$(7.3) \qquad S_N(t) = S_N(t_k^{(N)}) + \frac{t - t_k^{(N)}}{t_{k+1}^{(N)} - t_k^{(N)}}\left(S_N(t_{k+1}^{(N)}) - S_N(t_k^{(N)})\right)$$

on $t_k^{(N)} \le t < t_{k+1}^{(N)}$ for $k = 0, 1, \dots, N-1$, where $S_N(0) = 0$.

Problem 1.7.2 (PC-Exercise 1.5.7)   *Form a linearly interpolated random walk* $S_{100}(t)$ *on* $[0, 1]$ *using the two-point random number generator and plot* $S_{100}(t)$ *against t. Repeat this for other sequences corresponding to different initial seeds and compare the plotted paths.*

Program 1.7.2   Paths of a random walk

```
USES CRT,DOS,GRAPH,INIT,SETSCR,SERVICE,RANDNUMB;

CONST
  N=100;      { number of random variables used             }
  P=0.5;      { probability that Xk = X1                     }
  X1=-1.0;    { first value of the two-point distributed Xk  }
```

```
X2=+1.0;      { second value of the two-point distributed Xk }
ABSCMIN=0.0;  { left end point                              }
ABSCMAX=+1.0; { right end point                             }
ORDMIN=-2.0;  { minimum of the ordinate                     }
ORDMAX=+2.0;  { maximum of the ordinate                     }
ORDPOINT=0.0; { significant ordinate point                  }
   ...
VAR
   ...
DELTA:REAL;          { time step size                        }
SQDELTA:REAL;        { sqrt of the time step size            }
TK:REAL;             { subinterval end                       }
XK:REAL;             { two-point distributed random number   }
SK:REAL;             { value of the random walk S(N)(tk+1)   }
SNT:VECTOR;          { values of the trajectory              }
ABSCISSA:VECTOR;     { subinterval points                    }

PROCEDURE COORDSYS(CY,CX:STRING);
PROCEDURE PLOTGRAPH(LTN,NO,NN:INTEGER;ORDINATE,ABSCISSA:VECTOR);

{ Main program : }
   ...
{ Generation of the random walk : }

 DELTA:=(ABSCMAX-ABSCMIN)/N;
 SQDELTA:=SQRT(DELTA);
 SK:=0.0;SNT[0]:=0.0;
 TK:=ABSCMIN;ABSCISSA[0]:=TK;
 K:=0;
 REPEAT
  K:=K+1;
  TK:=TK+DELTA;
  GENERO1(P,X1,X2,XK);
  SK:=SK+XK;
  SNT[K]:=SK*SQDELTA;
  ABSCISSA[K]:=TK;
 UNTIL K=N;

{ Printout : }

 CLEARDEVICE;
 COORDSYS('S('+CHCR(N)+')(t)','t'); { draws the coordinate system }
 PLOTGRAPH(1,0,N,SNT,ABSCISSA);     { plots the trajectory        }
   ...
```

Notes 1.7.2

(i) As before, the procedure COORDSYS prepares the screen for plotting the sample path, which is carried out by the procedure PLOTGRAPH here. A unified version of COORDSYS, differing slightly from earlier versions, is used throughout this section with the x-axis in the middle of the screen.

(ii) In PLOTGRAPH the line thickness is denoted by LTN, while N0 and NN denote the lower and upper component indices used from the calculated trajectory vector SNT.

(iii) The vertical axis scaling is now prescribed and not adapted to fit the data. Consequently some trajectory points may lie outside the prepared screen area, but several trajectories can now be plotted together.

It can be shown by the Central Limit Theorem that S_N converges in distribution as $N \to \infty$ to a process with independent increments satisfying conditions (7.1), that is to a standard Wiener process.

Problem 1.7.3 (PC-Exercise 1.8.2) *Generate and plot linearly interpolated sample paths of the process $S_N(t)$, defined on $0 \le t \le 1$ by (7.3), for increasing values of $N = 10, 20, \ldots, 100$. To compare approximations of the same sample path with, say, $N = 50$ and $N = 100$, generate 100 random numbers $X_1, X_2, \ldots, X_{100}$ and use them to determine a sample path of $S_{100}(t)$. Then add successive pairs to form $\tilde{X}_1 = X_1 + X_2$, $\tilde{X}_2 = X_3 + X_4$, ..., $\tilde{X}_{50} = X_{99} + X_{100}$ and use $\tilde{X}_1, \tilde{X}_2, \ldots, \tilde{X}_{50}$ to determine the corresponding sample path of $S_{50}(t)$.*

Program 1.7.3   Random walks

```
   ...
{ Calculation of S(50)(t) from S(100)(t) : }

DELTA:=2.0*(ABSCMAX-ABSCMIN)/NUMBER;SQDELTA:=SQRT(DELTA);
SK:=0.0;S50T[0]:=SK;TK:=ABSCMIN;ABSCISSA[0]:=TK;
K:=0;
REPEAT { calculates S(50)(t) w.r.t. the last 100 random numbers Xk }
 K:=K+1;
 TK:=TK+DELTA;
 SK:=SK+X[2*K-1]+X[2*K];
 S50T[K]:=SK*SQDELTA;
 ABSCISSA[K]:=TK;
UNTIL K=TRUNC(NUMBER/2.0+0.1);
   ...
```

Notes 1.7.3
(i) The full program on the diskette first calculates different sample paths $S_N(t)$ for $N = 10, 20, \ldots, 100$ as in Program 1.7.2 and displays each on a separate screen.

(ii) Press any key to proceed to the next case and <ESC> to terminate the program at any time.

(iii) In the part of the program listed above, a sample path of $S_{50}(t)$ corresponding to that of $S_{100}(t)$ is reconstructed from the data for $S_{100}(t)$ and both are plotted together with different line thickness.

(iv) See the notes for Problem 1.7.2 regarding the graphics procedures used.

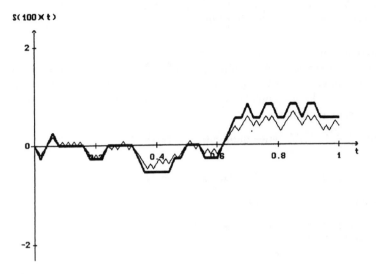

Figure 1.7.2 Sample paths of the random walk $S_{50}(t)$ and $S_{100}(t)$.

C. Some Properties of the Wiener Process

From the definition (7.1) of the Wiener process we have

$$\text{Var}(W(t)) = t$$

so the variance of the Wiener process grows without bound as time increases while the mean always remains zero. Consequently typical sample paths must attain larger and larger values, both positive and negative, as time increases. By a reinforcement of the Law of Large Numbers (see (1.4.8)) we find that

$$(7.4) \qquad \lim_{t \to \infty} \frac{W(t)}{t} = 0$$

in the mean square sense, which provides an asymptotic estimate for the growth rate of the Wiener process. A better impression of the rate of growth can be obtained from the law of the iterated logarithm which we will not discuss here.

Henceforth we shall usually write $W(t)$ as W_t which is conventional in stochastic calculus. The sample paths of the Wiener process are continuous, but not differentiable as the following problem indicates.

Problem 1.7.4 (PC-Exercise 1.8.6) *As in Problem 1.7.3 generate linearly interpolated random walks $S_N(t)$ on $[0,1]$ for increasing N to approximate the same sample path in the limiting process. For fixed N, say 50, evaluate the ratios*

$$\frac{S_N(h+0.5) - S_N(0.5)}{h}$$

*for successively smaller values of h and plot the linearly interpolated ratios
against h. Repeat this for larger values of N, say 100. What do these results
suggest about the differentiability of the limiting sample path at t = 0.5?*

The difference quotients in the above problem do not appear to converge. This
is because the derivative of a sample path of a Wiener process does not exist
at any fixed time. This can be proved theoretically and is a consequence of the
independence of the increments and their behaving like $\sqrt{\Delta t}$ rather than Δt
over small time increments.

Program 1.7.4   Difference quotients of a random walk

```
   ...
ABSCMAX:=0.5;ABSCMIN:=0.0;  { scaling parameters for the abscissa }

{ Calculation of the difference quotients : }

NUMBER:=TRUNC(N/2+0.1); { data index of 0.5 }
SK:=0.0;RATIOS[0]:=0.0;
H:=ABSCMIN;ABSCISSA[0]:=H;
K:=0;
REPEAT
 K:=K+1;
 H:=H+DELTA;
 RATIOS[K]:=(SNT[NUMBER+K]-SNT[NUMBER])/H;
 ABSCISSA[K]:=H;
UNTIL K=NUMBER;
   ...
```

Notes 1.7.4

(i)   The generation of the random walks and graphics is exactly the same as in
      Program 1.7.3, so only the part of the program calculating the difference
      quotient is listed above.

(ii) Attention is restricted to the right hand limit, that is to $h > 0$, with obvious
      changes for the left hand limit.

We denote by

$$\mathcal{A}_t = \sigma\{W_s, 0 \le s \le t\}$$

the σ-algebra of events generated by the values of the Wiener process until time
t. σ-algebras are important mathematical objects to characterize the available
information at a certain time, here about the Wiener process trajectory until
time t. It is not difficult to show that a Wiener process is a continuous time
Markov process with Gaussian transition distributions. It is also a *martingale*,
that is, the conditional expectation relation

$$E\left(W_t - W_s | \mathcal{A}_s\right) = 0$$

holds for all $0 \le s \le t < \infty$. Martingales satisfy powerful limit results and
inequalities which have significant practical and theoretical uses.

D. Karhunen-Loève Expansion

A standard Wiener process $W = \{W_t, t \geq 0\}$ consists of uncountably many random variables. It is, however, possible to represent it on any bounded interval $0 \leq t \leq T$ in terms of only countably many independent Gaussian random variables. This representation is similar to a Fourier series with the random variables as its coefficients and is called the *Karhunen-Loève expansion* of the process. In the sense of mean-square convergence we have

$$(7.5) \qquad W_t(\omega) = \sum_{n=0}^{\infty} Z_n(\omega)\phi_n(t) \quad \text{for} \quad 0 \leq t \leq T$$

where the $Z_0, Z_1, \ldots, Z_n, \ldots$ are independent standard Gaussian random variables and the $\phi_0, \phi_1, \ldots, \phi_n, \ldots$ are the nonrandom functions

$$(7.6) \qquad \phi_n(t) = \frac{2\sqrt{2T}}{(2n+1)\pi} \sin\left(\frac{(2n+1)\pi t}{2T}\right)$$

for $n = 0, 1, 2, \ldots$. These time dependent functions are themselves orthogonal with respect to the integral inner product

$$(\phi_i, \phi_j) = \int_0^T \phi_i(t)\phi_j(t)\, dt,$$

which is zero if $i \neq j$ for any $i, j = 0, 1, 2, \ldots$, and satisfy

$$\sum_{n=0}^{\infty} \phi_n(s)\phi_n(t) = \min\{s, t\},$$

the covariance function of the Wiener process. The random variables $Z_0, Z_1, Z_2, \ldots$ here are determined in an almost identical way to the coefficients of a Fourier series with

$$(7.7) \qquad Z_n(\omega) = \frac{2}{T}\left(\frac{(2n+1)\pi}{2\sqrt{2T}}\right)^2 \int_0^T W_t(\omega)\phi_n(t)\, dt$$

for $n = 0, 1, 2, \ldots$ which are independent standard Gaussian random variables. The random variables (7.7) and the time functions (7.6) clearly both depend on the particular time interval $[0, T]$.

We can truncate the series (7.5) to obtain an approximation for a Wiener process.

Problem 1.7.5 (PC-Exercise 2.4.2) *Generate 50 independent standard Gaussian pseudo-random numbers for use as realizations of the first 50 random coefficients $Z_0, Z_1, \ldots, Z_{49}$ in the series expansion (7.5). Then plot the graphs of the partial sums*

$$\sum_{n=0}^{49} Z_n(\omega)\phi_n(t)$$

against t on the interval $0 \leq t \leq T$ with $T = 1$.

Program 1.7.5   Truncated Karhunen-Loève expansion


```
USES CRT,DOS,GRAPH,INIT,SETSCR,SERVICE,RANDNUMB;

CONST
  T=1.0;          { interval end                       }
  NUMBER=64;      { number of time steps               }
  DELTA=T/NUMBER; { time step size                     }
  K=50;           { number of random coefficients Zn }
  ...
PROCEDURE COORDSYS(CY,CX:STRING);
PROCEDURE PLOTGRAPH(LTN,NO,NN:INTEGER;ORDINATE,ABSCISSA:VECTOR);

FUNCTION FI(NN:INTEGER;TI:REAL):REAL;
BEGIN
  FI:=2*SQRT(2*T)/((2*NN+1)*PI)*SIN((2*NN+1)*PI*TI/(2*T));
END;{ FI }

{ Main program : }
  ...
  FOR N:=1 TO K DO { generates the values of Zn }
    IF N MOD 2 = 1 THEN GENERATE(Z[N-1],U)
      ELSE Z[N-1]:=U;
  TI:=-DELTA;
  I:=-1;
  REPEAT
    I:=I+1;
    TI:=TI+DELTA; { time }
    XT:=0.0;
    N:=-1;
    REPEAT { calculates the current value of the Karhunen-Loeve expansion }
      N:=N+1;
      XT:=XT+Z[N]*FI(N,TI);
    UNTIL N=K-1;
    KL50T[I]:=XT;
    ABSCISSA[I]:=TI;
  UNTIL I=NUMBER;
  ...
```

Notes 1.7.5

(i)   As indicated in Notes 1.7.2 a selected segment of the calculated path can be plotted by an appropriate choice of indices N0 and NN in the PLOTGRAPH routine.

(ii) Since the axis scaling is not done automatically, the prescribed scaling parameters may have to be adjusted.

(iii) The function generating routine FI(NN,TI) can be changed to allow expansions in terms of other basis functions to be considered.

(iv) The functions are evaluated at time steps of length 2^{-6} here. Experiments with smaller time steps will show increasingly smoother looking paths.

Expansions of the Wiener process can also be formed using other orthogonal systems of functions.

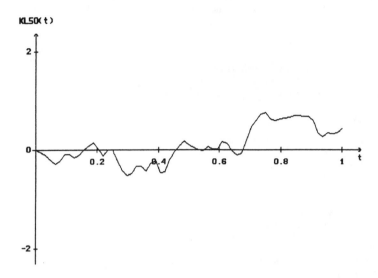

Figure 1.7.3   Approximate Wiener path by the Karhunen-Loève expansion.

E. Brownian Bridge Process

A useful modification of the Wiener process has sample paths which all pass through the same initial point x, not necessarily 0, and a given point y at a later time $t = T$. This process $B_{0,x}^{T,y}$ is defined sample pathwise for $0 \le t \le T$ by

(7.8) $$B_{0,x}^{T,y}(t,\omega) = x + W(t,\omega) - \frac{t}{T}\{W(T,\omega) - y + x\}$$

and is called a *Brownian bridge* or a *tied-down Wiener process*. It is a Gaussian process satisfying the constraints $B_{0,x}^{T,y}(0,\omega) = x$ and $B_{0,x}^{T,y}(T,\omega) = y$, so can be considered as a kind of conditional Wiener process. Since it is Gaussian it is determined uniquely by its means and covariances, which are

(7.9) $$\mu(t) = x - \frac{t}{T}(x - y) \quad \text{and} \quad C(s.t) = \min\{s,t\} - \frac{st}{T}$$

for $0 \le s, t \le T$, respectively.

Problem 1.7.6 *Generate and plot a linearly interpolated path of a Brownian bridge.*

Program 1.7.6   Brownian bridge

```
USES CRT,DOS,GRAPH,INIT,SETSCR,SERVICE,RANDNUMB,AAGRAPHS;

CONST
```

```
T=1.0;              { right end point                             }
DELTA=T/NUMINV; { minimum step size w.r.t. screen resolution }
X=0.0;              { initial value of the trajectory            }
Y=0.0;              { end value of the trajectory                }

VAR

  ...
WT:VECTOR;        { values of the Wiener trajectory            }
BXYT:VECTOR;      { values of the Brownian bridge trajectory }
ABSCISSA:VECTOR; { values of the subinterval points          }

{ Main program : }
  ...
{ Generation of the Wiener trajectory : }

SQDELTA:=SQRT(DELTA);
WT[0]:=0.0;U:=0.0;
K:=0;
REPEAT
 K:=K+1; { counter }
 IF K MOD 2 =1 THEN GENERATE(U1,U2)
   ELSE U1:=U2; { uses Polar Marsaglia }
 U:=U+U1;
 WT[K]:=U*SQDELTA; { value of the Wiener trajectory }
UNTIL K=NUMINV;

{ Generation of the Brownian bridge trajectory and the abscissa data : }

TK:=0.0-DELTA;
K:=-1;
REPEAT
 K:=K+1;          { counter }
 TK:=TK+DELTA; { time      }
 BXYT[K]:=X+WT[K]-(TK/T)*(WT[NUMINV]-Y+X); { value of the Brownian bridge }
 ABSCISSA[K]:=TK; { value of the X-axis }
UNTIL K=NUMINV;
  ...
```

Notes 1.7.6

(i)   The entire sample path of the Wiener process must be generated first since its final point is required for the calculation of the intermediate values of the corresponding sample path of the Brownian bridge process.

(ii) As in Program 1.7.1 the GRAPH111 graphics routine is used here, automatically scaling the vertical axis with respect to the sample path under consideration.

(iii) If several sample paths are required in the same picture, then the CO-ORDSYS and PLOTGRAPH routines from Programs 1.7.2 – 1.7.5 should be used.

We can also try to approximate a Brownian bridge process by a random walk fixed at the terminal time.

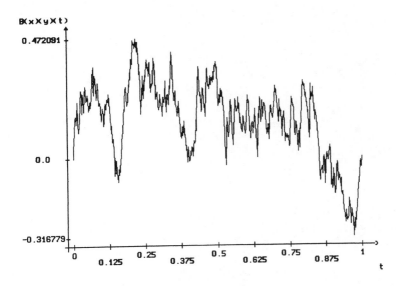

Figure 1.7.4 A path of a Brownian bridge process.

Problem 1.7.7 (PC-Exercise 1.8.8) *Modify the linearly interpolated random walks $S_N(t)$ on $[0,1]$ in Problem 1.7.3 to obtain approximations similar to the Brownian bridge $B_{0,0}^{1,0}$. For $N = 10, 20, \ldots, 100$ plot approximations to the same limiting sample path against time t for $0 \leq t \leq 1$.*

Notes 1.7.7 Program 1.7.7, which can be found on the diskette, is a simple modification to Program 1.7.3 including a modification of the Brownian bridge routine of Program 1.7.6.

Literature for Chapter 1

Karlin & Taylor (1970,1981), Chung (1975) and Shiryayev (1984) provide comprehensive introductions to probability theory. A computational approach to elementary probability and basic statistics can be found in Groeneveld (1979). Introductions and surveys to pseudo-random number generation are given in Rubinstein (1981), Ripley (1983), Morgan (1984), Marsaglia (1985) and Niederreiter (1992). Random number generation on supercomputers including lagged–Fibonacci generators is considered in Petersen (1988) and Anderson (1990). Markov chains are treated, e.g., in Chung (1975), Karlin & Taylor (1970) and Shiryayev (1984). For further information on the Wiener process and related questions see, e.g., Hida (1980), Karatzas & Shreve (1988).

Chapter 2

Stochastic Differential Equations

A brief, heuristic introduction into stochastic integration and stochastic differential equations is given in this chapter, as well a derivation of stochastic Taylor expansions.

2.1 Stochastic Integration

A. Introduction

An ordinary differential equation

$$(1.1) \qquad \dot{x} = \frac{dx}{dt} = a(t, x)$$

may be thought of as a degenerate form of a stochastic differential equation in the absence of randomness. It is therefore useful to review some of its basic properties. We could write (1.1) in the symbolic differential form

$$(1.2) \qquad dx = a(t, x)\, dt,$$

or more accurately as an integral equation

$$(1.3) \qquad x(t) = x_0 + \int_{t_0}^{t} a(s, x(s))\, ds$$

where $x(t) = x(t; x_0, t_0)$ is a solution satisfying the initial condition $x(t_0) = x_0$. Regularity assumptions, such as Lipschitz continuity, are usually made on a to ensure the existence of a unique solution $x(t; x_0, t_0)$ for each initial condition. These solutions are then related by the evolutionary property

$$(1.4) \qquad x(t; x_0, t_0) = x(t; x(s; x_0, t_0), s)$$

for all $t_0 \leq s \leq t$, which says that the future is determined completely by the present, with the past being involved only in that it determines the present. This is a deterministic version of the Markov property.

Following Einstein's explanation of observed Brownian motion during the first decade of this century, attempts were made by Langevin and others to formulate the dynamics of such motion in terms of differential equations. The resulting equations were written in the form as

$$(1.5) \qquad dX_t = a(t, X_t)\, dt + b(t, X_t)\, \xi_t\, dt$$

with a deterministic or averaged drift term as in (1.1) perturbed by a noisy, diffusive term $b(t, X_t) \xi_t$, where the ξ_t are standard Gaussian random variables for each t and $b(t, X_t)$ is a space-time dependent intensity factor. This symbolic differential was interpreted as an equation

$$(1.6) \qquad X_t(\omega) = X_{t_0}(\omega) + \int_{t_0}^{t} a(s, X_s(\omega))\, ds + \int_{t_0}^{t} b(s, X_s(\omega)) \xi_s(\omega)\, ds$$

for each sample path. When extrapolated to a limit, the observations of Brownian motion seemed to suggest that the covariance $C(t) = E(\xi_s \xi_{s+t})$ of the process ξ_t had a constant spectral density, that is with all time frequencies equally weighted in a Fourier transform of $C(t)$. Such a process became known as Gaussian white noise, particularly in the engineering literature. For the special case of (1.6) with $a \equiv 0$, $b \equiv 1$ we see that ξ_t should be the derivative of pure Brownian motion, that is the derivative of a Wiener process W_t, thus suggesting that we could write (1.6) alternatively as

$$(1.7) \qquad X_t(\omega) = X_{t_0}(\omega) + \int_{t_0}^{t} a(s, X_s(\omega))\, ds + \int_{t_0}^{t} b(s, X_s(\omega))\, dW_s(\omega).$$

The problem with this is, as mentioned in Section 1.7, that a Wiener process W_t is nowhere differentiable, so strictly speaking the white noise process ξ_t does not exist as a conventional function of t. The second integral in (1.7), thus, cannot be an ordinary integral.

B. Ito Integral

For constant $b(t, x) \equiv b$ we would expect the second integral in (1.7), however it is to be defined, to equal $b\{W_t(\omega) - W_{t_0}(\omega)\}$. This is the starting point for Ito's definition of a stochastic integral. We shall consider such an integral of a random function f over the unit time interval $0 \le t \le 1$, denoting it by $I(f)$ where

$$(1.8) \qquad I(f)(\omega) = \int_0^1 f(s, \omega)\, dW_s(\omega).$$

For a nonrandom step function $f(t, \omega) \equiv f_j$ on $t_j \le t < t_{j+1}$ for $j = 1, 2, \ldots, n$ where $0 = t_1 < t_2 < \cdots < t_{n+1} = 1$ we should obviously take, at least w.p.1,

$$(1.9) \qquad I(f)(\omega) = \sum_{j=1}^{n} f_j \left\{ W_{t_{j+1}}(\omega) - W_{t_j}(\omega) \right\}.$$

This random variable has zero mean since it is the sum of random variables with zero mean.

For random step functions appropriate measurability conditions must be imposed to ensure the nonanticipativeness of the integral with respect to the Wiener process. To be specific, suppose that $\{ \mathcal{A}_t, t \ge 0 \}$ is an increasing family of σ-algebras. $\mathcal{A}_t$ represents the information known to us until time t.

We assume that W_t is $\mathcal{A}_t$-measurable for each $t \geq 0$. Roughly speaking, the value of W_t is detectable by events given in $\mathcal{A}_t$. Then we consider a random step function $f(t, \omega) = f_j(\omega)$ on $t_j \leq t < t_{j+1}$ for $j = 1, 2, \ldots, n$ where $0 = t_1 < t_2 < \cdots < t_{n+1} = 1$ and f_j is $\mathcal{A}_{t_j}$-measurable, that is observable by events that can be detected at or before time t_j. We shall also assume that each f_j is mean-square integrable, hence $E(f_j^2) < \infty$ for $j = 1, 2, \ldots, n$. Since we have for the conditional expectation of $W_{t_{j+1}} - W_{t_j}$ under the events given in $\mathcal{A}_{t_j}$ the relation $E(W_{t_{j+1}} - W_{t_j} | \mathcal{A}_{t_j}) = 0$, w.p.1, it follows that the product $f_j\{W_{t_{j+1}} - W_{t_j}\}$ has expectation

$$E\left(f_j\left\{W_{t_{j+1}} - W_{t_j}\right\}\right) = E\left(f_j\, E\left(W_{t_{j+1}} - W_{t_j} | \mathcal{A}_{t_j}\right)\right) = 0$$

for each $j = 1, 2, \ldots, n$. Analogously to (1.9) we define the integral $I(f)$ by

(1.10) $$I(f)(\omega) = \sum_{j=1}^{n} f_j(\omega)\left\{W_{t_{j+1}}(\omega) - W_{t_j}(\omega)\right\}$$

w.p.1, which thus has mean zero. Since the jth-term in this sum is $\mathcal{A}_{t_{j+1}}$-measurable and hence $\mathcal{A}_1$-measurable, it follows that $I(f)$ is $\mathcal{A}_1$-measurable. In addition $I(f)$ is mean-square integrable with

(1.11) $$E\left(I(f)^2\right) = \sum_{j=1}^{n} E\left(f_j^2\, E\left(\left|W_{t_{j+1}} - W_{t_j}\right|^2 | \mathcal{A}_{t_j}\right)\right)$$

$$= \sum_{j=1}^{n} E\left(f_j^2\right)(t_{j+1} - t_j)$$

on account of the mean-square property of the increments $W_{t_{j+1}} - W_{t_j}$ for $j = 1, 2, \ldots, n$.

For a general integrand $f : [0, 1] \times \Omega \to \Re$ with $\mathcal{A}_t$-measurable $f(t, \cdot)$ for each t we shall define $I(f)$ as the limit of integrals $I(f^{(n)})$ of random step functions $f^{(n)}$ as above converging to f. To do this we need to specify conditions on f and determine an appropriate mode of convergence for which such an approximating sequence of step functions and limit exist. We form a partition $0 = t_1^{(n)} < t_2^{(n)} < \cdots < t_{n+1}^{(n)} = 1$ with

$$\delta^{(n)} = \max_{1 \leq j \leq n}\left\{t_{j+1}^{(n)} - t_j^{(n)}\right\} \to 0 \quad \text{as} \quad n \to \infty$$

and define a step function $f^{(n)}$ by $f^{(n)}(t, \omega) = f(t_j^{(n)}, \omega)$ on $t_j^{(n)} \leq t < t_{j+1}^{(n)}$ for the choice, where $j = 1, 2, \ldots, n$. In addition we assume that the step functions $f^{(n)}$ converge to the integrand f in an appropriate mode of convergence. Since the sample paths of a Wiener process do not have bounded variation the finite sums

(1.12) $$I\left(f^{(n)}\right)(\omega) = \sum_{j=1}^{n} f^{(n)}\left(t_j^{(n)}, \omega\right)\left\{W_{t_{j+1}^{(n)}}(\omega) - W_{t_j^{(n)}}(\omega)\right\}$$

will certainly not converge with probability one. On the other hand, a Wiener process has well-behaved mean-square properties and the right hand side of

$$(1.13) \qquad E\left(I\left(f^{(n)}\right)^2\right) = \sum_{j=1}^{n} E\left(f\left(t_j^{(n)},\cdot\right)^2\right)(t_{j+1} - t_j),$$

that is equality (1.11) for the step function $f^{(n)}$, converges to the integral $\int_0^1 E(f(s,\cdot)^2)\,ds$ for $n \to \infty$ provided $E(f(t,\cdot)^2)$ is assumed to be continuous in t. Together these suggest that we should use mean-square convergence. Now it is not hard to show that $E(|f^{(n)} - f|^2) \to 0$ as $n \to \infty$. Moreover, the mean-square limit of the $I(f^{(n)})$ exists and is unique, w.p.1. We shall denote it by $I(f)$ and call it the *Ito integral* of f on $0 \le t \le 1$. Obviously $I(f)$ is an $\mathcal{A}_1$-measurable random variable which is mean-square integrable that is

$$(1.14) \qquad E\left(I(f)^2\right) = \int_0^1 E\left(f(s,\cdot)^2\right)\,ds < \infty$$

with $E(I(f)) = 0$. We like to underline that the nonanticipativeness of the integrand was essential for the definition of the Ito integral.

The Ito integral is defined similarly on any bounded interval $[t_0, t]$, resulting in a random variable

$$(1.15) \qquad X_t(\omega) = \int_{t_0}^{t} f(s,\omega)\,dW_s(\omega),$$

which is $\mathcal{A}_t$-measurable and mean-square integrable with

$$E(X_t) = 0 \quad \text{and} \quad E\left(X_t^2\right) = \int_{t_0}^{t} E\left(f(s,\cdot)^2\right)\,ds.$$

From the independence of increments of a Wiener process on nonoverlapping time intervals in the step function sum (1.12), and in their mean-square limit, we have

$$(1.16) \qquad E\left(X_{t_2} - X_{t_1} | \mathcal{A}_{t_1}\right) = 0,$$

w.p.1, for any $t_0 \le t_1 \le t_2$, so $\{X_t, t \ge 0\}$ is a martingale analogous to what we pointed out for the Wiener process itself at the end of Subsection 1.7.C. However, the Ito integral also has the peculiar property, amongst others, that

$$(1.17) \qquad \int_0^t W_s(\omega)\,dW_s(\omega) = \frac{1}{2}W_t^2(\omega) - \frac{1}{2}t,$$

w.p.1, in contrast to

$$\int_0^t w_s\,dw_s = \frac{1}{2}w_t^2$$

of classical calculus for a differentiable function $w(t)$ with $w(0) = 0$.

We underline that in (1.12) we evaluated the integrand f at the left hand endpoint of the discretization intervals. This is crucial for the definition of the Ito integral. The choice of the evaluation point in the middle of the discretization intervals leads to the definition of the Stratonovich integral. We shall distinguish the Stratonovich integral by a " o " before the differential dW_t. Unlike the Ito integral, the Stratonovich integral enjoys some similar properties to those of classical calculus, in particular

$$\int_0^t W_s(\omega) \circ dW_s(\omega) = \frac{1}{2} W_t^2(\omega).$$

Unfortunately it does not have the martingale property which turns out to be crucial in stochastic analysis quite often. If one wants to compute the expectation of a Stratonovich integral it is better to translate it in Ito language as we will discuss in Subsection 2.2.D.

Problem 2.1.1 *Plot a linearly interpolated sample path over the interval $0 \leq t \leq 1$ of the Ito sums approximation (1.12) for the stochastic integral (1.17) calculated with equidistant steps $\Delta = 0.01$. Compare the results with the corresponding sample path of the exact solution.*

Program 2.1.1 Approximation by Ito sums

```
USES CRT,DOS,GRAPH,INIT,SETSCR,SERVICE,RANDNUMB;

CONST
N=100;              { number of time steps          }
T0=0.0;             { left end point                }
T=1.0;              { right end point               }
DELTA=(T-T0)/N;     { time step size                }
X0=0.0;             { initial value of the trajectory }
  ...

PROCEDURE COORDSYS(CY,CX:STRING);
PROCEDURE PLOTGRAPH(LTN,NO,NN:INTEGER;ORDINATE,ABSCISSA:VECTOR);

{ Main program : }
  ...
SQDELTA:=SQRT(DELTA);
U2:=0.0;WT:=0.0;RSTK:=0.0;
TK:=T0-DELTA;
K:=-1;
REPEAT
 K:=K+1;         { counter }
 TK:=TK+DELTA; { time     }
 IF K MOD 2 =1 THEN GENERATE(U1,U2)
   ELSE U1:=U2;    { uses Polar Marsaglia method            }
 DWT:=U1*SQDELTA; { increment of the Wiener trajectory     }
 WT:=WT+DWT;      { current value of the Wiener trajectory }
 XT[K]:=0.5*(WT*WT-TK);    { value of the explicit Ito solution }
 RST[K]:=RSTK+(WT-DWT)*DWT; { value of the Ito sum              }
 ABSCISSA[K]:=TK;          { value of the x-axis               }
 RSTK:=RST[K];
UNTIL K=N;
  ...
```

Notes 2.1.1

(i) The two sample paths are plotted on the same screen using the procedures COORDSYS and PLOTGRAPH, which were mentioned in Notes 1.7.2. Other, previously developed routines could be used to analyse the statistical properties of Ito sums approximations, such as estimating their moments and plotting empirical density functions.

(ii) The chosen step size is still too large for an accurate approximation. While better results can be obtained with smaller step sizes, Ito sums approximations remain an inefficient method for approximating stochastic integrals. Other methods will be considered in later chapters.

C. Ito Formula

The martingale property (1.16) of the Ito stochastic integral is one of its most advantageous features. There is a price to be paid for it, namely that stochastic differentials, which are interpreted in terms of stochastic integrals, do not transform according to the chain rule of classical calculus. Instead there is an additional term due, essentially, to the fact that the stochastic differential $(dW_t)^2$ behaves as dt in the mean-square sense. The resulting expression is called the Ito formula.

Let e and f be two functions from $[0,T] \times \Omega$ into $\Re$ with such measurability and integrability properties that the ordinary and stochastic integrals appearing in the following formula make sense. By a *stochastic differential* we mean an expression

$$dX_t(\omega) = e(t,\omega)\, dt + f(t,\omega)\, dW_t(\omega),$$

which is just a symbolical way of writing

$$(1.18) \qquad X_t(\omega) - X_s(\omega) = \int_s^t e(u,\omega)\, du + \int_s^t f(u,\omega)\, dW_u(\omega),$$

w.p.1, for any $0 \le s \le t \le T$. The first integral in (1.18) is an ordinary (Riemann or Lebesgue) integral for each $\omega \in \Omega$ and the second is an Ito integral.

Let $U : [0,T] \times \Re \to \Re$ have continuous partial derivatives $\frac{\partial U}{\partial t}$, $\frac{\partial U}{\partial x}$ and $\frac{\partial^2 U}{\partial x^2}$. A scalar transformation by U of the stochastic differential (1.18) results after some non trivial analysis in the *Ito formula*:

$$U(t,X_t) = U(s,X_s) + \int_s^t \left\{ \frac{\partial U}{\partial t} + e\frac{\partial U}{\partial x} + \frac{1}{2}f^2\frac{\partial^2 U}{\partial x^2} \right\} du + \int_s^t f\frac{\partial U}{\partial x}\, dW_u,$$

(1.19)

w.p.1, for any $0 \le s \le t \le T$, where the integrands are all evaluated at (u, X_u).

There is also a multi-dimensional version of the Ito formula for multi-dimensional stochastic differentials with multi-dimensional Wiener processes. We shall state it in Subsection 2.2.C.

2.2 Stochastic Differential Equations

A. Linear Stochastic Equations

The Ito stochastic integral provides us with the necessary means for formulating stochastic differential equations. Such equations describe the dynamics of many important continuous time stochastic systems. Let us consider a simple linear case which models the molecular bombardment of a speck of dust on a water surface responsible for Brownian motion. The intensity of this bombardment does not depend on the state variables, for instance the position and velocity of the speck. Taking X_t as one of the components of the velocity of the particle, Langevin wrote the equation

$$(2.1) \qquad \frac{dX_t}{dt} = -aX_t + b\xi_t$$

(see (1.5)) for the acceleration of the particle, that is, as the sum of a retarding frictional force depending on the velocity and the molecular forces represented by a white noise process ξ_t, with the intensity b independent of the velocity of the particle. Here a and b are positive constants. We now interpret the *Langevin equation* (2.1) symbolically as an Ito stochastic differential (equation)

$$(2.2) \qquad dX_t = -aX_t\,dt + b\,dW_t,$$

that is, as a stochastic integral equation

$$(2.3) \qquad X_t = X_0 - \int_0^t aX_s\,ds + \int_0^t b\,dW_s$$

where the second integral is an Ito stochastic integral. This process is also called *Ornstein-Uhlenbeck process*. Later we shall see that it does not matter in this case whether we choose the Ito or Stratonovich integral as both yield the same process here. We say that equation (2.3) has *additive noise* since the noise term does not depend on the state variable; it is also a linear equation. It can be shown that equation (2.3) has the explicit solution

$$(2.4) \qquad X_t = e^{-at}X_0 + e^{-at}\int_0^t e^{as}b\,dW_s$$

for $0 \leq t \leq T$.

With external fluctuations the intensity of the noise usually depends on the state of the system. For example, the growth coefficient in an exponential growth equation $\dot{x} = \alpha x$ may fluctuate on account of environmental effects, taking the form $\alpha = a + b\xi_t$ where a and b are positive constants and ξ_t is a white noise process. This results in the heuristically written equation

$$(2.5) \qquad \frac{dX_t}{dt} = aX_t + bX_t\,\xi_t,$$

which we shall interpret as the Ito stochastic differential equation

$$(2.6) \qquad dX_t = aX_t \, dt + bX_t \, dW_t,$$

that is, as the stochastic integral equation

$$(2.7) \qquad X_t = X_0 + \int_0^t aX_s \, ds + \int_0^t bX_s \, dW_s.$$

The second integral is again an Ito integral, but in contrast with (2.3) its integrand involves the unknown solution. We say that (2.7) has *multiplicative noise*; again it is a linear equation. Using the Ito formula (1.19) we can verify that it has the explicit solution

$$(2.8) \qquad X_t = X_0 \exp\left((a - \frac{1}{2} b^2)t + bW_t \right)$$

for $0 \le t \le T$. A Stratonovich integral in (2.7) now would yield a different solution process.

Problem 2.2.1 *Simulate with time step size $\Delta = 2^{-9}$ a linearly interpolated trajectory of the explicit solution (2.8) of the linear equation (2.7) with multiplicative noise, $a = b = 1.0$, starting at $X_0 = 1$ and plot it against time $t \in [0,1]$.*

In the following program we use the Polar Marsaglia method to generate the Gaussian random numbers required for the increments of the Wiener process in (2.8).

Program 2.2.1   Explicit solution of a linear SDE

```
USES CRT,DOS,GRAPH,INIT,SETSCR,SERVICE,RANDNUMB,AAGRAPHS;

CONST
   T0=0.0;                  { left end point                                    }
   T=1.0;                   { right end point                                   }
   DELTA=(T-T0)/NUMINV;     { minimum step size w.r.t. screen resolution        }
   X0=1.0;                  { initial value of the trajectory                   }
   A=1.0;                   { parameter of the drift coefficient of the SDE     }
   B=1.0;                   { parameter of the diffusion coefficient of the SDE }
   WPT0=0.0;                { initial value of the Wiener trajectory            }
   ...
FUNCTION EXPLSOL(TT,WPT:REAL):REAL; { can be replaced by other functions }
BEGIN
 EXPLSOL:=X0*EXP((A-0.5*B*B)*(TT-T0)+B*(WPT-WPT0));
END;{ EXPLSOL }

{ Main program : }
   ...
 SQDELTA:=SQRT(DELTA);
 U2:=0.0;WT:=WPT0;
 TK:=T0-DELTA;
```

```
K:=-1;
REPEAT
 K:=K+1;             { counter                }
 TK:=TK+DELTA;   { time                      }
 ABSCISSA[K]:=TK; { values of the x-axis }
 IF K MOD 2 =1 THEN GENERATE(U1,U2) ELSE U1:=U2; { uses Polar Marsaglia }
 WT:=WT+U1*SQDELTA;     { current value of the Wiener trajectory }
 XT[K]:=EXPLSOL(TK,WT); { values of the y-axis            }
UNTIL K=NUMINV;

GRAPH111(XT,ABSCISSA,'X(t)','t','');
 ...
```

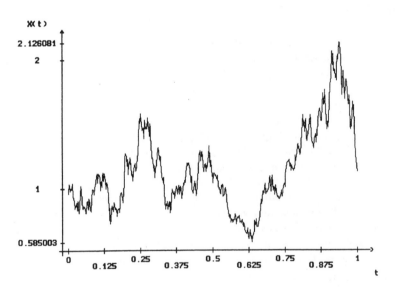

Figure 2.2.1   Solution path of a linear equation with multiplicative noise.

Notes 2.2.1

(i) Explicit solutions of other stochastic differential equations can be plotted
     with an appropriate change in the function EXPLSOL, provided they in-
     volve the Wiener process in a similar way. A nonzero initial time T0 can
     also be used.

(ii) The AAGRAPHS routine GRAPH111 from Program 1.1.1 is used, so a
     change in step size DELTA will require a corresponding change in the length
     of the time interval since the number of points allowed is bounded above
     by NUMINV+1. This restriction can be avoided by using the COORDSYS
     and PLOTGRAPH routines instead, but rescaling may then be necessary.

B. Ito Stochastic Differential Equations

When we have no explicit solution of an Ito stochastic equation

$$(2.9) \qquad X_t = X_{t_0} + \int_{t_0}^{t} a(X_s)\, ds + \int_{t_0}^{t} b(X_s)\, dW_s,$$

we need somehow to ensure the existence and uniqueness of a process $X = \{X_t, t \in [t_0, T]\}$ which satisfies it. We call $a(X_s)$ the drift coefficient and $b(X_s)$ the diffusion coefficient.

We shall say that the solutions of (2.9) are *pathwise unique* if any two such solutions $X = \{X_t, t \in [t_0, T]\}$ and $\tilde{X} = \{\tilde{X}_t, t \in [t_0, T]\}$ have, almost surely, the same sample paths on $[t_0, T]$, that is if

$$(2.10) \qquad P\left(\sup_{t_0 \le t \le T} \left| X_t - \tilde{X}_t \right| > 0 \right) = 0.$$

In this case we call X a unique strong solution of (2.9).

From a basic existence and uniqueness theorem it follows that equation (2.9) has a unique strong solution $X = \{X_t, t \in [t_0, T]\}$ on $[t_0, T]$ with

$$\sup_{t_0 \le t \le T} E\left(X_t^2 \right) < \infty$$

provided X_{t_0} is independent of $W = \{W_t, t \in [t_0, T]\}$ with $E(X_{t_0})^2 < \infty$ and the coefficients a, b satisfy Lipschitz conditions. We often call X an *Ito diffusion process* or, for brevity, just a *diffusion process*.

Sometimes equation (2.9) may have solutions which are unique in the weaker sense that only their probability laws coincide, but not necessarily their sample paths. We shall say then that we have a *unique weak solution*. Existence and uniqueness of weak solutions follow if, for instance, the coefficients a and b are bounded and continuous and b is nondegenerate $|b| \ge \epsilon > 0$. Of course, any unique strong solution is always a unique weak solution.

The nonlinear Ito equation

$$(2.11) \qquad X_t = X_0 + \int_{0}^{t} 1\, ds + \int_{0}^{t} 2\sqrt{X_s}\, dW_s.$$

has the explicit solution

$$(2.12) \qquad X_t = \left(W_t + \sqrt{X_0} \right)^2$$

for all $t \in [0, T]$, which can be checked with the Ito formula (1.19); it is unique in the strong sense even thought the diffusion coefficient is not a Lipschitz continuous function. Such a case is covered by a well–known Yamada–Watanabe strong existence and uniqueness theorem. Existence and uniqueness theorems usually involve sufficient rather than necessary conditions for their asserted results.

Exercise 2.2.1 (PC-Exercise 4.4.1) *Repeat Problem 2.2.1 to simulate a linearly interpolated trajectory on $[0, 1]$ of the explicit solution (2.12) which solves the stochastic equation (2.11) with $X_0 = 100$.*

C. Vector Stochastic Differential Equations

Many applications involving stochastic differential equations require the underlying dynamics to be modelled by a vector stochastic differential equation.

Here we introduce notation that will be used in the general vector case throughout the book. We shall interpret a vector as a column vector and its transpose as a row vector and consider an m-dimensional Wiener process $W = \{W_t, t \geq 0\}$ with components W_t^1, W_t^2, ..., W_t^m, which are independent scalar Wiener processes. Then, we take a d-dimensional vector valued function $a : [t_0, T] \times \Re^d \to \Re^d$, the drift coefficient, and a $d \times m$-matrix valued function $b : [t_0, T] \times \Re^d \to \Re^{d \times m}$, the diffusion coefficient, $t_0 \in [0, T]$, to form a d-dimensional *vector stochastic differential equation*

$$(2.13) \qquad dX_t = a(t, X_t)\, dt + b(t, X_t)\, dW_t.$$

We interpret this as a stochastic integral equation

$$(2.14) \qquad X_t = X_{t_0} + \int_{t_0}^t a(s, X_s)\, ds + \int_{t_0}^t b(s, X_s)\, dW_s$$

with initial value $X_{t_0} \in \Re^d$, where the Lebesgue and Ito integrals are determined component by component, with the ith component of (2.14) being

$$X_t^i = X_{t_0}^i + \int_{t_0}^t a^i(s, X_s)\, ds + \sum_{j=1}^m \int_{t_0}^t b^{i,j}(s, X_s)\, dW_s^j.$$

If the drift and diffusion coefficients do not depend on the time variable, that is if $a(t, x) \equiv a(x)$ and $b(t, x) \equiv b(x)$, then we say that the stochastic equation is *autonomous*. We can always write a nonautonomous equation as a vector autonomous equation of one dimension more by setting in the first component the drift coefficient equal to 1 and the diffusion coefficient as 0 to obtain as the first component of X_t the time variable $X_t^1 = t$.

There is a vector version of the Ito formula. For a sufficiently smooth transformation $U : [t_0, T] \times \Re^d \to \Re^k$ of the solution $X = \{X_t, t_0 \leq t \leq T\}$ of (2.13) we obtain a k-dimensional process $Y = \{Y_t = U(t, X_t), t_0 \leq t \leq T\}$ with the vector stochastic differential in component form

$$(2.15) \quad dY_t^p = \left(\frac{\partial U^p}{\partial t} + \sum_{i=1}^d a^i \frac{\partial U^p}{\partial x_i} + \frac{1}{2} \sum_{i,j=1}^d \sum_{l=1}^m b^{i,l} b^{j,l} \frac{\partial^2 U^p}{\partial x_i \partial x_j} \right) dt$$

$$+ \sum_{l=1}^m \sum_{i=1}^d b^{i,l} \frac{\partial U^p}{\partial x_i}\, dW_t^l$$

for $p = 1, 2, \ldots, k$, where the terms are all evaluated at (t, X_t). We can sometimes use this formula to determine the solutions of certain vector stochastic differential equations in terms of known solutions of other equations, for example linear equations.

The general form of a d-dimensional *linear stochastic differential equation* is

$$(2.16) \qquad dX_t = (A(t)X_t + a(t))\, dt + \sum_{l=1}^{m} \left(B^l(t)X_t + b^l(t)\right)\, dW_t^l$$

where $A(t)$, $B^1(t)$, $B^2(t)$, ..., $B^m(t)$ are $d \times d$-matrix functions and $a(t)$, $b^1(t)$, $b^2(t)$, ..., $b^m(t)$ are d-dimensional vector functions for $t \in [t_0, T]$. It can be shown that the solution of (2.16) has the form

$$(2.17) \qquad X_t = \Phi_{t,t_0}\left(X_{t_0} + \int_{t_0}^{t} \Phi_{s,t_0}^{-1}\left(a(s) - \sum_{l=1}^{m} B^l(s)b^l(s)\right) ds\right.$$
$$\left. + \sum_{l=1}^{m} \int_{t_0}^{t} \Phi_{s,t_0}^{-1} b^l(s)\, dW_s^l\right).$$

Here Φ_{t,t_0} is the $d \times d$ fundamental matrix satisfying $\Phi_{t_0,t_0} = I$, where I is the unit matrix, and the homogeneous *matrix stochastic differential equation*

$$(2.18) \qquad d\Phi_{t,t_0} = A(t)\Phi_{t,t_0}\, dt + \sum_{l=1}^{m} B^l(t)\Phi_{t,t_0}\, dW_t^l,$$

which we interpret column vector by column vector as vector stochastic differential equations. Unlike the scalar homogeneous linear equations, we cannot always solve (2.18) explicitly for the fundamental solution, even when all of the matrices are constant matrices. If, however, the matrices $A, B^1, B^2, \ldots, B^m$ are constants and commute, that is if

$$(2.19) \qquad AB^l = B^l A \quad \text{and} \quad B^l B^k = B^k B^l$$

for all $k, l = 1, 2, \ldots, m$, then we obtain the following explicit expression for the fundamental matrix solution

$$\Phi_{t,t_0} = \exp\left(\left(A - \frac{1}{2}\sum_{l=1}^{m}\left(B^l\right)^2\right)(t - t_0) + \sum_{l=1}^{m} B^l\left(W_t^l - W_{t_0}^l\right)\right).$$

It is possible to derive vector and matrix ordinary differential equations for the vector mean $m(t) = E(X_t)$ and the $d \times d$ matrix second moment $P(t) = E\left(X_t X_t^\top\right)$ of a general vector linear SDE (2.16). Here $X^\top$ is the transpose of a vector X, and the product of d-dimensional vectors X and Y, $XY^\top$, is a $d \times d$ matrix with ijth component $x_i y_j$. Thus $XX^\top$ is a symmetric matrix. We find that

$$(2.20) \qquad \frac{dm}{dt} = A(t)m + a(t)$$

and

$$(2.21) \quad \frac{dP}{dt} = A(t)P + PA(t)^\top + \sum_{l=1}^{m} B^l(t)PB^l(t)^\top$$

$$+ a(t)m(t)^\top + m(t)a(t)^\top$$

$$+ \sum_{l=1}^{m} \left(B^l(t)m(t)b^l(t)^\top + b^l(t)m(t)^\top B^l(t) + b^l(t)b^l(t)^\top \right)$$

with initial conditions $m(t_0) = E(X_{t_0})$ and $P(t_0) = E\left(X_{t_0} X_{t_0}^\top\right)$, respectively.

D. Stratonovich Stochastic Differential Equations

We saw in (2.13) and (2.14) that a d-dimensional stochastic differential equation

$$(2.22) \quad dX_t = a(t, X_t)\, dt + \sum_{j=1}^{m} b^j(t, X_t)\, dW_t^j$$

for $t \in [t_0, T]$ with initial value $X_{t_0} \in \Re^d$ can be interpreted mathematically as a vector stochastic integral equation

$$(2.23) \quad X_t = X_{t_0} + \int_{t_0}^{t} a(s, X_s)\, ds + \sum_{j=1}^{m} \int_{t_0}^{t} b^j(s, X_s)\, dW_s^j$$

where $a = (a^i)_{i=1}^d$ and $b^j = (b^{i,j})_{i=1}^d$ are d-dimensional vectors. So far we have taken the stochastic integral to be an Ito stochastic integral, calling (2.22) an *Ito stochastic differential equation*.

With different choices of stochastic integrals we would generally obtain different solutions for the above integral equation, which would justify our saying that we have different stochastic differential equations, even though they all have the same coefficients. Of all these integrals, the one proposed by Stratonovich, in which the integrand is, essentially, evaluated at the midpoint $\frac{1}{2}(t_j^{(n)} + t_{j+1}^{(n)})$ of each partition subinterval $[t_j^{(n)}, t_{j+1}^{(n)}]$, is the most appealing because it alone satisfies the usual transformation rules of classical calculus. If we use the Stratonovich stochastic integral, it is appropriate to say that we have a *Stratonovich stochastic differential equation*. We shall write it as

$$(2.24) \quad dX_t = \underline{a}(t, X_t)\, dt + \sum_{j=1}^{m} b^j(t, X_t) \circ dW_t^j$$

for $t \in [t_0, T]$ with initial value X_{t_0} or

$$(2.25) \quad X_t = X_{t_0} + \int_{t_0}^{t} \underline{a}(s. X_s)\, ds + \sum_{j=1}^{m} \int_{t_0}^{t} b^j(s, X_s) \circ dW_s^j,$$

and reserve (2.22) and (2.23) for the Ito case. With the choice of a corrected drift

$$(2.26) \qquad \underline{a}(t,x) = a(t,x) - \frac{1}{2} \sum_{j=1}^{d} \sum_{k=1}^{m} b^{j,k}(t,x) \frac{\partial b^k}{\partial x_j}(t,x)$$

it turns out that the processes satisfying the Ito stochastic equation (2.23) and the Stratonovich stochastic equation (2.25) are equivalent. Thus in general every Ito stochastic equation can be converted in an equivalent Stratonovich one and vice versa.

For a sufficiently smooth transformation $U : [0,T] \times \Re^d \to \Re^k$ of the solution of (2.25) the k-dimensional process $Y_t = U(t, X_t)$ has the Stratonovich stochastic differential

$$(2.27) \qquad dY_t^p = \left(\frac{\partial U^p}{\partial t} + \sum_{i=1}^{d} \underline{a}^i \frac{\partial U^p}{\partial x_i} \right) dt + \sum_{l=1}^{m} \sum_{i=1}^{d} b^{i,l} \frac{\partial U^p}{\partial x_i} \circ dW_t^l$$

for $p = 1, 2, \ldots, k$, where all the terms are evaluated at (t, X_t). This chain rule for Stratonovich equations is equivalent to the deterministic chain rule and is clearly simpler than its counterpart (2.15) for Ito equations. This has advantages in certain theoretical and practical investigations, but is at the expense of the direct link to martingale theory that the Ito integral provides. Nevertheless we can always switch, via the drift correction (2.26), between equivalent Ito and Stratonovich equations when this is advantageous.

2.3   Stochastic Taylor Expansions

A. Deterministic Taylor Expansion

Deterministic Taylor expansions are well known. We shall review them here using terminology that will facilitate our presentation of their stochastic counterparts. To begin we consider the solution $X = \{X_t, t \in [t_0, T]\}$ of a 1-dimensional ordinary differential equation

$$(3.1) \qquad \frac{d}{dt} X_t = a(X_t),$$

with initial value X_{t_0}, for $t \in [t_0, T]$ where $0 \leq t_0 < T$, which we can write in the equivalent integral equation form as

$$(3.2) \qquad X_t = X_{t_0} + \int_{t_0}^{t} a(X_s) \, ds.$$

To justify the following constructions we require that the function a satisfies appropriate properties, for instance to be sufficiently smooth with a linear growth bound. Let $f : \Re \to \Re$ be a continuously differentiable function. Then, by the chain rule, we have

$$(3.3) \qquad \frac{d}{dt} f(X_t) = a(X_t) \, f'(X_t),$$

which, using the operator

(3.4)
$$Lf = a f'$$

where $'$ denotes differentiation with respect to x, we can express (3.3) as the integral relation

(3.5)
$$f(X_t) = f(X_{t_0}) + \int_{t_0}^t Lf(X_s)\, ds$$

for all $t \in [t_0, T]$. When $f(x) \equiv x$ we have $Lf = a$, $L^2 f = La$, ... and (3.5) reduces to

(3.6)
$$X_t = X_{t_0} + \int_{t_0}^t a(X_s)\, ds,$$

that is, to equation (3.2).

If we now apply the relation (3.5) to the function $f = a$ in the integral in (3.6), we obtain

(3.7)
$$
\begin{aligned}
X_t &= X_{t_0} + \int_{t_0}^t \left(a(X_{t_0}) + \int_{t_0}^s La(X_z)\, dz \right) ds \\
&= X_{t_0} + a(X_{t_0}) \int_{t_0}^t ds + \int_{t_0}^t \int_{t_0}^s La(X_z)\, dz\, ds,
\end{aligned}
$$

which is the simplest nontrivial Taylor expansion for X_t. We can apply (3.5) again to the function $f = La$ in the double integral of (3.7) to derive

(3.8)
$$X_t = X_{t_0} + a(X_{t_0}) \int_{t_0}^t ds + La(X_{t_0}) \int_{t_0}^t \int_{t_0}^s dz\, ds + R_3$$

with remainder

(3.9)
$$R_3 = \int_{t_0}^t \int_{t_0}^s \int_{t_0}^z L^2 a(X_u)\, du\, dz\, ds,$$

for $t \in [t_0, T]$. For a general $r + 1$ times continuously differentiable function $f : \Re \to \Re$ this method gives the classical *Taylor formula* in integral form

(3.10)
$$
\begin{aligned}
f(X_t) &= f(X_{t_0}) + \sum_{l=1}^r \frac{(t - t_0)^l}{l!} L^l f(X_{t_0}) \\
&\quad + \int_{t_0}^t \cdots \int_{t_0}^{s_2} L^{r+1} f(X_{s_1})\, ds_1 \ldots ds_{r+1}
\end{aligned}
$$

for $t \in [t_0, T]$ and $r = 1, 2, 3, \ldots$ since

$$\int_{t_0}^t \int_{t_0}^{s_1} \cdots \int_{t_0}^{s_{l-1}} ds_1 \ldots ds_l = \frac{1}{l!}(t - t_0)^l$$

for $l = 1, 2, \ldots$.

The Taylor formula (3.10) has proven to be a very useful tool in both theoretical and practical investigations, particularly in numerical analysis. It allows the approximation of a sufficiently smooth function in a neighbourhood of a given point to any desired order of accuracy. The expansion depends on the values of the function and some of its higher derivatives at the expansion point, weighted by corresponding multiple time integrals. In addition, there is a remainder term which contains the next following multiple time integral, but now with a time dependent integrand.

B. The Ito-Taylor Expansion   see P.142

A stochastic counterpart of the deterministic Taylor formula for the expansion of smooth functions of an Ito process about a given value has many potential applications in stochastic analysis, for instance in the derivation of numerical methods for stochastic differential equations. There are several possibilities for such a stochastic Taylor formula. One is based on the iterated application of the Ito formula (1.19), which we shall call the *Ito-Taylor expansion*. We shall indicate it here for the solution X_t of the 1-dimensional Ito stochastic differential equation in integral form

$$(3.11) \qquad X_t = X_{t_0} + \int_{t_0}^t a(X_s)\,ds + \int_{t_0}^t b(X_s)\,dW_s$$

for $t \in [t_0, T]$, where the second integral in (3.11) is an Ito stochastic integral and the coefficients a and b are sufficiently smooth real valued functions satisfying a linear growth bound. For any twice continuously differentiable function $f : \Re \to \Re$, the Ito formula (1.19) then gives

$$(3.12) \quad \begin{aligned} f(X_t) &= f(X_{t_0}) + \int_{t_0}^t \left(a(X_s)f'(X_s) + \frac{1}{2}b^2(X_s)f''(X_s) \right) ds \\ &\quad + \int_{t_0}^t b(X_s)\,f'(X_s)\,dW_s \\ &= f(X_{t_0}) + \int_{t_0}^t L^0 f(X_s)\,ds + \int_{t_0}^t L^1 f(X_s)\,dW_s, \end{aligned}$$

for $t \in [t_0, T]$. Here we have introduced the operators

$$(3.13) \qquad L^0 f = a\,f' + \frac{1}{2}b^2\,f''$$

and
$$(3.14) \qquad L^1 f = b\,f'.$$

Obviously for $f(x) \equiv x$ we have $L^0 f = a$ and $L^1 f = b$, in which case (3.12) reduces to the original Ito equation for X_t, that is to

$$(3.15) \qquad X_t = X_{t_0} + \int_{t_0}^t a(X_s)\,ds + \int_{t_0}^t b(X_s)\,dW_s.$$

In analogy with the deterministic expansions above, if we apply the Ito formula (3.12) to the functions $f = a$ and $f = b$ in (3.15) we obtain

$$(3.16) \quad X_t = X_{t_0}$$
$$+ \int_{t_0}^t \left(a(X_{t_0}) + \int_{t_0}^s L^0 a(X_z) \, dz + \int_{t_0}^s L^1 a(X_z) \, dW_z \right) ds$$
$$+ \int_{t_0}^t \left(b(X_{t_0}) + \int_{t_0}^s L^0 b(X_z) \, dz + \int_{t_0}^s L^1 b(X_z) \, dW_z \right) dW_s$$
$$= X_{t_0} + a(X_{t_0}) \int_{t_0}^t ds + b(X_{t_0}) \int_{t_0}^t dW_s + R$$

with remainder

$$R = \int_{t_0}^t \int_{t_0}^s L^0 a(X_z) \, dz \, ds + \int_{t_0}^t \int_{t_0}^s L^1 a(X_z) \, dW_z \, ds$$
$$+ \int_{t_0}^t \int_{t_0}^s L^0 b(X_z) \, dz \, dW_s + \int_{t_0}^t \int_{t_0}^s L^1 b(X_z) \, dW_z \, dW_s.$$

This is the simplest nontrivial Ito-Taylor expansion of X_t. It involves integrals with respect to both the time variable and the Wiener process, with multiple integrals with respect to both in the remainder.

We can repeat the above procedure, for instance by applying the Ito formula (3.12) to $f = L^1 b$ in (3.16), in which case we get

$$(3.17) \quad X_t = X_{t_0} + a(X_{t_0}) \int_{t_0}^t ds + b(X_{t_0}) \int_{t_0}^t dW_s$$
$$+ L^1 b(X_{t_0}) \int_{t_0}^t \int_{t_0}^s dW_z \, dW_s + \bar{R}$$

with remainder

$$\bar{R} = \int_{t_0}^t \int_{t_0}^s L^0 a(X_z) \, dz \, ds + \int_{t_0}^t \int_{t_0}^s L^1 a(X_z) \, dW_z \, ds$$
$$+ \int_{t_0}^t \int_{t_0}^s L^0 b(X_z) \, dz \, dW_s + \int_{t_0}^t \int_{t_0}^s \int_{t_0}^z L^0 L^1 b(X_u) \, du \, dW_z \, dW_s$$
$$+ \int_{t_0}^t \int_{t_0}^s \int_{t_0}^z L^1 L^1 b(X_u) \, dW_u \, dW_z \, dW_s.$$

It is possible to express the Ito-Taylor expansion for a general function f and arbitrarily many expansion terms in a succinct way. Nevertheless, its main properties are already apparent in the preceding example, with the multiple Ito integrals

$$\int_{t_0}^t ds, \qquad \int_{t_0}^t dW_s, \qquad \int_{t_0}^t \int_{t_0}^s dW_z \, dW_s$$

multiplied by certain constants and a remainder term involving the next following multiple Ito integrals, but now with nonconstant integrands. The Ito-Taylor
expansion can thus be considered as a generalization of both the Ito formula
and the deterministic Taylor formula.

C. The Stratonovich-Taylor Expansion

Analogous expansions, called *Stratonovich-Taylor expansions*, hold for Ito processes satisfying Stratonovich stochastic differential equations. To illustrate
them we consider

$$(3.18) \qquad X_t = X_{t_0} + \int_{t_0}^t \underline{a}(X_s)\, ds + \int_{t_0}^t b(X_s) \circ dW_s$$

for $t \in [t_0, T]$, where the second integral in (3.18) is a Stratonovich stochastic
integral and the coefficients $\underline{a}$ and b are sufficiently smooth real valued functions
satisfying a linear growth bound. We know from (2.27) that the solution of a
Stratonovich SDE transforms according to the deterministic chain rule, so for
any twice continuously differentiable function $f : \Re \to \Re$ we have

$$(3.19) \qquad f(X_t) = f(X_{t_0}) + \int_{t_0}^t \underline{L}^0 f(X_s)\, ds + \int_{t_0}^t \underline{L}^1 f(X_s) \circ dW_s,$$

for $t \in [t_0, T]$, with the operators

$$(3.20) \qquad \underline{L}^0 f = \underline{a}\, f'$$

and
$$(3.21) \qquad \underline{L}^1 f = b\, f'.$$

For $f(x) \equiv x$ we have $\underline{L}^0 f = \underline{a}$ and $\underline{L}^1 f = b$, in which case (3.19) reduces to
the original Stratonovich equation

$$(3.22) \qquad X_t = X_{t_0} + \int_{t_0}^t \underline{a}(X_s)\, ds + \int_{t_0}^t b(X_s) \circ dW_s.$$

As in the Ito case just considered, we can now apply (3.19) to the integrand
functions $f = \underline{a}$ and $f = b$ in (3.18) to obtain

$$X_t = X_{t_0} + \underline{a}(X_{t_0}) \int_{t_0}^t ds + b(X_{t_0}) \int_{t_0}^t \circ dW_s + R$$

with remainder

$$R = \int_{t_0}^t \int_{t_0}^s \underline{L}^0 \underline{a}(X_z)\, dz\, ds + \int_{t_0}^t \int_{t_0}^s \underline{L}^1 \underline{a}(X_z) \circ dW_z\, ds$$
$$+ \int_{t_0}^t \int_{t_0}^s \underline{L}^0 b(X_z)\, dz \circ dW_s + \int_{t_0}^t \int_{t_0}^s \underline{L}^1 b(X_z) \circ dW_z \circ dW_s.$$

This is the simplest nontrivial Stratonovich-Taylor expansion of X_t. We can continue expanding, for instance by applying (3.19) to the integrand $f = \underline{L}^1 b$ in the above equation to obtain

$$(3.23) \qquad X_t = X_{t_0} + \underline{a}(X_{t_0}) \int_{t_0}^t ds + b(X_{t_0}) \int_{t_0}^t \circ dW_s$$

$$+ \underline{L}^1 b(X_{t_0}) \int_{t_0}^t \int_{t_0}^s \circ dW_z \circ dW_s + \bar{R}$$

with an appropriate remainder $\bar{R}$.

It is also possible to formulate the Stratonovich-Taylor expansion for a general function f and arbitrarily many terms. This is similar to the corresponding Ito-Taylor expansion, but with multiple Stratonovich stochastic integrals. While formally similar to the Ito-Taylor expansion, the Stratonovich-Taylor expansion in fact has a simpler structure which makes it a more natural generalization of the deterministic Taylor formula and more convenient to use in some questions of stochastic numerical analysis. Convergence proofs, however, make extensive use of martingale theory, so the Ito-Taylor expansion has advantages here. It is nevertheless possible to switch between Ito and Stratonovich versions based on equivalent versions of stochastic differential equations.

Similar expansions hold for multi-dimensional Ito processes satisfying nonautonomous stochastic differential equations.

D. Approximate Multiple Stratonovich Integrals

We see from the above stochastic Taylor expansions that multiple stochastic integrals are the building blocks for representations of solutions of stochastic differential equations and their functionals. In other words, they are the basic elements of the Wiener chaos. This fact is widely used to construct efficient higher order numerical schemes for the approximate solution of stochastic equation as we shall see later. Consequently, we need to be able to generate or at least to approximate multiple stochastic integrals.

Multiple stochastic integrals of higher multiplicity cannot always be expressed in terms of simpler stochastic integrals, especially when the Wiener process is multi-dimensional. Nevertheless it is still possible to represent them in a reasonably efficient way. Here we present such a method for multiple Stratonovich integrals based on a Kahunen-Loève or random Fourier series expansion of the Wiener process similar to (1.7.5). Our starting point is the Brownian bridge process

$$\left\{ W_t - \frac{t}{\Delta} W_\Delta, 0 \leq t \leq \Delta \right\}$$

formed from the given m-dimensional Wiener process $W_t = (W_t^1, \ldots, W_t^m)$ on the time interval $[0, \Delta]$. The componentwise Fourier expansion of this process

$$(3.24) \quad W_t^j - \frac{t}{\Delta} W_\Delta^j = \frac{1}{2} a_{j,0} + \sum_{r=1}^\infty \left(a_{j,r} \cos\left(\frac{2r\pi t}{\Delta}\right) + b_{j,r} \sin\left(\frac{2r\pi t}{\Delta}\right) \right)$$

has random coefficients

$$(3.25) \qquad a_{j,r} = \frac{2}{\Delta} \int_0^{\Delta} \left(W_s^j - \frac{s}{\Delta} W_{\Delta}^j \right) \cos\left(\frac{2r\pi s}{\Delta} \right) ds$$

and

$$(3.26) \qquad b_{j,r} = \frac{2}{\Delta} \int_0^{\Delta} \left(W_s^j - \frac{s}{\Delta} W_{\Delta}^j \right) \sin\left(\frac{2r\pi s}{\Delta} \right) ds,$$

which are Gaussian random variables, for $j = 1, \ldots, m$ and $r = 0, 1, 2, \ldots$. The series in (3.24) is understood to converge in the mean-square sense. We can truncate it to obtain an approximation of the Brownian bridge process. For each $p = 1, 2, \ldots$ we obtain the process

$$(3.27) \quad W_t^{j,p} = \frac{t}{\Delta} W_{\Delta}^j + \frac{1}{2} a_{j,0} + \sum_{r=1}^p \left(a_{j,r} \cos\left(\frac{2r\pi t}{\Delta} \right) + b_{j,r} \sin\left(\frac{2r\pi t}{\Delta} \right) \right)$$

which has differentiable sample paths on $[0, \Delta]$. Since Riemann-Stieltjes integrals with respect to such a process converge to Stratonovich stochastic integrals, we can use such integrals to approximate multiple Stratonovich integrals.

We shall denote a multiple Stratonovich integral

$$(3.28) \qquad J_{(j_1, j_2, \ldots, j_l), t} = \int_0^t \int_0^{s_l} \cdots \int_0^{s_2} dW_{s_1}^{j_1} \circ \ldots \circ dW_{s_{l-1}}^{j_{l-1}} \circ dW_{s_l}^{j_l}$$

for a multi-index $\alpha = (j_1, j_2, \ldots, j_l) \in \{0, 1, \ldots, m\}^l$ and a given m-dimensional Wiener process $W = (W^1, \ldots, W^m)$, setting $W_t^0 = t$ so that we can include integration with respect to t in the same formalism. In addition, we write

$$(3.29) \qquad J_{(j_1, j_2, \ldots, j_l), t}^p = \int_0^t \int_0^{s_l} \cdots \int_0^{s_2} dW_{s_1}^{j_1, p} dW_{s_2}^{j_2, p} \cdots dW_{s_l}^{j_l, p}$$

for the corresponding Riemann-Stieltjes integrals of the smooth functions defined in (3.27).

For each $j = 1, \ldots, m$ and $r = 1, \ldots, p$ with $p = 1, 2, \ldots$ we define independent standard Gaussian random variables $\xi_j, \zeta_{j,r}, \eta_{j,r}, \mu_{j,p}$ and $\phi_{j,p}$ by

$$(3.30) \quad \xi_j = \frac{1}{\sqrt{\Delta}} W_{\Delta}^j, \qquad \zeta_{j,r} = \sqrt{\frac{2}{\Delta}} \pi r a_{j,r}, \qquad \eta_{j,r} = \sqrt{\frac{2}{\Delta}} \pi r b_{j,r},$$

$$\mu_{j,p} = \frac{1}{\sqrt{\Delta \rho_p}} \sum_{r=p+1}^{\infty} a_{j,r}, \qquad \phi_{j,p} = \frac{1}{\sqrt{\Delta \alpha_p}} \sum_{r=p+1}^{\infty} \frac{1}{r} b_{j,r}$$

where

$$\rho_p = \frac{1}{12} - \frac{1}{2\pi^2} \sum_{r=1}^p \frac{1}{r^2}, \qquad \alpha_p = \frac{\pi^2}{180} - \frac{1}{2\pi^2} \sum_{r=1}^p \frac{1}{r^4}.$$

Using these random variables it turns out after lengthy computations that we can approximate a multiple Stratonovich integral $J_{(j_1, \ldots, j_l), \Delta}$ by $J_{(j_1, \ldots, j_l), \Delta}^p$ for

$p = 1, 2, \ldots$ in the manner described below. For brevity we omit the time subscript Δ from the J_α^p. We have for $j, j_1, j_2, j_3 \in \{1, \ldots, m\}$

(3.31)
$$J_{(0)}^p = \Delta, \quad J_{(j)}^p = \sqrt{\Delta}\,\xi_j, \quad J_{(0,0)}^p = \frac{1}{2}\Delta^2,$$

$$J_{(j,0)}^p = \frac{1}{2}\Delta\left(\sqrt{\Delta}\,\xi_j + a_{j,0}\right), \qquad J_{(0,j)}^p = \frac{1}{2}\Delta\left(\sqrt{\Delta}\,\xi_j - a_{j,0}\right)$$

where

$$a_{j,0} = -\frac{1}{\pi}\sqrt{2\Delta}\,\sum_{r=1}^{p}\frac{1}{r}\,\zeta_{j,r} - 2\sqrt{\Delta\rho_p}\,\mu_{j,p};$$

(3.32)
$$J_{(j_1,j_2)}^p = \frac{1}{2}\Delta\xi_{j_1}\xi_{j_2} - \frac{1}{2}\sqrt{\Delta}\,\left(a_{j_2,0}\xi_{j_1} - a_{j_1,0}\xi_{j_2}\right) + \Delta\,A_{j_1,j_2}^p$$

with

$$A_{j_1,j_2}^p = \frac{1}{2\pi}\sum_{r=1}^{p}\frac{1}{r}\left(\zeta_{j_1,r}\eta_{j_2,r} - \eta_{j_1,r}\zeta_{j_2,r}\right);$$

(3.33)
$$J_{(0,0,0)}^p = \frac{1}{3!}\Delta^3, \qquad J_{(0,j,0)}^p = \frac{1}{3!}\Delta^{5/2}\xi_j - \frac{1}{\pi}\Delta^2 b_j,$$

$$J_{(j,0,0)}^p = \frac{1}{3!}\Delta^{5/2}\xi_j + \frac{1}{4}\Delta^2 a_{j,0} + \frac{1}{2\pi}\Delta^2 b_j,$$

$$J_{(0,0,j)}^p = \frac{1}{3!}\Delta^{5/2}\xi_j - \frac{1}{4}\Delta^2 a_{j,0} + \frac{1}{2\pi}\Delta^2 b_j$$

with

$$b_j = \sqrt{\frac{\Delta}{2}}\,\sum_{r=1}^{p}\frac{1}{r^2}\eta_{j,r} + \sqrt{\Delta\alpha_p}\,\phi_{j,p};$$

(3.34)
$$J_{(j_1,0,j_2)}^p = \frac{1}{3!}\Delta^2\xi_{j_1}\xi_{j_2} + \frac{1}{2}a_{j_1,0}J_{(0,j_2)}^p + \frac{1}{2\pi}\Delta^{3/2}\xi_{j_2}b_{j_1}$$

$$\qquad\qquad - \Delta^2 B_{j_1,j_2}^p - \frac{1}{4}\Delta^{3/2}a_{j_2,0}\xi_{j_1} + \frac{1}{2\pi}\Delta^{3/2}\xi_{j_1}b_{j_2},$$

$$J_{(0,j_1,j_2)}^p = \frac{1}{3!}\Delta^2\xi_{j_1}\xi_{j_2} - \frac{1}{\pi}\Delta^{3/2}\xi_{j_2}b_{j_1} + \Delta^2 B_{j_1,j_2}^p$$

$$\qquad\qquad - \frac{1}{4}\Delta^{3/2}a_{j_2,0}\xi_{j_1} + \frac{1}{2\pi}\Delta^{3/2}\xi_{j_1}b_{j_2} + \Delta^2 C_{j_1,j_2}^p$$

$$\qquad\qquad + \frac{1}{2}\Delta^2 A_{j_1,j_2}^p,$$

with

$$B_{j_1,j_2}^p = \frac{1}{4\pi^2}\sum_{r=1}^{p}\frac{1}{r^2}\left(\zeta_{j_1,r}\zeta_{j_2,r} + \eta_{j_1,r}\eta_{j_2,r}\right)$$

and

$$C^p_{j_1,j_2} = -\frac{1}{2\pi^2} \sum_{\substack{r,l=1 \\ r\neq l}}^{p} \frac{r}{r^2 - l^2} \left(\frac{1}{l} \zeta_{j_1,r} \zeta_{j_2,l} - \frac{l}{r} \eta_{j_1,r} \eta_{j_2,l} \right);$$

$$(3.35) \quad J^p_{(j_1,j_2,0)} = \frac{1}{2} \Delta^2 \xi_{j_1} \xi_{j_2} - \frac{1}{2} \Delta^{3/2} \left(a_{j_2,0} \xi_{j_1} - a_{j_1,0} \xi_{j_2} \right) + \Delta^2 A^p_{j_1,j_2}$$
$$- J^p_{(j_1,0,j_2)} - J^p_{(0,j_1,j_2)};$$

$$(3.36) \quad J^p_{(j_1,j_2,j_3)} = \frac{1}{\sqrt{\Delta}} \xi_{j_1} J^p_{(0,j_2,j_3)} + \frac{1}{2} a_{j_1,0} J^p_{(j_2,j_3)}$$
$$+ \frac{1}{2\pi} \Delta b_{j_1} \xi_{j_2} \xi_{j_3} - \Delta^{3/2} \xi_{j_2} B^p_{j_1,j_3}$$
$$+ \Delta^{3/2} \xi_{j_3} \left(\frac{1}{2} A^p_{j_1,j_2} - C^p_{j_2,j_1} \right) + \Delta^{3/2} D^p_{j_1,j_2,j_3}$$

with

$$D^p_{j_1,j_2,j_3} = -\frac{1}{\pi^2 2^{5/2}} \sum_{r,l=1}^{p} \frac{1}{l(l+r)} \left[\zeta_{j_2,l} \left(\zeta_{j_3,l+r} \eta_{j_1,r} - \zeta_{j_1,r} \eta_{j_1,l+r} \right) \right.$$
$$\left. + \eta_{j_2,l} \left(\zeta_{j_1,r} \zeta_{j_3,l+r} + \eta_{j_1,r} \eta_{j_3,l+r} \right) \right]$$
$$+ \frac{1}{\pi^2 2^{5/2}} \sum_{l=1}^{p} \sum_{r=1}^{l-1} \frac{1}{r(l-r)} \left[\zeta_{j_2,l} \left(\zeta_{j_1,r} \eta_{j_3,l-r} + \zeta_{j_3,l-r} \eta_{j_1,r} \right) \right.$$
$$\left. - \eta_{j_2,l} \left(\zeta_{j_1,r} \zeta_{j_3,l-r} - \eta_{j_1,r} \eta_{j_3,l-r} \right) \right]$$
$$+ \frac{1}{\pi^2 2^{5/2}} \sum_{l=1}^{p} \sum_{r=l+1}^{2p} \frac{1}{r(r-l)} \left[\zeta_{j_2,l} \left(\zeta_{j_3,r-l} \eta_{j_1,r} - \zeta_{j_1,r} \eta_{j_3,r-l} \right) \right.$$
$$\left. + \eta_{j_2,l} \left(\zeta_{j_1,r} \zeta_{j_3,r-l} + \eta_{j_1,r} \eta_{j_3,r-l} \right) \right],$$

where for $r > p$ we set $\zeta_{j,r} = 0$ and $\eta_{j,r} = 0$ for all j=1,...,m.

The most sensitive approximation is $J^p_{(j_1,j_2),\Delta}$ because the other $J^p_{\alpha,\Delta}$ are either identical to $J_{\alpha,\Delta}$ or their mean-square error can be estimated by a constant times Δ^γ for some $\gamma \geq 3$. It can be shown that

$$(3.37) \quad E\left(\left| J^p_{(j_1,j_2),\Delta} - J_{(j_1,j_2),\Delta} \right|^2 \right) = \frac{\Delta^2}{2\pi^2} \sum_{r=p+1}^{\infty} \frac{1}{r^2} = \Delta^2 \rho_p,$$

where

$$(3.38) \qquad \rho_p = \frac{1}{2\pi^2} \sum_{r=p+1}^{\infty} \frac{1}{r^2} \leq \frac{1}{2\pi^2} \int_p^{\infty} \frac{1}{u^2} \, du = \frac{1}{2\pi^2 p},$$

so

$$(3.39) \qquad E\left(\left|J_{\alpha,\Delta}^p - J_{\alpha,\Delta}\right|^2\right) \leq \Delta^2 \rho_p$$

for multi-indices α of length $l(\alpha) \leq 3$ provided Δ is sufficiently small.

E. Generation of Multiple Stratonovich Integrals

Here we generate some of the approximate multiple Stratonovich integrals that we shall later use in the numerical schemes. In the following problems we shall test whether these random variables are close to the theoretical values.

Let us start with multiple integrals that involve only a single Wiener process. We do not have to generate every integral directly since some of them follow from others by relationships like

$$(3.40) \qquad J_{(0,1),\Delta} + J_{(1,0),\Delta} = \Delta \, J_{(1),\Delta},$$

$$J_{(1,1,0),\Delta} + J_{(1,0,1),\Delta} + J_{(0,1,1),\Delta} = \Delta \, J_{(1,1),\Delta},$$

$$J_{(1,1),\Delta} = \frac{1}{2} \left(J_{(1),\Delta}\right)^2 \quad \text{and} \quad J_{(1,1,1),\Delta} = \frac{1}{6} \left(J_{(1),\Delta}\right)^3.$$

Problem 2.3.1 *Generate and plot linearly interpolated sample paths of the multiple Stratonovich integral approximants $J_{(1),t}^p$, $J_{(0,1),t}^p$, $J_{(1,0),t}^p$, $J_{(1,1),t}^p$, $J_{(0,1,1),t}^p$, $J_{(1,0,1),t}^p$ and $J_{(1,1,0),t}^p$ over the interval $0 \leq t \leq 1$ for $p = 5$ using 2^9 subintervals of equal length. Repeat for $p = 50$ and 100.*

Program 2.3.1 Approximating Stratonovich integrals

```
USES CRT,DOS,GRAPH,INIT,SETSCR,SERVICE,RANDNUMB,AAGRAPHS;

CONST
  N=512;              { number of the subintervals                    }
  T0=0.0;             { begin of the time interval                    }
  T=1.0;              { end of the time interval                      }
  DELTA=(T-T0)/N;     { time step size                                }
  P=20;               { truncation index in the approximate multiple integrals }
  ...
VAR
  ...
  ORDMIN:REAL;        { minimum of the ordinate                       }
  ORDMAX:REAL;        { maximum of the ordinate                       }
  SQDELTA:REAL;       { square root of the time step size             }
  TI:REAL;            { subinterval end                               }
  ALPHAP,ROP:REAL;    { approximation parameters                      }
  J1,J11:REAL;        { exact multiple Stratonovich integrals         }
  J01,J10:REAL;       { multiple Stratonovich integrals approximated  }
  J011,J101,J110:REAL; { multiple Stratonovich integrals approximated  }
```

```
 DWT:VECTOR;              { Gaussian random numbers                        }
  ...

PROCEDURE COORDSYS(CY,CX:STRING);
PROCEDURE PLOTGRAPH(LTN,NO,NN:INTEGER;ORDINATE,ABSCISSA:VECTOR);
PROCEDURE CONTROLSCR(CL,BOTTOMCR:STRING;ORD1,ORD2:REAL;XT:VECTOR);

PROCEDURE SETUPPARAMULTIINT(P:INTEGER;VAR ALPHAP,ROP:REAL);
VAR
 R:INTEGER;
 INCRE:REAL;
BEGIN
 ALPHAP:=0.0;ROP:=0.0;
 FOR R:=1 TO P DO
  BEGIN
   INCRE:=1/(R*R);ROP:=ROP+INCRE;ALPHAP:=ALPHAP+INCRE*INCRE;
  END;
 ALPHAP:=(PI*PI*PI*PI/90-ALPHAP)/(2*PI*PI);
 ROP:=PI*PI/6-ROP;ROP:=ROP/(2*PI*PI);
END;{ SETUPPARAMULTIINT }

PROCEDURE MULTIINT(P:INTEGER;DELTA,DWT:REAL;
          VAR J1,J01,J10,J11,J011,J101,J110:REAL);
VAR
 R,L:INTEGER;
 A10,B11P,B1P,C11P,FI1P,GSI,MUE1P:REAL;
 ETA1,FI1:VECTOR;
BEGIN
 GENERATE(FI1P,MUE1P);
 A10:=0.0;B1P:=0.0;B11P:=0.0;
 FOR R:=1 TO P DO
  BEGIN
   GENERATE(FI1[R],ETA1[R]);
   A10:=A10+FI1[R]/R;
   B1P:=B1P+ETA1[R]/(R*R);
   B11P:=B11P+(FI1[R]*FI1[R]+ETA1[R]*ETA1[R])/(R*R);
  END;
 A10:=-(1./PI)*A10*SQRT(2.0)*SQDELTA;
 A10:=A10-2.*SQDELTA*SQRT(ROP)*MUE1P;
 B1P:=B1P*SQDELTA/SQRT(2.0)+SQDELTA*SQRT(ALPHAP)*FI1P;
 B11P:=B11P/(4.*PI*PI);
 GSI:=DWT;
 J1:=GSI*SQDELTA;              { increment of the Wiener trajectory = J1 }
 J10:=0.5*DELTA*(J1+A10);      { multiple integral J10                  }
 J01:=J1*DELTA-J10;            { multiple integral J01                  }
 J11:=0.5*J1*J1;               { multiple integral J11                  }
 C11P:=0.0;
 R:=0;
 REPEAT
  R:=R+1;
  L:=0;
  REPEAT
   L:=L+1;
   IF L<>R THEN
    C11P:=C11P+(R/(R*R-L*L))*(FI1[R]*FI1[L]/L-ETA1[R]*ETA1[L]*L/R);
  UNTIL L=P;
 UNTIL R=P;
 C11P:=-C11P/(2.*PI*PI);
```

```
      J101:=SQR(DELTA*GSI)/6-DELTA*A10*A10/4;
      J101:=J101+SQDELTA*DELTA*GSI*B1P/PI;
      J101:=J101-DELTA*DELTA*B11P; { multiple integral J101 }
      J110:=SQR(DELTA*GSI)/6+DELTA*A10*A10/4;
      J110:=J110-SQDELTA*DELTA*GSI*B1P/(2.*PI);
      J110:=J110+SQDELTA*DELTA*A10*GSI/4;
      J110:=J110-DELTA*DELTA*C11P; { multiple integral J110 }
      J011:=J11*DELTA-J101-J110;   { multiple integral J011 }
    END;{ MULTIINT }

  { Main program : }
      ...
    SQDELTA:=SQRT(DELTA);
    SETUPPARAMULTIINT(P,ALPHAP,ROP); { sets approximation parameters }

    I:=0;       { time step index                          }
    TI:=T0;     { initial time                             }
    WT[0]:=0.0; { starting value of the Wiener trajectory }
    J1:=0.0;J01:=0.0;J10:=0.0;J11:=0.0;J011:=0.0;J101:=0.0;J110:=0.0;
      ...
    WHILE (TI < T) DO
      BEGIN
        I:=I+1;
        TI:=TI+DELTA;
        IF I MOD 2 = 1 THEN GENERATE(DWT[I-1],DWT[I]);
        MULTIINT(P,DELTA,DWT[I-1],J1,J01,J10,J11,J011,J101,J110);
          ...
      END;{ WHILE }
      ...
```

Notes 2.3.1

(i) In each case the appropriate formula from (3.31)–(3.35) provides the increment over a given subinterval. Not all need to be generated in this way since exact expressions or relationships such as (3.40) with others are known in some cases. The MULTINT procedure requires the approximation index P, the step size DELTA and a Gaussian random number DWT as its inputs. It also assumes that SETUPPARAMULTINT has already been called.

(ii) The GRAPH111 routine should be used first to check on scaling magnitudes, which can then be used in the COORDSYS–PLOTGRAPH routines to enable several paths to be plotted on the screen at once, for example corresponding to different P values.

To investigate the convergence rates of the approximations (3.31)–(3.35) of multiple Stratonovich integrals we examine the statistical properties of a sample of $J^p_{(1,0),\Delta}$ realizations on a single subinterval $[0, \Delta]$ for different values of p. The program, which is omitted, uses routines from Program 2.3.1 for the approximations and from Program 1.3.1 for the calculation of sample statistics, namely COMPSAMPLEPARA. The multiple integral under consideration can be changed easily.

Problem 2.3.2 *Estimate the sample averages and variances of samples of 10^3 realizations each of the approximation $J^p_{(1,0),\Delta}$ with $\Delta = 0.05$ for $p = 2, 5$ and 10.*

Approximating multiple Stratonovich integrals which involve different independent Wiener processes is more complicated than the above computations. Here we consider the simplest example $J_{(1,2),\Delta}$ and its approximation $J^p_{(1,2),\Delta}$ given by (3.32). In particular, we investigate the relationship between $J_{(1,2),\Delta}$ and the product $\frac{1}{2}\Delta W^1\Delta W^2$ to convince the reader that mixed multiple integrals cannot be evaluated in terms of the simple increments $\Delta W^1 = J_{(1),\Delta}$ and $\Delta W^2 = J_{(2),\Delta}$ of the involved Wiener processes W^1 and W^2.

Problem 2.3.3   *Generate corresponding sample path values of $J^p_{(1,2),\Delta}$ on the interval $[0,\Delta]$ where $\Delta = 1.0$ for $p = 5, 20$, and 100. Compare the results with the corresponding sample path of $\frac{1}{2}\Delta W^1\Delta W^2$.*

Program 2.3.3   Approximations of J12

```
USES CRT,DOS,GRAPH,INIT,SETSCR,SERVICE,RANDNUMB;

CONST
 DELTA=1.0; { time step size        }
 P1=5;      { truncation parameter }
 P2=20;     { truncation parameter }
 P3=100;    { truncation parameter }

TYPE
 VECTOR=ARRAY[1..3] OF REAL;

VAR
 CR:STRING;        { help string                                        }
 SQDELTA:REAL;     { square root of the time step size                  }
 ROP:REAL;         { current approximation parameter                    }
 SUMVAL:REAL;      { summation help variable                            }
 MU1,MU2:REAL;     { Gaussian random numbers                            }
 GSI1,GSI2:REAL;   { Gaussian random numbers generating Wiener increments }
 HALFDW1DW2:REAL;  { dW1*dW2/2                                           }
 J12P:VECTOR;      { multiple Stratonovich integral J12 approximated    }

PROCEDURE SETTABLETOSCR;

PROCEDURE SETUPPARAMULTIINTJ12(P:INTEGER;VAR ROP:REAL);
VAR
 R:INTEGER;
BEGIN
 ROP:=0.0;
 FOR R:=1 TO P DO ROP:=ROP+1/(R*R);
 ROP:=PI*PI/6-ROP;ROP:=ROP/(2*PI*PI);
END;{ SETUPPARAMULTIINTJ12 }

PROCEDURE MULTIINTJ12(PL,PR:INTEGER;ROP,DELTA,GSI1,GSI2,MU1,MU2:REAL;
                      VAR SUMVAL,J12:REAL);
VAR
 R:INTEGER;
 ETA,ZETA:ARRAY[1..2,1..100] OF REAL;
BEGIN
 FOR R:=PL TO PR DO
  BEGIN
   GENERATE(ZETA[1,R],ZETA[2,R]);GENERATE(ETA[1,R],ETA[2,R]);
```

```
SUMVAL:=SUMVAL+(1/R)*(ZETA[1,R]*(SQRT(2)*GSI2+ETA[2,R])
              -ZETA[2,R]*(SQRT(2)*GSI1+ETA[1,R])));
END;
J12:=SUMVAL*DELTA/PI;
J12:=J12+DELTA*(GSI1*GSI2/2+SQRT(ROP)*(MU1*GSI2-MU2*GSI1));
END;{ MULTIINTJ12 }

{ Main program : }
    ...
GENERATE(GSI1,GSI2);
HALFDW1DW2:=GSI1*GSI2*DELTA/2;
GENERATE(MU1,MU2); { generates the Gaussian random values MU1 and MU2 }
SETUPPARAMULTIINTJ12(P1,ROP); { sets the current parameter ROP        }
SUMVAL:=0.0; { help value for the summation }
MULTIINTJ12(1,P1,ROP,DELTA,GSI1,GSI2,MU1,MU2,SUMVAL,J12P[1]);
SETUPPARAMULTIINTJ12(P2,ROP); { sets the current parameter ROP        }
MULTIINTJ12(P1+1,P2,ROP,DELTA,GSI1,GSI2,MU1,MU2,SUMVAL,J12P[2]);
SETUPPARAMULTIINTJ12(P3,ROP); { sets the current parameter ROP        }
MULTIINTJ12(P2+1,P3,ROP,DELTA,GSI1,GSI2,MU1,MU2,SUMVAL,J12P[3]);
    ...
```

Notes 2.3.3

(i) An approximation is saved as SUMVAL to begin the calculation of its successor. The accuracy parameter P takes preassigned values $P1 \leq P2 \leq P3$ with an upper bound of 100 in the program.

(ii) Each step in the summation of an approximation involves four Gaussian random numbers GS11, GS12, MU1 and MU2, with the first two being used to form the corresponding increments of the Wiener processes W^1 and W^2.

To further convince the reader that mixed multiple integrals cannot be expressed as simple formulas involving the corresponding increments of the Wiener processes, we can compare the sample statistics of approximations $J^p_{(1,2),\Delta}$ with the known statistics of the product $\frac{1}{2} \Delta W^1 \Delta W^2$, namely

$$E\left(\frac{1}{2} \Delta W^1 \Delta W^2\right) = 0, \quad \text{Var}\left(\frac{1}{2} \Delta W^1 \Delta W^2\right) = \frac{1}{4} \Delta^2.$$

The required program combines procedures from Programs 2.3.2 and 2.3.3.

Exercise 2.3.1 *Estimate the sample mean and variance of a sample of 10^3 realizations of $J^p_{(1,2),\Delta}$ on the interval $[0, \Delta]$ with $\Delta = 1.0$ and $p = 5, 20$, and 100. Compare the results with the mean and variance of $\frac{1}{2} \Delta W^1 \Delta W^2$.*

F. Relations between Multiple Ito and Stratonovich Integrals

Multiple Ito integrals $I_{(j_1,...,j_l),\Delta} = \int_0^\Delta \ldots \int_0^{s_2} dW^{j_1}_{s_1} \ldots dW^{j_l}_{s_l}$ are analogous defined as multiple Stratonovich integrals in (3.28). There are simple relationships between multiple Ito and Stratonovich integrals which allow the computation of Ito integrals from their Stratonovich counterparts or other Ito integrals. For all $j_1, j_2, j_3 \in \{0, 1, \ldots, m\}$ we have

$$(3.41) \qquad\qquad I_{(j_1),\Delta} = J_{(j_1),\Delta},$$

$$(3.42) \qquad\qquad I_{(j_1,j_2),\Delta} = J_{(j_1,j_2),\Delta} - \frac{1}{2} I_{\{j_1=j_2\neq 0\}} \Delta$$

$$(3.43) \qquad I_{(j_1,j_2,j_3),\Delta} = J_{(j_1,j_2,j_3),\Delta} - \frac{1}{2} I_{\{j_1=j_2\neq 0\}} J_{(0,j_3),\Delta}$$
$$- \frac{1}{2} I_{\{j_2=j_3\neq 0\}} J_{(j_1,0),\Delta},$$

where I_A is the indicator function with value 1 if A is true and 0 otherwise. Further we mention that

$$(3.44) \qquad\qquad I_{(j,j),\Delta} = \frac{1}{2} \left((\Delta W^j)^2 - \Delta \right)$$

and

$$(3.45) \qquad\qquad I_{(j,j,j),\Delta} = \frac{1}{3!} \left((\Delta W^j)^2 - 3\Delta \right) \Delta W^j$$

for $j \in \{1, \ldots, m\}$.

Finally we remark that there are also the following relations between multiple Ito integrals

$$(3.46) \qquad\qquad \Delta W^j \Delta = I_{(j,0),\Delta} + I_{(0,j),\Delta}$$

$$(3.47) \qquad\qquad I_{(j,j),\Delta} \Delta = I_{(j,j,0),\Delta} + I_{(j,0,j),\Delta} + I_{(0,j,j),\Delta}$$

$$(3.48) \qquad\qquad I_{(0,j),\Delta} \Delta W^j = 2 I_{(0,j,j),\Delta} + I_{(j,0,j),\Delta} + \frac{1}{2} \Delta^2$$

$$(3.49) \qquad\qquad I_{(j,0),\Delta} \Delta W^j = I_{(j,0,j),\Delta} + 2 I_{(j,j,0),\Delta} + \frac{1}{2} \Delta^2$$

for $j \in \{1, \ldots, m\}$. We gave analogous relations for Stratonovich integrals already in (3.40).

Literature for Chapter 2

There are many books which introduce the theory of stochastic differential equations such as Gikhman & Skorokhod (1972), Arnold (1974), Øksendal (1985) and Gard (1988). More advanced treatments are provided e.g. by Ikeda & Watanabe (1981, 89), Karatzas & Shreve (1988) or Protter (1990). First results on the Ito-Taylor formula can be found in Wagner & Platen (1978) and Platen & Wagner (1982). Azencott (1982) generalized this. The Stratonovich-Taylor formula was given in Kloeden & Platen (1991a). Both are derived in Kloeden & Platen (1992a). Approximations on multiple stochastic integrals are considered in Liske, Platen & Wagner (1982), Milstein (1988a) and Kloeden, Platen & Wright (1992). Relations between multiple Ito- and Stratonovich integrals are considered in Kloeden & Platen (1991b).

Chapter 3

Introduction to
Discrete Time Approximation

In this chapter we introduce the reader to the basic issues concerning the discrete time approximation of solutions of stochastic differential equations. First, we review the use and properties of numerical methods for deterministic ordinary differential equations. Then we examine the stochastic Euler scheme in some detail, introducing the concepts of strong and weak convergence for discrete time approximations.

3.1 Numerical Methods for Ordinary Differential Equations

Here we shall review some basic facts and properties of numerical methods for ordinary differential equations which will also be important in the stochastic context.

A. Euler Method

Consider the solution $x = x(t; t_0, x_0)$ of an *initial value problem* (IVP)

$$(1.1) \qquad \frac{dx}{dt} = a(t, x), \qquad x(t_0) = x_0$$

for a deterministic ordinary differential equation. The most widely applicable and commonly used numerical methods to solve the IVP (1.1) are the *discrete time approximation* or *difference methods*.

The simplest difference method for the IVP (1.1) is the *Euler method*

$$(1.2) \qquad y_{n+1} = y_n + a(t_n, y_n)\, \Delta_n$$

for a given time discretization $t_0 < t_1 < t_2 < \cdots < t_n < \cdots$ with increments $\Delta_n = t_{n+1} - t_n$ where $n = 0, 1, 2, \ldots$. Once the initial value y_0 has been specified, usually $y_0 = x_0$, the approximations $y_1, y_2, \ldots, y_n, \ldots$ can be calculated by applying formula (1.2) recursively. It is derived by freezing the right hand side of the differential equation over the time interval $t_n \leq t < t_{n+1}$ at the value $a(t_n, y_n)$ and then integrating to obtain the tangent to the solution $x(t; t_n, y_n)$ of the differential equation with the initial value $x(t_n) = y_n$. The difference

$$(1.3) \qquad l_{n+1} = x(t_{n+1}; t_n, y_n) - y_{n+1},$$

which is generally not zero, is called the *local discretization error* for the nth time step. It is usually not the same as the *global discretization error*

$$(1.4) \qquad e_{n+1} = x(t_{n+1}; t_0, x_0) - y_{n+1}$$

for the same time step, which is the error with respect to the solution $x(t; t_0, x_0)$ of the original IVP (1.1).

Problem 3.1.1 (PC-Exercise 8.1.1) *Apply the Euler method (1.2) to the IVP*

$$\frac{dx}{dt} = -5\,x, \qquad x(0) = 1,$$

with time steps of equal length $\Delta_n = \Delta = 2^{-3}$ *and* 2^{-5} *over the time interval* $0 \le t \le 1$. *Plot the results and the exact solution* $x(t) = e^{-5t}$ *against* t.

Program 3.1.1   Exact solution and Euler approximation

```
USES CRT,DOS,GRAPH,INIT,SETSCR,SERVICE;

CONST
  NUM=NUMINV;           { maximum number of time steps used         }
  T0=0.0;               { left end point                            }
  T=1.0;                { right end point                           }
  DELTA0=(T-T0)/NUM;    { time step size for plotting the exact solution }
  DELTA1=(T-T0)/8;      { time step size of the Euler approximation }
  DELTA2=(T-T0)/32;     { time step size of the Euler approximation }
  X0=1.0;               { initial value                             }
  ALPHA=-5.0;           { parameter of the function A(t,x)          }
  ABSCMIN=T0;           { left end point                            }
  ABSCMAX=T;            { right end point                           }
  ORDMIN=0.0;           { minimum of the ordinate                   }
  ORDMAX=+1.0;          { maximum of the ordinate                   }
  ORDPOINT=0.5;         { significant ordinate point                }

TYPE
  VECTOR=ARRAY[0..NUM] OF REAL;

VAR
  CR:STRING;                { help string                           }
  AXISX,AXISY:INTEGER;      { location of the axes                  }
  DISTX,DISTY:INTEGER;      { scale parameters                      }
  N0,N1,N2:INTEGER;         { numbers of time steps                 }
  I:INTEGER;                { time step                             }
  TI:REAL;                  { subinterval point                     }
  YTOLD:REAL;               { last value of the approximation       }
  XT:VECTOR;                { values of the exact solution X(t)     }
  YT1:VECTOR;               { values of the Euler approximation using DELTA1 }
  YT2:VECTOR;               { values of the Euler approximation using DELTA2 }
  ABSCISSA0:VECTOR;         { values of the subinterval points      }
  ABSCISSA1:VECTOR;         { values of the subinterval points      }
  ABSCISSA2:VECTOR;         { values of the subinterval points      }

FUNCTION A(TI,XI:REAL):REAL;
BEGIN
```

```
 A:=ALPHA*XI;
END;{ A }

FUNCTION EXPLSOL(TI:REAL):REAL;
BEGIN
 EXPLSOL:=X0*EXP(ALPHA*(TI-T0));
END;{ EXPLSOL }

PROCEDURE COORDSYS(CY,CX:STRING);
PROCEDURE PLOTGRAPH1(LTN,NO,NN:INTEGER;ORDINATE,ABSCISSA:VECTOR);

{ Main program : }
   ...
 NO:=TRUNC((T-T0)/DELTA0+0.1);N1:=TRUNC((T-T0)/DELTA1+0.1);
 N2:=TRUNC((T-T0)/DELTA2+0.1);
 TI:=T0-DELTA0;
 I:=-1;
 REPEAT
  I:=I+1;               { time step                       }
  TI:=TI+DELTA0;        { time                            }
  XT[I]:=EXPLSOL(TI); { exact value of the solution }
  ABSCISSA0[I]:=TI;   { value of the x-axis          }
 UNTIL I=NO;
 YT1[0]:=X0;ABSCISSA1[0]:=T0;
 TI:=T0;
 I:=0;
 REPEAT
  I:=I+1; { time step }
  YTOLD:=YT1[I-1];
  YT1[I]:=YTOLD+A(TI,YTOLD)*DELTA1; { Euler scheme }
  TI:=TI+DELTA1;    { time                }
  ABSCISSA1[I]:=TI; { value of the x-axis }
 UNTIL I=N1;
 YT2[0]:=X0;ABSCISSA2[0]:=T0;
 TI:=T0;
 I:=0;
 REPEAT
  I:=I+1; { time step }
  YTOLD:=YT2[I-1];
  YT2[I]:=YTOLD+A(TI,YTOLD)*DELTA2; { Euler scheme }
  TI:=TI+DELTA2;    { time                }
  ABSCISSA2[I]:=TI; { value of the x-axis }
 UNTIL I=N2;

{ Printout : }
   ...
```

Notes 3.1.1

(i) The procedures COORDSYS and PLOTGRAPH1, which prepare the screen and plot the trajectories, are similar to those in Program 1.7.2. A difference is that PLOTGRAPH1 puts the x-axis at the bottom of the screen whereas PLOTGRAPH places it in the centre.

(ii) Numerical experiments with different equations are facilitated by the use of the routines FUNCTION and EXPLSOL for the function on the right hand side of the differential equation and the corresponding explicit solution. The parameter ALPHA is the coefficient of the linear function under consideration. Computationally more efficient programs are possible, but at the expense of the flexibility and structural transparency of the one here.

(iii) The stepsize DELTA0 = 1/NUMINV allows the highest possible screen resolution for plotting the explicit solution, this being 2^{-9} for an EGA or VGA graphics card. This and other parameters in the program can be changed as desired, but the effect on the graphics must be borne in mind. Numerical instabilities can be observed when the step size is large.

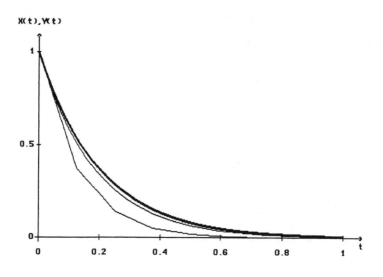

Figure 3.1.1   Exact solution and Euler approximation.

In Figure 3.1.1 we see that the lowest curve is for the Euler approximation with $\Delta = 2^{-3}$. The global discretization error here is smaller for smaller time step sizes.

We shall examine the dependence of the global truncation error on the step size. For this we recall that a function $f(\Delta) = A \Delta^{\gamma}$ becomes linear in logarithmic coordinates, that is

$$\log_a f(\Delta) = \log_a A + \gamma \log_a \Delta$$

for logarithms to the base $a \neq 1$. In comparative studies it is convenient to take time steps of the form $\Delta = a^{-k}$ for $k = 1, 2, \ldots$ and $a > 1$. We shall usually halve the time step successively, in which case logarithms to the base $a = 2$ will be appropriate.

Problem 3.1.2 *Plot the linearly interpolated values of* $\log_2 f(\Delta)$ *against* $\log_2 \Delta$ *for the two functions* $f(\Delta) = \Delta$ *and* $f(\Delta) = \Delta^2$ *with* $\Delta = 2^{-k}$ *for* $k = 0, 1, \ldots, 5$.

We omit the program as it is elementary, though such logarithmic scaling of the axes will be used frequently in what follows.

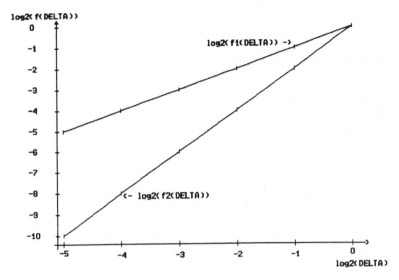

Figure 3.1.2 $\log_2 f(\Delta)$ versus $\log_2 \Delta$.

Problem 3.1.3 (PC-Exercise 8.1.2)   *For the IVP in Problem 3.1.1 calculate the global discretization error at time $t = 1$ for the Euler method with time steps of equal length $\Delta = 1, 2^{-1}, 2^{-2}, \ldots, 2^{-12}$, rounding off to 4 significant digits. Plot the logarithm to the base 2 of these errors against $\log_2 \Delta$ and determine the slope of the resulting curve.*

Program 3.1.3   Global discretization error

```
USES CRT,DOS,GRAPH,INIT,SETSCR,SERVICE,AAGRAPHS;

CONST
  TO=0;        { left end point                                       }
  T=1;         { right end point                                      }
  DELTA=1;     { largest time step size                               }
  NUM=13;      { number of different time step sizes                  }
  X0=1.0;      { initial value                                        }
  ALPHA=-5.0;  { parameter of the function A(t,x)                     }
  R=4;         { parameter for the simulation of the roundoff error   }

VAR
  CR:STRING;      { help string                     }
  I:INTEGER;      { time step                       }
  K:INTEGER;      { index of the time step size used }
  TI:REAL;        { subinterval point               }
  DELTA_Y:REAL;   { time step size                  }
  Q0:REAL;        { constant for rounding off       }
  XT:REAL;        { exact solution X(t) at time t=T }
```

```
YT:VECTOR1;    { approximations                        }
EPS:VECTOR1;   { global discretization errors          }
DEL:VECTOR1;   { time step sizes                        }

FUNCTION A(TI,XI:REAL):REAL;
BEGIN
 A:=ALPHA*XI;
END;{ A }

FUNCTION EXPLSOL(TI:REAL):REAL;

BEGIN
 EXPLSOL:=X0*EXP(ALPHA*(TI-T0));
END;{ EXPLSOL }

{ Main program : }
  ...
 XT:=EXPLSOL(T); { exact value of the solution }

 Q0:=EXP(R*LN(10));
 DELTA_Y:=DELTA*2;
 K:=0;
 REPEAT
  K:=K+1;              { index of the time step size DELTA used }
  I:=0;                { number of time steps                   }
  TI:=T0;              { initial time                           }
  YT[K]:=X0;           { initial value of the approximation     }
  DELTA_Y:=DELTA_Y/2;  { current time step size                 }
  WHILE (TI < T) DO
   BEGIN
    I:=I+1; { time step }
    TI:=TI+DELTA_Y; { right subinterval end point }
    YT[K]:=YT[K]+A(TI-DELTA_Y,YT[K])*DELTA_Y; { Euler scheme }
    YT[K]:=ROUND(Q0*YT[K])/Q0; { simulates the roundoff error }
   END;
 UNTIL K=NUM;

 FOR K:=1 TO NUM DO
  BEGIN
   EPS[K]:=ABS(XT-YT[K]); { global discretization error at time T }
   IF K>1 THEN DEL[K]:=DEL[K-1]/2 ELSE DEL[K]:=DELTA;
  END;
  ...
 CR:='EPS(Euler)';
 GRAPH313(NUM,EPS,DEL,'log2(EPS)','log2(DELTA)',CR);
  ...
```

Notes 3.1.3

(i) The vectors EPS and DEL of the abscissa values in powers of 2 and the corresponding discretization error values, respectively, have dimension NUM $\leq$ 50. These vectors are the inputs to the automatically adapted graphics routine GRAPH313 which plots the linearly interpolated error trajectory on the screen in logarithmic coordinates. Linear interpolation is used to highlight the slope of different parts of the curve.

(ii) The integer parameter R denotes the number of significant digits to be re-
tained in the simulated rounding off process. Repeating the calculations for
increasing values of R provides some insight into the actual role of roundoff
error in computational work. As in Program 3.1.3 the differential equation
function and corresponding explicit solution are defined in subroutines at
the beginning of the program for ease of change to allow the effect of different equations to be investigated.

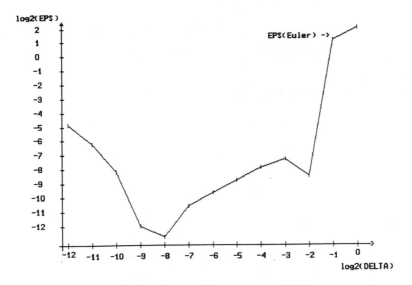

Figure 3.1.3   Global discretization error for the Euler method.

We see from Figure 3.1.3 that the calculated global discretization error is proportional to the step size Δ provided Δ is not too large; $\Delta \leq 2^{-3}$ here. For
$\Delta \leq 2^{-8}$ the global error begins to increase here as Δ is further decreased.
This effect is due to the roundoff error which influences the calculated global
error. Here we have artificially enhanced the influence of the roundoff error by
rounding off to 4 significant digits in each step.

Exercise 3.1.3   *Repeat Problem 3.1.3 with rounding off to 3, 5 and 6 significant digits.*

B. Higher Order Methods

We say that a method converges with order γ if there exists a constant $K < \infty$
such that the absolute global discretization error $|e_{n+1}|$ can be bounded from
above by $K\Delta^\gamma$ for all $\Delta \in (0, \delta_0)$ for some $\delta_0 > 0$. The Euler method has order
$\gamma = 1.0$.

The influence of the roundoff error is neglected here. In practice, its presence means that there is a minimum step size $\Delta_{\min}$ for each initial value problem, below which we cannot improve the accuracy of the approximations calculated by means of the given method such as the Euler method. To obtain a more accurate approximation we need to use another method with a higher order discretization error.

For instance, we could use the *trapezoidal method*

$$(1.5) \qquad y_{n+1} = y_n + \frac{1}{2} \left\{ a(t_n, y_n) + a(t_{n+1}, y_{n+1}) \right\} \Delta,$$

where Δ is the equidistant stepsize. This is called an *implicit* scheme because the unknown quantity y_{n+1} appears in both sides of (1.5) and, in general, cannot be solved for algebraically. To circumvent this difficulty we could use the Euler method (1.2) to approximate the y_{n+1} term on the right hand side of (1.5). Then we obtain the *modified trapezoidal method*

$$\bar{y}_{n+1} \;\; = \;\; y_n + a(t_n, y_n)\,\Delta$$

$$y_{n+1} \;\; = \;\; y_n + \frac{1}{2} \left\{ a(t_n, y_n) + a(t_{n+1}, \bar{y}_{n+1}) \right\} \Delta,$$

or, written together,

$$(1.6) \qquad y_{n+1} = y_n + \frac{1}{2} \left\{ a(t_n, y_n) + a\left(t_{n+1}, y_n + a(t_n, y_n)\,\Delta \right) \right\} \Delta,$$

which is also known as the *improved Euler* or *Heun method*. It is a simple example of a *predictor-corrector method* with the predictor $\bar{y}_{n+1}$ inserted into the corrector equation to give the next iterate y_{n+1}.

Both the trapezoidal and the modified trapezoidal methods have local discretization errors of order three in Δ.

Problem 3.1.4 (PC-Exercise 8.1.3) *Repeat Problem 3.1.3 with the usual arithmetic of your PC for the modified trapezoidal method (1.6). Compare the results with those for the Euler method.*

Notes 3.1.4 We omit Program 3.1.4 because it is similar to Program 3.1.3, using the modified trapezoidal method (1.6), which can be rearranged algebraically here as an explicit method, instead of the Euler scheme and the simulated rounding procedure.

The corresponding printout from the screen for Problem 3.1.4 should show that the global error for the modified trapezoidal method is proportional to Δ^2 for $2^{-12} \leq \Delta \leq 2^{-3}$ whereas that of the Euler method in Figure 3.1.3 was proportional to Δ within some smaller interval of stepsizes.

There are different methods, called *multistep methods*, which achieve higher accuracy by using information from previous discretization intervals. An example is the 3-*step Adams-Bashford method*

$$(1.7) \quad y_{n+1} = y_n + \frac{1}{12} \left\{ 23\, a(t_n, y_n) - 16\, a(t_{n-1}, y_{n-1}) + 5\, a(t_{n-2}, y_{n-2}) \right\} \Delta$$

which turns out to have third order global discretization error. A one-step method of the same order is usually used to generate the values required to start a multistep method.

Problem 3.1.5 (PC-Exercise 8.1.5) *Repeat Problem 3.1.4 using the 3-step Adams-Bashford method (1.7) with the Heun method (1.6) as its starting routine.*

Program 3.1.5 Adams-Bashford method

```
USES CRT,DOS,GRAPH,INIT,SETSCR,SERVICE,AAGRAPHS;

CONST
  T0=0;       { left end point                         }
  T=1;        { right end point                        }
  DELTA=1;    { largest time step size                 }
  NUM=13;     { number of different time step sizes    }
  X0=1.0;     { initial value                          }
  ALPHA=-5.0; { parameter of the function A(t,x)       }

VAR
  CR:STRING;         { help string                                     }
  I:INTEGER;         { time step                                       }

  K:INTEGER;         { index of the time step size used                }
  TI:REAL;           { subinterval point                               }
  DELTA_Y:REAL;      { time step size                                  }
  AA1,AA2,AA3:REAL;  { values of A at the left subinterval end points  }
  XT:REAL;           { exact solution X(t) at time t=T                 }
  YT:VECTOR1;        { approximations                                  }
  EPS:VECTOR1;       { global discretization errors                    }
  DEL:VECTOR1;       { time step sizes                                 }

FUNCTION A(TI,XI:REAL):REAL;
BEGIN
 A:=ALPHA*XI;
END;{ A }

FUNCTION EXPLSOL(TI:REAL):REAL;
BEGIN
 EXPLSOL:=X0*EXP(ALPHA*(TI-T0));
END;{ EXPLSOL }

{ Main program : }
  ...
  XT:=EXPLSOL(T); { exact value of the solution }

  DELTA_Y:=DELTA*2;
  K:=0;
  REPEAT
   K:=K+1;                    { index of the time step size DELTA used }
```

```
I:=0;                    { number of time steps            }
TI:=T0;                  { initial time                    }
YT[K]:=X0;               { initial value of the approximation }
DELTA_Y:=DELTA_Y/2; { current time step size          }

{ Heun method as the starting routine : }

TI:=TI+DELTA_Y;
AA1:=A(TI-DELTA_Y,YT[K]);
YT[K]:=YT[K]+0.5*(AA1+A(TI,YT[K]+AA1*DELTA_Y))*DELTA_Y;
TI:=TI+DELTA_Y;
IF TI<=T THEN
  BEGIN
    AA2:=A(TI-DELTA_Y,YT[K]);
    YT[K]:=YT[K]+0.5*(AA2+A(TI,YT[K]+AA2*DELTA_Y))*DELTA_Y;
    I:=2;
  END;

{ Iteration : }

WHILE (TI < T) DO
  BEGIN
    I:=I+1; { time step }
    TI:=TI+DELTA_Y; { right subinterval end point }

  { 3-step Adams-Bashford scheme : }

    AA3:=A(TI-DELTA_Y,YT[K]);
    YT[K]:=YT[K]+(1/12)*(23*AA3-16*AA2+5*AA1)*DELTA_Y;

    AA1:=AA2;AA2:=AA3;
  END;
UNTIL K=NUM;

FOR K:=1 TO NUM DO
  BEGIN
    EPS[K]:=ABS(XT-YT[K]); { global discretization error at time T }
    IF K>1 THEN DEL[K]:=DEL[K-1]/2 ELSE DEL[K]:=DELTA;
  END;

CR:='EPS(3-step Adams-Bashford)';
GRAPH313(NUM,EPS,DEL,'log2(EPS)','log2(DELTA)',CR);
  ...
```

Notes 3.1.5   We list only the starting routine and multistep iteration parts of the program here, the rest being the same as in Program 3.1.3.

Sometimes we can obtain higher order accuracy from a one-step scheme by an *extrapolation method*. For example, suppose we use the Euler scheme (1.2) with N equal time steps $\Delta = T/N$ on the interval $0 \leq t \leq T$ to determine the value $y_N(\Delta)$. We can then repeat this calculation with $2N$ equal time steps of length $\frac{1}{2}\Delta$ to evaluate $y_{2N}\left(\frac{1}{2}\Delta\right)$. Then the Romberg or Richardson extrapolation formula

$$(1.8) \qquad Z_N(\Delta) = 2y_{2N}\left(\frac{1}{2}\Delta\right) - y_N(\Delta),$$

provides a second order approximation. It is possible to generalize this to higher order extrapolations.

Problem 3.1.6 (PC-Exercise 8.1.7) *Compare the error of the Euler and Richardson extrapolation approximations of $x(1)$ for the solution of the initial value problem*

$$\frac{dx}{dt} = -x, \qquad x(0) = 1$$

for equal time steps $\Delta = 2^{-3}, 2^{-4}, \dots, 2^{-10}$. Plot the logarithm to the base 2 of the errors against $\log_2 \Delta$.

Notes 3.1.6 Program 3.1.6 can be found on the diskette. It is a simple variation of Programs 3.1.1 - 3.1.3 with the graphics routine GRAPH312 from Program 3.1.2 which allows two error curves to be plotted on the screen at the same time.

C. Further Higher Order Schemes

It is easy to derive higher order methods by truncating Taylor expansions. These are usually more of theoretical than practical interest because they involve higher order derivatives of the function a, which may be complicated.

The 1st order truncated Taylor method is just the Euler method (1.2). The *2nd order truncated Taylor method* is

$$(1.9) \qquad y_{n+1} = y_n + a\,\Delta + \frac{1}{2!}\left\{\frac{\partial}{\partial t}a + a\frac{\partial}{\partial x}a\right\}\Delta^2$$

and the *3rd order truncated Taylor method*

$$(1.10) \quad y_{n+1} = y_n + a\,\Delta + \frac{1}{2!}\left\{\frac{\partial}{\partial t}a + a\frac{\partial}{\partial x}a\right\}\Delta^2$$

$$+ \frac{1}{3!}\left\{\frac{\partial^2}{\partial t^2}a + 2a\frac{\partial^2}{\partial t\,\partial x}a + a^2\frac{\partial^2}{\partial x^2}a + \left(\frac{\partial}{\partial t}a\right)\frac{\partial}{\partial x}a + a\left(\frac{\partial}{\partial x}a\right)^2\right\}\Delta^3,$$

where a and its partial derivatives are evaluated at (t_n, y_n).

Problem 3.1.7 (PC-Exercise 8.2.1) *Use the 2nd order truncated Taylor method (1.9) with equal length time steps $\Delta = 2^{-3}, \dots, 2^{-10}$ to calculate approximations to the solution $x(t) = 2/(1 + e^{-t^2})$ of the initial value problem*

$$\frac{dx}{dt} = t\,x\,(2 - x), \qquad x(0) = 1$$

over the interval $0 \leq t \leq 0.5$. Repeat the calculations using the 3rd order truncated Taylor method (1.10). Plot $\log_2$ of the global discretization errors at time $t = 0.5$ against $\log_2 \Delta$.

Program 3.1.7   Truncated Taylor scheme

```
   ...
FUNCTION A(TI,XI:REAL):REAL;
BEGIN
 A:=TI*XI*(2.-XI);
END;{ A }

FUNCTION DAT(TI,XI:REAL):REAL;
BEGIN
 DAT:=XI*(2.-XI);
END;{ DAT }

FUNCTION DAX(TI,XI:REAL):REAL;
BEGIN
 DAX:=TI*(2.-2.*XI);
END;{ DAX }

FUNCTION DDATT(TI,XI:REAL):REAL;
BEGIN
 DDATT:=0.0;
END;{ DDATT }

FUNCTION DDATX(TI,XI:REAL):REAL;
BEGIN
 DDATX:=2.*(1.-XI);
END;{ DDATX }

FUNCTION DDAXX(TI,XI:REAL):REAL;
BEGIN
 DDAXX:=-2.*TI;
END;{ DDAXX }

FUNCTION EXPLSOL(TI:REAL):REAL;
VAR
 C:REAL;
BEGIN
 C:=X0/(2.-X0);
 EXPLSOL:=2.*(C/(C+EXP(-TI*TI)));
END;{ EXPLSOL }

{ Main program : }
   ...
XT:=EXPLSOL(T); { exact value of the solution }

DELTA_Y:=DELTA*2;
K:=0;
REPEAT
 K:=K+1;              { index of the time step size DELTA used         }
 I:=0;                { number of time steps                           }
 TI:=T0;              { initial time                                   }
 Y1T[K]:=X0;          { initial value of the 2nd order Taylor approximation }
 Y2T[K]:=X0;          { initial value of the 3rd order Taylor approximation }
 DELTA_Y:=DELTA_Y/2;  { current time step size                         }
 WHILE (TI < T) DO
  BEGIN
   I:=I+1;            { time step                                      }
   TI:=TI+DELTA_Y;    { right subinterval end point                    }
```

```
{ 2nd order Taylor scheme : }

  AA:=A(TI-DELTA_Y,Y1T[K]);
  D1:=DAT(TI-DELTA_Y,Y1T[K]);
  D2:=DAX(TI-DELTA_Y,Y1T[K]);
  Y1T[K]:=Y1T[K]+AA*DELTA_Y+0.5*(D1+D2*AA)*DELTA_Y*DELTA_Y;

{ 3rd order Taylor scheme : }

  AA:=A(TI-DELTA_Y,Y2T[K]);
  D1:=DAT(TI-DELTA_Y,Y2T[K]);
  D2:=DAX(TI-DELTA_Y,Y2T[K]);
  DD11:=DDATT(TI-DELTA_Y,Y2T[K]);
  DD12:=DDATX(TI-DELTA_Y,Y2T[K]);
  DD22:=DDAXX(TI-DELTA_Y,Y2T[K]);
  H1:=DD11+2*DD12*AA+DD22*AA*AA+D1*D2+D2*D2*AA;
  H1:=H1*DELTA_Y*DELTA_Y*DELTA_Y;
  Y2T[K]:=Y2T[K]+AA*DELTA_Y+(1/2)*(D1+D2*AA)*DELTA_Y*DELTA_Y+H1/6;

  END;
UNTIL K=NUM;

FOR K:=1 TO NUM DO
  BEGIN
  EPS1[K]:=ABS(XT-Y1T[K]); { global discretization error(2nd Taylor) at t=T }
  EPS2[K]:=ABS(XT-Y2T[K]); { global discretization error(3rd Taylor) at t=T }
  IF K>1 THEN DEL[K]:=DEL[K-1]/2 ELSE DEL[K]:=DELTA;
  END;
  ...
```

Notes 3.1.7

(i) The essential change in Program 3.1.7 from the previous ones is in its iterative core listed above for the 2nd and 3rd order truncated Taylor methods together.

(ii) The problem can be repeated for other differential equations by changing the function $A(T,X)$ on the right hand side, the required partial derivatives of A and the corresponding explicit solution EXPLSOL.

(iii) The partial derivatives $\frac{\partial A}{\partial t}$ and $\frac{\partial^2 A}{\partial t \, \partial x}$ are denoted by DAT and DDATX, respectively. with analogous notation for the other partial derivatives of A.

In the corresponding printout a curve with slope 2 should be obtained for the 2nd order truncated Taylor method and one with slope 3 for the 3rd order truncated Taylor method.

The classical 4*th order Runge-Kutta method* is an important explicit method which avoids the use of derivatives. It has the form

$$(1.11) \qquad y_{n+1} = y_n + \frac{1}{6} \left\{ k_n^{(1)} + 2\, k_n^{(2)} + 2\, k_n^{(3)} + k_n^{(4)} \right\} \Delta$$

where

$$k_n^{(1)} \;=\; a\left(t_n, y_n\right),$$

$$k_n^{(2)} = a\left(t_n + \frac{1}{2}\Delta, y_n + \frac{1}{2}k_n^{(1)}\Delta\right),$$

$$k_n^{(3)} = a\left(t_n + \frac{1}{2}\Delta, y_n + \frac{1}{2}k_n^{(2)}\Delta\right),$$

$$k_n^{(4)} = a\left(t_{n+1}, y_n + k_n^{(3)}\Delta\right).$$

Problem 3.1.8 (PC-Exercise 8.2.2) *Repeat Problem 3.1.7 using the classical 4th order Runge-Kutta method (1.11) with equal length time steps $\Delta = 2^{-2}$, ..., 2^{-7}.*

Notes 3.1.8 Program 3.1.8, listed on the diskette, is an obvious variation of Program 3.1.7; the corresponding curve would show a 4th order rate of convergence.

Multistep methods often do not require as many evaluations of the function a per time step as one-step methods of the same order. They may, however, suffer from serious numerical instabilities. An example for which this occurs is the simple *midpoint method*

$$(1.12) \qquad\qquad y_{n+1} = y_{n-1} + 2\,a(t_n, y_n)\,\Delta,$$

which is a two-step method.

Problem 3.1.9 (PC-Exercise 8.2.3) *Calculate the discretization errors in using the Euler method (1.2) and the midpoint method (1.12) started with the Euler method to approximate the solution $x(t) = \frac{2}{3}e^{-3t} + \frac{1}{3}$ of the initial value problem*

$$\frac{dx}{dt} = -3x + 1, \qquad x(0) = 1$$

over the interval $0 \le t \le 1$. Use time steps of equal length $\Delta = 0.1$ and plot the results against t.

Program 3.1.9   Euler and midpoint method

```
   ...
FUNCTION A(TI,XI:REAL):REAL;
BEGIN
 A:=-3.*XI+1.;
END;{ A }

FUNCTION EXPLSOL(TI:REAL):REAL;
BEGIN
 EXPLSOL:=X0*((2/3)*EXP(-3.*(TI-T0))+1/3);
END;{ EXPLSOL }

PROCEDURE COORDSYS(CY,CX:STRING);
```

```
PROCEDURE PLOTGRAPH(LTN,NO,NN:INTEGER;ORDINATE,ABSCISSA:VECTOR);

{ Main program : }
  ...
  I:=0;              { number of time steps                    }
  TI:=T0;            { initial time                            }
  Y1T[I]:=X0;        { initial value of the Euler approximation }
  Y2T[I]:=X0;        { initial value of the midpoint approximation }
  EPS1[0]:=0.0;      { initial error of the Euler approximation }
  EPS2[0]:=0.0;      { initial error of the midpoint approximation }
  ABSCISSA[0]:=T0;   { initial value of the x-axis             }

  I:=I+1;                        { new time step                    }
  TI:=T0+DELTA;                  { new subinterval point            }
  Y1T[I]:=X0+A(TI-DELTA,X0)*DELTA; { one Euler step for the midpoint scheme }
  Y2T[I]:=Y1T[I];
  ABSCISSA[I]:=TI;   { value of the x-axis          }
  XT:=EXPLSOL(TI);   { exact solution               }
  EPS1[I]:=XT-Y1T[I]; { error of the Euler method    }
  EPS2[I]:=XT-Y2T[I]; { error of the midpoint method }

  WHILE (TI < T) DO
    BEGIN
    I:=I+1; { time step }
    TI:=TI+DELTA; { right subinterval end point }

    { Euler scheme : }

    Y1T[I]:=Y1T[I-1]+A(TI-DELTA,Y1T[I-1])*DELTA;

    { Midpoint scheme : }

    Y2T[I]:=Y2T[I-2]+2.*A(TI-DELTA,Y2T[I-1])*DELTA;

    { Calculation of the global discretization errors at time t = TI : }

    XT:=EXPLSOL(TI);    { exact solution               }
    EPS1[I]:=XT-Y1T[I]; { error of the Euler method    }
    EPS2[I]:=XT-Y2T[I]; { error of the midpoint method }

    ABSCISSA[I]:=TI;    { value of the x-axis          }
    END;
  ...
```

Notes 3.1.9

(i) The new feature in this problem is that it examines the time evolution of the global discretization error for a fixed time step size. The errors of the midpoint method errors oscillate with increasing amplitude, so the PLOTGRAPH routine is used since it places the x-axis in the centre of the screen.

(ii) The influence of the particular differential equation on the results can be investigated by simply changing the declared equation function A(T,X) and the corresponding explicit solution EXPLSOL. Care must be taken with the graphics scaling. Repetition of the calculations for different step sizes suggests that the midpoint method is inherently numerically unstable.

The printout here should show a reasonable approximation by the Euler method, but the path generated by the midpoint method may be unacceptable due to numerical instabilities.

D. Roundoff Error

In practice arithmetic operations can only be carried out to a finite number of significant decimal places of accuracy on a digital computer and roundoff errors are not avoidable. Some compilers round off randomly but most do it according to a simple deterministic roundoff rule.

Quite realistic estimates for the accumulated roundoff error can be determined from a statistical analysis, assuming that the local roundoff errors are independent, identically distributed random variables. It is commonly assumed that they are uniformly distributed over the interval

$$(1.13) \qquad\qquad \left[-5 \times 10^{-(s+1)}, \; 5 \times 10^{-(s+1)}\right],$$

where s is the number of significant decimal places used. To check the appropriateness of this distribution we will repeat some calculations using an arithmetic to s decimal places, say $s = 4$, and compare it with the results obtained under the usual precision of the computer.

Problem 3.1.10 (PC-Exercise 8.4.1)   *Calculate* 300 *iterates of*

$$y_{k+1} = \frac{\pi}{3}\, y_k$$

with initial value $y_0 = 0.1$ *using the prescribed arithmetic of the PC, at each step rounding the value of* y_{k+1} *obtained to four significant figures. Plot the relative frequencies of the "roundoff errors" in a histogram on the interval*

$$[-5 \times 10^{-5}, 5 \times 10^{-5}]$$

using 40 *equal subintervals.*

Program 3.1.10   Local roundoff error

```
USES CRT,DOS,GRAPH,INIT,SETSCR,SERVICE;

CONST
  N=300;                              { number of iterations             }
  S=4;                               { number of the significant figures }
  Y0=0.1;                            { initial value for the iteration   }
  ABSCMIN=-0.00005;                  { left end point                   }
  ABSCMAX=+0.00005;                  { right end point                  }
  INVLENGTH=(ABSCMAX-ABSCMIN)/40;    { length of subinterval            }
  ORDMIN=0.0;                        { minimum of the ordinate          }
  ORDMAX=1/(10*INVLENGTH);           { maximum of the ordinate          }
  ORDPOINT=ORDMAX/4;                 { significant ordinate point       }

TYPE
```

```
VECTOR=ARRAY[1..N] OF REAL;

VAR
  CR:STRING;                { help string                    }
  I:INTEGER;                { counter                        }
  AXISX,AXISY:INTEGER;      { location of the axes           }
  DISTX,DISTY:INTEGER;      { scale parameters               }
  QO:REAL;                  { constant 10^S for rounding off }
  YI:REAL;                  { current iteration value        }
  AVERAGE,VARIANCE:REAL;    { statistical parameters         }
  LROE:VECTOR;              { local roundoff errors          }

FUNCTION ITSTEP(YI:REAL):REAL;
BEGIN
 ITSTEP:=PI*YI/3;
END;{ ITSTEP }


PROCEDURE COORDSYS;
PROCEDURE QSORT(DOWN,UP:INTEGER;VAR F:VECTOR);
PROCEDURE HISTOGRAM(NN:INTEGER;F:VECTOR);
PROCEDURE COMPSAMPLEPARA(NN:INTEGER;X:VECTOR;VAR SAVERAGE,SVARIANCE:REAL);
PROCEDURE STATDATATOSCR(AVERAGE,VARIANCE:REAL);


{ Main program : }
   ...
 QO:=EXP(S*LN(10)); { roundoff constant     }
 YI:=YO;  { initial value for the iteration }
 FOR I:=1 TO N DO
   BEGIN
     YI:=ITSTEP(YI);                    { iteration step       }
     LROE[I]:=YI-ROUND(YI*QO)/QO;  { local roundoff error }
   END;
 COMPSAMPLEPARA(N,LROE,AVERAGE,VARIANCE); { computes the sample parameters }

{ Printout : }

 QSORT(1,N,LROE); { sorts the vector LROE }
 CLEARDEVICE;
 COORDSYS; { draws the coordinate system }
 HISTOGRAM(N,LROE); { plots the histogram }
 STATDATATOSCR(AVERAGE,VARIANCE);
   ...
```

Notes 3.1.10

(i) The procedures COORDSYS and HISTOGRAM from Program 1.3.1 are used here. The STATDATATOSCR now displays the sample standard deviation on the screen rather than the sample variance, which can be very small and not visible, even when the printout accuracy parameter ACCUR (in the CHCR procedure of the SERVICES unit) is changed from 6 to 10 to allow a 10^{-10} printout accuracy.

(ii) The FUNCTION subroutine for $\pi y/3$, defined at the beginning of the program, is first evaluated using the usual arithmetic of the PC and the result is then rounded to the desired number of significant digits to give the next iterate. The total number N of iterates can be extended by replacing the divisor 3 by a rational number which is closer to π, since the ratio is then closer to 1 and more powers are possible before arithmetic overflow occurs.

If one considers the worst possible case for an accumulation of roundoff errors lying in an interval like (1.13), then in some linear problem after approximately $N = 10^s$ calculations an accumulated roundoff error of the form

$$R_N = \sum_{k=1}^{N} r_k$$

could cause the loss of all decimal places of accuracy. That means we can only ensure that R_N lies in the interval

(1.14) $$\left[-5N \times 10^{-(s+1)}, 5N \times 10^{-(s+1)}\right].$$

It is more realistic to suppose that the r_k are uniformly distributed over the interval (1.13) having mean and variance

$$\mu = E\left(r_k\right) = 0, \qquad \sigma^2 = \text{Var}\left(r_k\right) = \frac{1}{12} 10^{-2s}.$$

If they are also independent, the accumulated roundoff error has mean and variance

$$E\left(R_N\right) = 0, \qquad \text{Var}\left(R_N\right) = N\sigma^2.$$

By the Central Limit Theorem (see Subsection 1.4.C) the normalized random variables $Z_N = R_N/\sigma\sqrt{N}$ are approximately standard Gaussian for large N. From this, as in Section 1.5, we can conclude that the values of R_N lie with probability 0.95 in the interval

(1.15) $$\left[-1.96 \times 10^{-s}\sqrt{N/12},\ 1.96 \times 10^{-s}\sqrt{N/12}\right]$$

when N is large.

We observe that the interval length increases in (1.14) proportional to N compared with $\sqrt{N}$ in (1.15), which leads in the latter case to considerably smaller and usually more realistic intervals for the accumulated roundoff errors.

Let us study the accumulative roundoff error for different initial conditions in the following linear problem.

Problem 3.1.11 (PC-Exercise 8.4.2) *Use the Euler method with equal time steps* $\Delta = 2^{-2}$ *for the differential equation*

$$\frac{dx}{dt} = x$$

over the interval $0 \leq t \leq 1$ *with* $L = 10^3$ *different initial values* $x(0)$ *between 0.4 and 0.6. Use both four significant figure arithmetic and the prescribed arithmetic of the PC and determine the final accumulative roundoff error* $R_{1/\Delta}$ *in*

each case, plotting them in a histogram on the interval $[-5 \times 10^{-4}, 5 \times 10^{-4}]$
*with 40 equal subintervals. In addition, calculate the sample mean and sample
variance of the* $R_{1/\Delta}$ *values.*

Program 3.1.11   Accumulative roundoff error

```
   ...
QO:=EXP(S*LN(10)); { roundoff constant }

K:=0;
REPEAT
  K:=K+1;                        { index of the initial value used     }
  XO:=X1+(K/(N+1))*(X2-X1); { current initial value                }
  YT:=XO;YTWR:=XO;               { initialization of the approximations }
  AROE[K]:=0.0;                  { initial accumulative error           }
  TI:=TO;
  I:=0;
  WHILE TI<T DO
    BEGIN
      I:=I+1;                              { time step                          }
      TI:=TI+DELTA;                        { time                               }
      YT:=YT+A(TI-DELTA,YT)*DELTA;    { Euler scheme without rounding off }
      AROE[K]:=AROE[K]+YT-ROUND(YT*QO)/QO; { accumulative roundoff error }
    END;
  AROE[K]:=AROE[K]*QO*SQRT(12/NUM);
UNTIL K=N;
COMPSAMPLEPARA(N,AROE,AVERAGE,VARIANCE); { computes the sample parameters }
   ...
```

Notes 3.1.11
(i) The rounding routine from Program 3.1.3 and histogram graphics from
    Program 3.1.10 are used here. Rounding is done within each iteration of
    the Euler method after all calculations have been performed using the usual
    arithmetic of the PC. Essentially, each initial value of the Euler method
    provides a new sample path of the random variable Z_N.

(ii) The choice of differential equations has little effect on the result, as can
     be seen by changing the equation function $A(T,X)$ or its linear coefficient
     ALPHA at the beginning of the program.

Finally we repeat the above problem with different time step sizes.

Problem 3.1.12 (PC-Exercise 8.4.3)   *Repeat Problem 3.1.11 with* $L = 200$
and with time steps $\Delta = 2^{-2}$, 2^{-3}, 2^{-4} *and* 2^{-5}, *determining* $R_{1/\Delta}$ *in each
case. Plot the 90% confidence intervals for the mean value of the error against*
Δ.

A comparison of the confidence intervals for the mean of the accumulated
roundoff error here will show no significant differences for different time step
sizes. Thus the roundoff error can be considered to be in some sense independent of the discretization.

3.2 A Stochastic Discrete Time Simulation

Here we introduce an elementary discrete time approximation called the Euler approximation. We consider a simple example of a simulation where this approximation is used to generate an approximate sample path of a solution of a stochastic differential equation.

A. Euler Approximation

We shall consider an Ito process $X = \{X_t, t_0 \leq t \leq T \}$ satisfying the scalar stochastic differential equation

$$(2.1) \qquad\qquad dX_t = a(t, X_t)\, dt + b(t, X_t)\, dW_t$$

on $t_0 \leq t \leq T$ with the initial value

$$(2.2) \qquad\qquad X_{t_0} = X_0.$$

For a given discretization $t_0 = \tau_0 < \tau_1 < \cdots < \tau_n < \cdots < \tau_N = T$ of the time interval $[t_0, T]$, an *Euler approximation* is a continuous time stochastic process $Y = \{Y(t), t_0 \leq t \leq T\}$ satisfying the iterative scheme

$$(2.3) \quad Y_{n+1} = Y_n + a(\tau_n, Y_n)\, (\tau_{n+1} - \tau_n) + b(\tau_n, Y_n)\left(W_{\tau_{n+1}} - W_{\tau_n}\right),$$

for $n = 0, 1, 2, \ldots, N - 1$ with initial value

$$(2.4) \qquad\qquad Y_0 = X_0,$$

where we have written
$$(2.5) \qquad\qquad Y_n = Y(\tau_n)$$

for the value of the approximation at the discretization time τ_n. In most parts of the book we shall consider equidistant discretization times

$$(2.6) \qquad\qquad \tau_n = t_0 + n\, \Delta$$

with stepsize $\Delta = (T - t_0)/N$ for some integer N. Between the discretization times we may find it convenient to interpolate the approximation, for example, as discussed in the following subsection. When the diffusion coefficient is identically zero, that is when $b \equiv 0$, the stochastic iterative scheme (2.3) reduces to the deterministic Euler scheme (1.2).

The sequence $\{Y_n, n = 0, 1, \ldots, N\}$ of values of the Euler approximation (2.3) at the instants τ_n, $n = 0, 1, \ldots, N$ of the time discretization can be computed in a similar way to those of the deterministic case. The main difference is that we now need to generate the random increments

$$(2.7) \qquad\qquad \Delta W_n = W_{\tau_{n+1}} - W_{\tau_n},$$

for $n = 0, 1, \ldots, N - 1$, of the Wiener process $W = \{W_t, t \geq 0\}$. From Section 1.7 we know that these increments are independent Gaussian random variables with mean
$$(2.8) \qquad\qquad E\left(\Delta W_n\right) = 0$$

and variance
$$(2.9) \qquad E\left((\Delta W_n)^2\right) = \Delta.$$

The increments (2.7) of the Wiener process can be generated by one of the random number generators described in Section 1.2 for independent Gaussian pseudo-random numbers, for example the Box-Muller or Polar Marsaglia generator. Later in Subsection C we shall apply the *Euler scheme*

$$(2.10) \qquad Y_{n+1} = Y_n + a\,\Delta + b\,\Delta W_n,$$

for $n = 0, 1, \ldots, N-1$ to approximate a specific stochastic differential equation. In (2.10) we have omitted the arguments (τ_n, Y_n) in the drift and diffusion coefficients, as well as the initial condition, a practice we shall often follow for other schemes in the sequel.

B. Interpolation of Discrete Time Approximations

The recursive structure of the Euler scheme, which evaluates approximate values of the Ito process at the discretization instants only, is the key to its successful implementation on a digital computer. In this book we shall focus on discrete time approximations with such a recursive structure. We shall use the term *scheme* to denote a recursive algorithm which provides the values of a discrete time approximation at the given discretization instants. We emphasize that we shall always consider a discrete time approximation to be a continuous time stochastic process defined on the whole interval $[t_0, T]$, although we shall mainly be interested in its values at the discretization times. This facilitates many theoretical proofs and the visualization of the resulting sample paths.

The Euler scheme (2.10), as other schemes given later, determines values of the approximating process at the discretization times only. If required, values can then be determined at the intermediate instants by an appropriate interpolation method. The simplest is the *piecewise constant interpolation* with

$$(2.11) \qquad Y(t) = Y_{n_t}$$

for $t \in [0, \infty)$, where n_t is the integer defined by

$$(2.12) \qquad n_t = \max\{n = 0, 1, \ldots, N : \tau_n \le t\},$$

that is the largest integer n for which τ_n does not exceed t. However, the *linear interpolation*

$$(2.13) \qquad Y(t) = Y_{n_t} + \frac{t - \tau_{n_t}}{\tau_{n_t+1} - \tau_{n_t}}\left(Y_{n_t+1} - Y_{n_t}\right)$$

is often used because it is continuous and also simple.

We shall concentrate on the values of a discrete time approximation at the given discretization instants in this book. It will not be possible to reproduce the finer structure of sample paths of an Ito process as these inherit the irregularity of the sample paths of the driving Wiener process, in particular their nondifferentiability.

C. Example of a Simulation

Here we shall examine a simple example of a simulation of a discrete time approximation in some detail. We consider the Ito process $X = \{X_t,\ t \geq 0\}$ satisfying the linear stochastic differential equation

$$(2.14) \qquad\qquad dX_t = aX_t\,dt + bX_t\,dW_t$$

for $t \in [0, T]$ with an initial value $X_0 \in (-\infty, \infty)$. From (2.2.8) we know that (2.14) has the explicit solution

$$(2.15) \qquad X_t = X_0 \exp\left(\left(a - \frac{1}{2}b^2\right)t + bW_t\right)$$

for $t \in [0, T]$ and the given Wiener process $W = \{W_t, t \geq 0\}$. This will allow us to compare the Euler approximation with the exact solution and to calculate the error.

To simulate a linearly interpolated trajectory of the Euler approximation for a given time discretization we start from the initial value $Y_0 = X_0$, and proceed recursively to generate the value

$$(2.16) \qquad\qquad Y_{n+1} = Y_n + aY_n\,\Delta + bY_n\,\Delta W_n$$

at the next instant for $n = 0, 1, 2, \ldots$ according to the Euler scheme (2.10). Here ΔW_n is the $N(0; \Delta)$ distributed Gaussian increment of the Wiener process W over the subinterval $\tau_n \leq t \leq \tau_{n+1}$ which we generate below by the Polar Marsaglia method. For comparison, we can use (2.15) to determine the corresponding values of the exact solution using the same sample path of the Wiener process, namely

$$(2.17) \qquad X_{\tau_n} = X_0 \exp\left(\left(a - \frac{1}{2}b^2\right)\tau_n + b\sum_{i=1}^{n}\Delta W_{i-1}\right).$$

Problem 3.2.1 (PC-Exercise 9.2.1)   *Generate equidistant Euler approximations on the time interval $[0, 1]$ with equal step size $\Delta = 2^{-2}$ for the Ito process X satisfying (2.14) with $X_0 = 1.0$, $a = 1.5$ and $b = 1.0$. Plot both the linearly interpolated approximation and the exact solution for the same sample path of the Wiener process.*

Program 3.2.1   Explicit solution and Euler approximation

```
    . . .
FUNCTION A(TI,XI:REAL):REAL;
BEGIN
 A:=ALPHA*XI;
END;{ A }

FUNCTION B(TI,XI:REAL):REAL;
BEGIN
```

```
 B:=BETA*XI;
END;{ B }

FUNCTION EXPLSOL(TI,WT:REAL):REAL;
BEGIN
 EXPLSOL:=X0*EXP((ALPHA-0.5*BETA*BETA)*(TI-T0)+BETA*WT);
END;{ EXPLSOL }

{ Main program : }
    ...
 APPROXSTEP:=TRUNC(NUMINV*DELTA+0.0000001); { ratio of step sizes  }
 DELTA_X:=(T-T0)/NUMINV;   { time step size for the exact solution }
 SQDELTA_X:=SQRT(DELTA_X); { square root of DELTA_X                }
 TI:=T0;                   { initial time                          }
 XT[0]:=X0;                { initial value of the exact solution   }
 ABSCISSA[0]:=T0;          { value of the x-axis                   }
 WT:=0.0;                  { value of the Wiener process at t = T0 }
 FOR I:=1 TO NUMINV DO
  BEGIN
   TI:=TI+DELTA_X; { time }
   IF I MOD 2 = 1 THEN GENERATE(G1,G2) { uses Polar Marsaglia method }
    ELSE G1:=G2;
   DWT[I-1]:=G1*SQDELTA_X; { Wiener process increment  W(t(i+1)) - W(ti) }
   WT:=WT+DWT[I-1];        { value of the Wiener process W(t) at t = TI  }
   XT[I]:=EXPLSOL(TI,WT);  { exact value of the solution                }
   ABSCISSA[I]:=TI;        { value of the x-axis                        }
  END;

{ Generation of the Euler approximation : }

 I:=0;
 TI:=T0;    { initial time                          }
 YT[0]:=X0; { initial value of the approximation }
 WHILE TI<T DO
  BEGIN
   I:=I+1;  { time step }
   TI:=TI+DELTA; { time }
   YTOLD:=YT[(I-1)*APPROXSTEP]; { saves the old value }
   DWTI:=0.0; { calculates Wiener process increment for Euler time step }
   FOR J:=1 TO APPROXSTEP DO DWTI:=DWTI+DWT[(I-1)*APPROXSTEP+J-1];

   { Euler scheme : }

   YT[I*APPROXSTEP]:=YTOLD+A(TI-DELTA,YTOLD)*DELTA+B(TI-DELTA,YTOLD)*DWTI;

   { Interpolation for the other values : }

    IF APPROXSTEP>1 THEN
     FOR J:=1 TO APPROXSTEP-1 DO
      YT[(I-1)*APPROXSTEP+J]:=YTOLD+J*(YT[I*APPROXSTEP]-YTOLD)/APPROXSTEP;

   END;{ WHILE }

{ Printout : }

 CLEARDEVICE;
 GRAPH321(XT,YT,ABSCISSA,'X(t), Y(t)','t','X(t)','Y(t)'); { plots the paths }
   ...
```

Notes 3.2.1

(i)   Wiener increments are generated by the Polar Marsaglia method for the
step size DELTA_X of the explicit solution, taken to be the smallest possible
for the given screen resolution. These increments are added to provide the
corresponding Wiener process increments for the larger time step DELTA
of the Euler approximation.

(ii) The exact and approximate solutions are plotted on the same screen by the
AAGRAPHS routine GRAPH311.

(iii) The program is modifiable to other stochastic differential equations with the
Wiener process appearing in a similar way in the explicit solution. Simply
change the drift A(T,X), diffusion B(T,X) and explicit solution EXPLSOL
expressions at the beginning of the program.

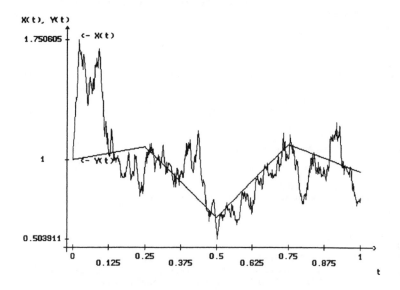

Figure 3.2.1   Euler approximation and exact solution.

In Figure 3.2.1 we see a typical output for Problem 3.2.1. The linearly interpolated Euler approximation obviously differs from the exact solution. However,
we may expect a closer resemblence if we use a smaller step size. We can observe this improvement if we repeat Problem 3.2.1 with a smaller step size.

Exercise 3.2.1 (PC-Exercise 9.2.2)   *Repeat Problem 3.2.1 with step size*
$\Delta = 2^{-4}$.

For nonlinear stochastic differential equations care must be taken in writing
down a scheme to ensure that the resulting expressions are meaningful. For
instance, a variable in a square root may become negative for an approximation,
so the root of its absolute value should be used in the corresponding scheme.

Exercise 3.2.2 *Repeat Problem 3.2.1 with the nonlinear stochastic differential equation (2.2.11) and explicit solution (2.2.12).*

3.3 Pathwise Approximation and Strong Convergence

To judge the quality of a discrete time approximation a criterion needs to be specified. Such a criterion should reflect the main goal of practical simulations. There are two basic types of tasks connected with the simulation of solutions of stochastic differential equations. The first occurs in situations where a good *pathwise approximation* is required, for instance in direct simulations, filtering or testing statistical estimators. In the second interest focuses on *approximating expectations* of functionals of the Ito process, such as its probability distribution and its moments. This is relevant in many practical problems because, often, such functionals cannot be determined analytically.

In the following we shall introduce the absolute error criterion which is appropriate for a situation where a pathwise approximation is required.

A. Absolute Error Criterion

The *absolute error criterion* is just the expectation of the absolute value of the difference between the approximation and the Ito process at the time T, that is

$$(3.1) \qquad \epsilon = E\left(|X_T - Y(T)|\right),$$

and gives a measure of the pathwise closeness at the end of the time interval $[0, T]$.

For a deterministic initial value problem (3.1.1), that is with $b \equiv 0$, the absolute error criterion coincides with the usual deterministic error criterion defined as absolute global discretization error (3.1.4).

We shall use the example of the preceding section to examine the absolute error criterion more closely. Rather than derive a theoretical estimate for the absolute error here, we shall try to estimate it statistically using computer experiments. To this end we shall repeat N different simulations of sample paths of the Ito process and their Euler approximation corresponding to the same sample paths of the Wiener process. We denote the values at time T of the kth simulated trajectories by $X_{T,k}$ and $Y_{T,k}$, respectively, and estimate the absolute error by the statistic

$$(3.2) \qquad \hat{\epsilon} = \frac{1}{N} \sum_{k=1}^{N} |X_{T,k} - Y_{T,k}|.$$

Problem 3.3.1 (PC-Exercise 9.3.1) *Simulate $N = 25$ trajectories of the Ito process X satisfying (2.14) with $X_0 = 1.0$, $a = 1.5$, $b = 1.0$ and their Euler approximations with equidistant time steps of step size $\Delta = 2^{-4}$ corresponding* *to the same sample paths of the Wiener process on the time interval* $[0, T]$ *for* $T = 1$. *Evaluate the statistic* $\hat{e}$ *defined by (3.2). Repeat for step sizes* $\Delta = 2^{-5}$, 2^{-6} *and* 2^{-7}, *and form a table of the corresponding* Δ *and* $\hat{e}$ *values.*

Program 3.3.1   Absolute error for different time steps

```
FUNCTION A(TI,XI:REAL):REAL;
BEGIN
 A:=ALPHA*XI;
END;{ A }

FUNCTION B(TI,XI:REAL):REAL;
BEGIN
 B:=BETA*XI;
END;{ B }

FUNCTION EXPLSOL(TI,WT:REAL):REAL;
BEGIN
 EXPLSOL:=XO*EXP((ALPHA-0.5*BETA*BETA)*(TI-TO)+BETA*WT);
END;{ EXPLSOL }

FUNCTION ABSERR(XT,YT:REAL):REAL;
BEGIN
 ABSERR:=ABS(XT-YT); { absolute error }
END;{ ABSERR }

PROCEDURE SETTABLETOSCR;

{ Main program : }
   ...
 DELTA_Y:=2.*DELTA;
 G:=0;
 REPEAT
  G:=G+1;                    { index of the time step size used    }
  DELTA_Y:=DELTA_Y/2;        { time step size of the approximation }
  SQDELTA_Y:=SQRT(DELTA_Y);  { square root of DELTA_Y              }
  EPS[G]:=0.0;               { initial error                       }

  { Generation of different trajectories : }

  K:=0;
  REPEAT
   K:=K+1;  { index of the trajectory used        }
   WT:=0.0; { value of the Wiener process at t = TO }

   { Generation of the Euler approximation and its absolute error : }

   I:=0;
   TI:=TO; { initial time                          }
   YT:=XO; { initial value of the approximation }
   WHILE TI<T DO
     BEGIN
      I:=I+1;         { time step }
      TI:=TI+DELTA_Y; { time      }
      IF I MOD 2 = 1 THEN GENERATE(G1,G2) { uses Polar Marsaglia method }
       ELSE G1:=G2;
      DWTI:=G1*SQDELTA_Y; { Wiener process increment  W(t(i+1)) - W(ti) }
```

```
    WT:=WT+DWTI;          { value of the Wiener process W(t) at t = TI  }

  { Euler scheme : }

    YT:=YT+A(TI-DELTA_Y,YT)*DELTA_Y+B(TI-DELTA_Y,YT)*DWTI;

    END;{ WHILE }

  { Summation of the absolute errors : }

    XT:=EXPLSOL(TI,WT); { exact value of the solution }
    EPS[G]:=EPS[G]+ABSERR(XT,YT);

  UNTIL K=N;{ REPEAT for different samples }
  EPS[G]:=EPS[G]/N;DEL[G]:=DELTA_Y;
 UNTIL G=NUM;{ REPEAT for different time step sizes }
  ...
```

Notes 3.3.1

(i) Large fluctuations in the absolute errors may occur due to the small sample size. These can be reduced by taking a smaller noise intensity parameter BETA or drift parameter ALPHA. Note that different sample paths are used for the different step sizes.

(ii) Other error criteria can be investigated by appropriate changes in the ABSERR expression. For different stochastic differential equations we refer to the Notes 3.2.1.

The printout should show the absolute error $\hat\epsilon$ decreasing with decreasing step size. But how reliable is this tendency? Let us repeat the above numerical experiment.

Exercise 3.3.1 (PC-Exercise 9.3.2) *Repeat Problem 3.3.1 using a different seed, that is initial value, for the random number generator.*

The program is exactly the same as above and the different seed is usually chosen by the PC by simply repeating the run without having reset the computer. In the printout we should see that the estimate of the absolute error decreases in both cases for decreasing step size. However, these estimates are random variables and take different values in the two runs.

In the next subsection we shall derive confidence intervals which provide more reliable estimates than those here.

B. Confidence Intervals for the Absolute Error

For large N we know from the Central Limit Theorem, mentioned in Section 1.4, that the error $\hat\epsilon$ behaves asymptotically like a Gaussian random variable and converges in distribution to the nonrandom expectation ϵ of the absolute value of the error as $N \to \infty$.

It is impossible to generate an infinite number of trajectories. However, we can estimate the variance $\hat\sigma_\epsilon^2$ of $\hat\epsilon$ and then use it to construct a confidence interval for the absolute error ϵ. To do this we arrange the simulations into M batches of N simulations each and estimate the variance of $\hat{\epsilon}$ in the following way. We denote by $Y_{T,k,j}$ the value of the kth generated Euler trajectory in the jth batch at time T and by $X_{T,k,j}$ the corresponding value of the Ito process. The average errors

$$(3.3) \qquad \hat{\epsilon}_j = \frac{1}{N} \sum_{k=1}^{N} |X_{T,k,j} - Y_{T,k,j}|$$

of the M batches $j = 1, 2, \ldots, M$ are then independent and approximately Gaussian for large N. We have arranged the errors into batches because, as explained in Section 1.5, we can then use the Student t-distribution to construct confidence intervals for a sum of independent Gaussian (or in this case approximately Gaussian) random variables with unknown variance. In particular, we estimate the mean of the batch averages

$$(3.4) \qquad \hat{\epsilon} = \frac{1}{M} \sum_{j=1}^{M} \hat{\epsilon}_j = \frac{1}{NM} \sum_{j=1}^{M} \sum_{k=1}^{N} |X_{T,k,j} - Y_{T,k,j}|$$

and then use the formula

$$(3.5) \qquad \hat{\sigma}_\epsilon^2 = \frac{1}{M-1} \sum_{j=1}^{M} (\hat{\epsilon}_j - \hat{\epsilon})^2$$

to estimate the variance σ_ϵ^2 of the batch averages. Experience has shown that the batch averages can be interpreted as being Gaussian for batch sizes $N \geq 15$; we shall usually take $N = 100$. For the Student t-distribution with $M-1$ degrees of freedom an $100(1-\alpha)\%$ confidence interval for ϵ has the form

$$(3.6) \qquad (\hat{\epsilon} - \Delta\hat{\epsilon}, \; \hat{\epsilon} + \Delta\hat{\epsilon})$$

with

$$(3.7) \qquad \Delta\hat{\epsilon} = t_{1-\alpha,M-1} \sqrt{\frac{\hat{\sigma}_\epsilon^2}{M}},$$

where $t_{1-\alpha,M-1}$ is determined from the Student t-distribution with $M-1$ degrees of freedom. For $M = 20$ and $\alpha = 0.1$ we have $t_{1-\alpha,M-1} \approx 1.73$ from Table 1.5.1. In this case the absolute error ϵ will lie in the corresponding confidence interval (3.6) with probability $1 - \alpha = 0.9$.

Problem 3.3.2 (PC-Exercise 9.3.3) *Simulate $M = 10$ batches each with $N = 100$ trajectories of the Ito process X satisfying (2.14) with $X_0 = 1.0$, $a = 1.5$, $b = 0.1$ and their Euler approximations with equidistant time steps of step size $\Delta = 2^{-4}$ corresponding to the same sample paths of the Wiener process on the time interval $[0, T]$ for $T = 1$. Evaluate the 90% confidence interval for the absolute error ϵ. Repeat this for $M = 20$, 40 and 100 batches, and plot the confidence intervals on ϵ versus M axes.*

Program 3.3.2   Confidence intervals for different batch sizes

```
   ...
FUNCTION A(TI,XI:REAL):REAL;
FUNCTION B(TI,XI:REAL):REAL;
FUNCTION EXPLSOL(TI,WT:REAL):REAL;
FUNCTION ABSERR(XT,YT:REAL):REAL;
PROCEDURE COMPSAMPLEPARA(NN:INTEGER;X:VECTORO;VAR SAVERAGE,SVARIANCE:REAL);


{ Main program : }
   ...
 EPS[1]:=0.0;EPS[2]:=0.0;EPS[3]:=0.0;EPS[4]:=0.0; { initial errors }
 SQDELTA:=SQRT(DELTA); { square root of DELTA }
 J:=0;
 REPEAT
  J:=J+1;           { index of the batch used                        }
  EPSYLON[J]:=0.0; { sum of the absolute errors within the batch used }

 { Generation of different trajectories : }

  K:=0;
  REPEAT
   K:=K+1;  { index of the trajectory used     }
   WT:=0.0; { value of the Wiener process at t = TO }

  { Generation of the Euler approximation and its absolute error : }

   I:=0;
   TI:=TO; { initial time                        }
   YT:=XO; { initial value of the approximation }
   WHILE TI<T DO
    BEGIN
     I:=I+1;      { time step }
     TI:=TI+DELTA; { time      }
     IF I MOD 2 = 1 THEN GENERATE(G1,G2) { uses Polar Marsaglia method }
      ELSE G1:=G2;
     DWTI:=G1*SQDELTA; { Wiener process increment  W(t(i+1)) - W(ti) }
     WT:=WT+DWTI;      { value of the Wiener process W(t) at t = TI  }

    { Euler scheme : }

     YT:=YT+A(TI-DELTA,YT)*DELTA+B(TI-DELTA,YT)*DWTI;

    END;{ WHILE }

  { Summation of the absolute errors : }

   XT:=EXPLSOL(TI,WT); { exact value of the solution }
   EPSYLON[J]:=EPSYLON[J]+ABSERR(XT,YT);

  UNTIL K=N;              { REPEAT for different samples             }
  EPSYLON[J]:=EPSYLON[J]/N; { estimate of the absolute error of the batch }
 UNTIL J=MAXM;            { REPEAT for different batches            }

{ Calculation of the confidence intervals and initialization of data : }
```

```
FOR J:=1 TO 4 DO
  BEGIN
    CASE J OF { initializes batch numbers and percentage points }
      1 : BEGIN M[J]:=10;QUANTILE:=1.83 END;
      2 : BEGIN M[J]:=20;QUANTILE:=1.73 END;
      3 : BEGIN M[J]:=40;QUANTILE:=1.68 END;
      4 : BEGIN M[J]:=100;QUANTILE:=1.66 END;
    END;
    COMPSAMPLEPARA(TRUNC(M[J]),EPSYLON,AVERAGE,VARIANCE);
    EPS[J]:=AVERAGE;              { midpoint of the confidence interval }
    DIFFER[J]:=QUANTILE*SQRT(VARIANCE/M[J]); { half the interval length }
  END;

{ Printout of the results : }

CLEARDEVICE;
CONFINV(4,EPS,DIFFER,M,'EPS','number of batches');
  ...
```

Notes 3.3.2

(i) The sample statistics are calculated with the routine COMPSAM-PLEPARA from Program 1.3.1 and plotted on the screen with CONFINV from Program 3.1.12. Time could be saved by including values from previous batches in the larger runs, but would require the program to be restructured.

(ii) As before, many parameters, the equation and its solution, and the error criterion can be changed easily. If changing the confidence interval significance level, one should not forget to change the Student t-test quantile $t_{1-\alpha,M-1}$, see Table 1.5.1.

Our results in the printout indicate that the length of the confidence interval for the absolute error decreases as the number of batches increases. In fact, this is predicted by formula (3.7), which says that we need to increase the number of batches fourfold in order to halve the length of the confidence interval. This provides us with a method for determining the number of simulations needed to obtain a confidence interval of specified length for the absolute error ϵ. Since the length $2\Delta\hat{\epsilon}$ of the confidence interval is inversely proportional to the square root of the number of batches M only, the required number of batches for a chosen confidence interval of sufficiently small length may be very large. Consequently, some thought should be given to decide just how much accuracy is really needed in the answer of a given problem.

C. Dependence of the Absolute Error on the Step Size

We now examine the relationship between the absolute error and the step size.

Problem 3.3.3 (PC-Exercise 9.3.4) *Simulate $M = 20$ batches each with $N = 100$ trajectories of the Ito process X satisfying (2.14) with $X_0 = 1.0$, $a = 1.5$, $b = 0.1$ and their Euler approximations with equidistant time steps of step size $\Delta = 2^{-2}$ corresponding to the same sample paths of the Wiener process* *on the time interval $[0, T]$ for $T = 1$. Evaluate the 90% confidence interval for the absolute error ϵ. Repeat this for step sizes $\Delta = 2^{-3}$, 2^{-4} and 2^{-5}, and plot the confidence intervals on ϵ versus Δ axes.*

Program 3.3.3   Confidence intervals for increasing step size

```
...
FUNCTION A(TI,XI:REAL):REAL;
FUNCTION B(TI,XI:REAL):REAL;
FUNCTION EXPLSOL(TI,WT:REAL):REAL;
FUNCTION ABSERR(XT,YT:REAL):REAL;
PROCEDURE COMPSAMPLEPARA(NN:INTEGER;X:VECTORO;VAR SAVERAGE,SVARIANCE:REAL);

{ Main program : }
 ...
 DELTA_Y:=2.*DELTA;
 G:=0;
 REPEAT
  G:=G+1;                   { index of the batch used }
  DELTA_Y:=DELTA_Y/2;       { current time step size  }
  SQDELTA_Y:=SQRT(DELTA_Y); { square root of DELTA_Y  }

 { Generation for different batches : }

  J:=0;
  REPEAT
   J:=J+1;             { batch index                              }
   EPSYLON[J]:=0.0; { sum of the absolute errors of the batch used }

  { Generation of different trajectories : }
   ...

   { Generation of the Euler approximation and its absolute error : }
   ...

{ Calculation of the confidence interval and initialization of data : }

  DEL[G]:=DELTA_Y; { current time step size }
  COMPSAMPLEPARA(M,EPSYLON,AVERAGE,VARIANCE);
  EPS[G]:=AVERAGE;               { midpoint of the confidence interval }
  DIFFER[G]:=QUANTILE*SQRT(VARIANCE/M); { half the interval length }

 UNTIL G=NUM;{ REPEAT for different time step sizes }
 ...
```

Notes 3.3.3

(i) The program takes some time to run. Some parts coincide with those of Program 3.3.2. It also plots the results in logarithmic coordinates as is required in Problem 3.3.4 below. After the first picture has been examined press the <RETURN> key for the corresponding picture in logarithmic coordinates.

(ii) While 4 different step sizes provide a good indication of the slope of the curve, the automatic scaling makes the confidence intervals appear small. It is worthwhile to compare with the results for only 3 different step sizes. (Exercise 3.3.3)

Exercise 3.3.3   *Repeat Problem 3.3.3 for the step sizes* $\Delta = 2^{-4}$, 2^{-5} *and* 2^{-6}.

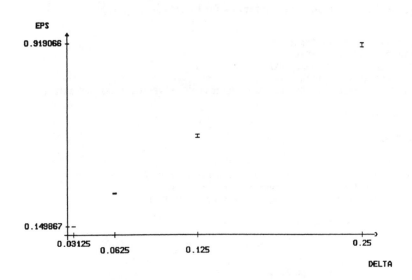

Figure 3.3.1   Confidence intervals for increasing step size.

As we can see in Figure 3.3.1 the step size Δ certainly has an influence on the magnitude of the absolute error ϵ.

For the Euler scheme the absolute error turns out to be proportional to $\sqrt{\Delta}$. We can see this more clearly if we replot the results using $\log_2$ versus $\log_2$ coordinates as in Problem 3.1.2 and 3.1.3.

Problem 3.3.4 (PC-Exercise 9.3.5)   *Replot the results of Problem 3.3.3 on* $\log_2 \epsilon$ *versus* $\log_2 \Delta$ *axes.*

The confidence intervals in Figure 3.3.2 closely follow a straight line with slope 1/2. This slope represents the experimentally obtained order of strong convergence of the Euler scheme for the given stochastic differential equation.

D. Order of Strong Convergence

In Subsection A we considered the pathwise approximation of an Ito process X by an Euler approximation Y and introduced the absolute error criterion (3.1).

We shall say that a discrete time approximation Y with maximum time step size δ *converges strongly* to X at time T if

$$(3.8) \qquad\qquad \lim_{\delta \downarrow 0} E\left(|X_T - Y(T)|\right) = 0.$$

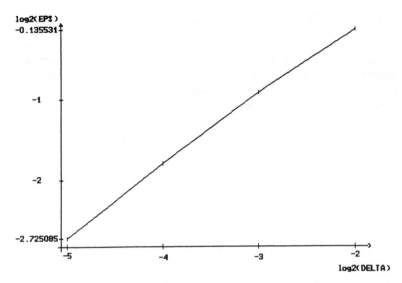

Figure 3.3.2 $\log_2$ of the absolute error versus $\log_2 \Delta$.

While the Euler approximation is the simplest useful discrete time approx-imation, it is, generally, not particularly efficient numerically. We shall thus investigate other discrete time approximations in the following chapters. In order to assess and compare different discrete time approximations, we need to know their rates of strong convergence.

We shall say that a discrete time approximation Y *converges strongly with order* $\gamma > 0$ at time T if there exists a positive constant C, which does not depend on δ, and a $\delta_0 > 0$ such that

$$(3.9) \qquad\qquad \epsilon(\delta) = E\left(|X_T - Y(T)|\right) \le C\,\delta^\gamma$$

for each $\delta \in (0, \delta_0)$.

The criterion (3.9) is a straightforward generalization of the usual deter-ministic convergence criterion and reduces to it when the diffusion coefficient vanishes and the initial value is deterministic.

In Chapter 4 we shall investigate the strong convergence of a number of different discrete time approximations experimentally. We shall see, in partic-ular, that the Euler approximation has strong order of convergence $\gamma = 0.5$, as suggested by the computer experiments in Subsection 3.3.C.

3.4 Approximation of Moments and Weak Convergence

As we mentioned already at the beginning of the previous section there also exists a second type of problems that does not require a good pathwise approximation. For instance, we may only be interested in the computation of moments, probabilities or other functionals of the Ito process. Since the requirements for their simulation are not as demanding as for pathwise approximations, it is natural and convenient to classify them as a separate class of problems with respect to a weaker convergence criterion than that considered above.

A. Mean Error

To introduce this weaker type of convergence, we shall carry out some computer experiments to investigate the mean error

$$(4.1) \qquad \mu = E(Y(T)) - E(X_T)$$

for the same linear stochastic differential equation as in Section 2

$$(4.2) \qquad dX_t = a\,X_t\,dt + b\,X_t\,dW_t,$$

for $t \in [0,T]$, and its Euler approximation

$$Y_{n+1} = Y_n + a\,Y_n\Delta_n + b\,Y_n\,\Delta W_n$$

for $n = 0, 1, 2, \ldots, N-1$. Here, as before, $\Delta_n = \tau_{n+1} - \tau_n$ denotes the step size and $\Delta W_n = W_{\tau_{n+1}} - W_{\tau_n}$ the increment of the Wiener process. In Subsection 3.4.D we shall generalize (4.1) to the approximation of polynomial and more general functionals of the process, including all its higher moments. Note that μ can take negative as well as positive values.

For equation (4.2) the mean of this Ito process is

$$(4.3) \qquad E(X_T) = E(X_0)\,\exp(aT).$$

As in the previous section, we shall arrange the simulated trajectories of the Euler approximation into M batches with N trajectories in each. We shall then estimate the mean error of the jth batch by the statistic

$$(4.4) \qquad \hat{\mu}_j = \frac{1}{N}\sum_{k=1}^{N} Y_{T,k,j} - E(X_T),$$

for $j = 1, 2, \ldots, M$, and their average by the statistic

$$(4.5) \qquad \hat{\mu} = \frac{1}{M}\sum_{j=1}^{M}\hat{\mu}_j = \frac{1}{MN}\sum_{j=1}^{M}\sum_{k=1}^{N} Y_{T,k,j} - E(X_T).$$

Similarly, we shall estimate the variance of the batch averages $\hat{\mu}_j$ by

(4.6)
$$\hat{\sigma}^2_\mu = \frac{1}{M-1} \sum_{j=1}^{M} (\hat{\mu}_j - \hat{\mu})^2 .$$

The $100(1-\alpha)\%$ confidence interval of the Student t-distribution with $M-1$ degrees of freedom for the mean error μ is then

(4.7)
$$(\hat{\mu} - \Delta\hat{\mu}, \hat{\mu} + \Delta\hat{\mu})$$

where

(4.8)
$$\Delta\hat{\mu} = t_{1-\alpha,M-1} \sqrt{\frac{\hat{\sigma}^2_\mu}{M}}.$$

The mean error μ will lie in this confidence interval with at least probability $1 - \alpha$.

Problem 3.4.1 (PC-Exercise 9.4.1) *Generate $M = 10$ batches of $N = 100$ trajectories of the Euler approximation for the Ito process (4.2) with $X_0 = 1.0$, $a = 1.5$, $b = 0.1$ for step length $\Delta = 2^{-4}$ and terminal time $T = 1$. Determine the 90% confidence interval for the mean error μ. Then repeat this for $M = 20$, 40 and 100 batches using the batches already simulated and plot the intervals on μ versus M axes.*

Program 3.4.1 Mean error for different batches

```
...
FUNCTION MEANGXT(TI:REAL):REAL;
BEGIN
 MEANGXT:=XO*EXP(ALPHA*(TI-TO));
END;
...
FUNCTION GXT(X:REAL):REAL;
BEGIN
 GXT:=X;
END;
...
```

Notes 3.4.1
(i) The program is similar to Program 3.3.2 except that the explicit solution need no longer be evaluated since the desired mean value is known theoretically.
(ii) To allow other weak error criteria to be considered a general function GXT and its theoretically known mean MEANGXT have been used.

The printout should show that a fourfold increase in the number of batches halves the length of the confidence interval, just as was needed in the previous section for pathwise approximations.

B. Dependence of the Mean Error on the Step Size

We shall now study the dependence of the mean error on the time step size.

Problem 3.4.2 (PC-Exercise 9.4.2) *Generate $M = 20$ batches of $N = 100$ trajectories of the Euler approximation as in Problem 3.4.1. Determine the 90% confidence interval for the mean error μ. Then repeat this for step sizes $\Delta = 2^{-3}$, 2^{-4} and 2^{-5}, and plot the intervals on μ versus Δ axes.*

The program is similar to Program 3.3.3 with the obvious change mentioned in Notes 3.4.1 above. When the picture has been examined, press the <RETURN> key and the corresponding picture in logarithmic coordinates, needed for Problem 3.4.3 below, will appear.

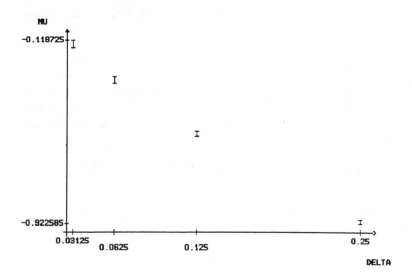

Figure 3.4.1   Confidence intervals for increasing time step size.

We see in Figure 3.4.1 that the choice of the time step size Δ influences significantly the mean error. When Δ is very small the variance may be again large due to roundoff errors and can distroy the result. Nevertheless it appears that $\hat{\mu}$ is nearly proportional to Δ. As in the last section it is useful here to plot the results in $\log_2$ versus $\log_2$ coordinates. We shall also call $|\mu|$ the mean error for convenience.

Problem 3.4.3 (PC-Exercise 9.4.3) *Replot the results of Problem 3.4.2 on $\log_2 |\mu|$ versus $\log_2 \Delta$ axes.*

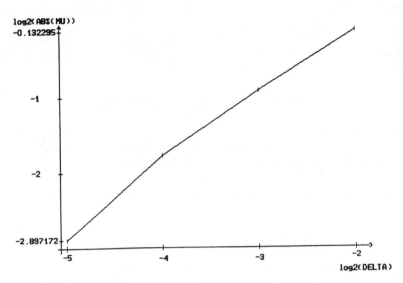

Figure 3.4.2 $\log_2$ of the mean error versus $\log_2 \Delta$.

From Figure 3.4.2 we see that the experimental mean error for the Euler scheme follows closely a straight line of slope 1 in logarithmic coordinates. This contrasts with the slope 0.5 for the strong pathwise Euler approximations in the previous section

In Chapter 5 we shall also consider other schemes with respect to the more general, weak convergence criterion which we shall specify below.

C. Systematic and Statistical Error

We can decompose the random estimate $\hat{\mu}$ for the mean error μ into a *systematic error* μ_{sys} and a *statistical error* μ_{stat}, with

(4.9)
$$\hat{\mu} = \mu_{sys} + \mu_{stat}$$

where
(4.10)
$$\mu_{sys} = E\left(\hat{\mu}\right).$$

Then

(4.11)
$$\mu_{sys} = E\left(\frac{1}{MN}\sum_{j=1}^{M}\sum_{k=1}^{N}Y_{T,k,j}\right) - E(X_T)$$

$$= E(Y(T)) - E(X_T)$$

$$= \mu,$$

so the systematic error is the same as the mean error. In Chapter 5 we shall introduce more complicated numerical schemes with the objective of decreasing the mean error for a given step size.

For a large number MN of independent simulations we conclude from the Central Limit Theorem that the statistical error $\hat{\mu}_{stat}$ becomes asymptotically Gaussian with mean zero and variance

$$(4.12) \qquad \mathrm{Var}\,(\mu_{stat}) = \mathrm{Var}\,(\hat{\mu}) = \frac{1}{MN}\,\mathrm{Var}(Y(T)).$$

This depends on the total number MN of simulations and not separately on the number M of batches or number N of simulations in each batch.

A fourfold increase in the number of simulations is also required here to halve the length of a confidence interval. The successful implementation of weak approximations is often in direct conflict with the size of this variance of the estimated functional.

D. Order of Weak Convergence

In the Subsections A and B we examined the approximation of the first moment of a particular Ito process X by the Euler approximation with respect to the mean error

$$(4.13) \qquad \mu = E\,(Y(T)) - E\,(X_T)\,.$$

In particular, we saw that this criterion differs in its properties from the strong convergence criterion. To some extent (4.13) is special and not appropriate for applications where the approximation of some higher moment

$$E\,(|X_T|^q)$$

with $q = 2, 3, \ldots$ or of some functional

$$(4.14) \qquad\qquad E\,(g\,(X_T))$$

is of interest. Like (4.13) these do not require a good pathwise approximation of the Ito process, but only an approximation of the probability distribution of X_T.

We shall say that a general discrete time approximation Y with maximum time step size δ *converges weakly* to X at time T as $\delta \downarrow 0$ with respect to a class $\mathcal{C}$ of test functions $g\colon \Re^d \to \Re$ if we have

$$(4.15) \qquad \lim_{\delta \downarrow 0} |E\,(g\,(X_T)) - E\,(g\,(Y(T)))| = 0$$

for all $g \in \mathcal{C}$. If $\mathcal{C}$ contains all polynomials this definition implies the convergence of all moments. so investigations involving it will require the existence of all moments. In the deterministic case with a zero diffusion coefficient and a nonrandom initial value, (4.15) with $g(x) \equiv x$ reduces to the usual deterministic convergence criterion, just as the strong convergence criterion does.

We shall say that a time discrete approximation Y *converges weakly with order* $\beta > 0$ to X at time T as $\delta \downarrow 0$ if for each polynomial g there exists a positive constant C, which does not depend on δ, and a finite $\delta_0 > 0$ such that

$$(4.16) \qquad |E\left(g\left(X_T\right)\right) - E\left(g\left(Y(T)\right)\right)| \le C\,\delta^\beta$$

for each $\delta \in (0, \delta_0)$.

In Chapter 5 we shall investigate the order of weak convergence of various discrete time approximations experimentally. In particular, we shall see that the Euler approximation usually converges with weak order $\beta = 1$, in contrast with the strong order $\gamma = 0.5$. We shall also see that the strong and weak convergence criteria lead to the development of different discrete time approximations which are only efficient with respect to one of the two criteria. This fact makes it important to clarify the aim of a simulation before choosing an approximation scheme. The following question should be answered:

Is a good pathwise approximation of the Ito process required or is the approximation of some functional of the Ito process the real objective?

3.5 Numerical Stability

The convergence of a numerical scheme alone is no guarantee that the scheme can be used efficiently in a practical simulation. It may be numerically unstable, which makes it useless for the problem at hand. Here we shall discuss such situations first for deterministic examples and then in the stochastic context. The concepts of stiff stochastic differential equations, stochastic A-stability and implicit schemes will be considered.

A. Numerical Stability in the Deterministic Case

It is well known in deterministic numerical analysis that difficulties may be encountered when trying to implement a difference method which is known to be convergent. For example, the differential equation

$$(5.1) \qquad \frac{dx}{dt} = -16\,x$$

has exact solutions $x(t) = x_0 e^{-16t}$, which all converge very rapidly to zero for different $x_0 > 0$. For this equation the Euler method with constant time step size Δ,

$$y_{n+1} = (1 - 16\Delta)\,y_n,$$

has exact iterates $y_n = (1 - 16\Delta)^n\,y_0$. If we choose $\Delta > 2^{-3}$ these iterates oscillate with increasing amplitude instead of converging to zero like the exact solutions of (5.1).

Problem 3.5.1   *Solve numerically equation (5.1) on the interval* $[0,1]$ *with* $x_0 = 1$ *by the Euler scheme and plot the linearly interpolated results against t for step sizes* $\Delta = 2^{-3}$ *and* 2^{-5} *together with the exact solution.*

Notes 3.5.1   The program is exactly the same as Program 3.1.1 except that the parameter ALPHA is now -16 instead of -5. To avoid future confusion this change should be made in a copy of Program 3.1.1 called Program 3.5.1 rather than in the original version.

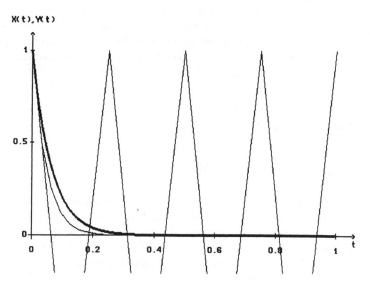

Figure 3.5.1   Euler approximation for $\Delta = 2^{-3}, 2^{-5}$ and exact solution.

We have just seen a simple example of *numerical instability*. In this particular case, we can overcome this instability by taking a time step $\Delta < 2^{-3}$, but for other multidimensional cases the numerical instabilities may persist no matter how small we take Δ.

We say that a one-step method is *numerically stable* if for each interval $[t_0, T]$ and differential equation with drift satisfying a Lipschitz condition there exist positive constants Δ_0 and M such that

$$(5.2) \qquad\qquad |y_n - \tilde{y}_n| \leq M |y_0 - \tilde{y}_0|$$

for $n = 0, 1, \ldots, n_T$ and two solutions $y_n, \tilde{y}_n$ of the one-step method starting at any $y_0, \tilde{y}_0$, respectively, for any time discretizations with $\max_n \Delta_n < \Delta_0$. The constants Δ_0 and M here may also depend on the particular time interval $t_0 \leq t \leq T$ in addition to the differential equation under consideration.

The commonly used one-step methods are numerically stable for Δ small enough. However, the constant M in (5.2) may be quite large. For example, if we replace the minus sign by a plus sign in the differential equation (5.1), we obtain

$$y_n - \tilde{y}_n = (1 + 16\Delta)^n (y_0 - \tilde{y}_0)$$

for the Euler method. The numerical stability condition (5.2) requires a bound like $e^{16(T-t_0)}$ for M, in contrast with $M \leq 1$, when $\Delta_0 < 2^{-3}$, for the original differential equation. The difference is due to the fact that the solutions of the modified differential equation are diverging exponentially fast, whereas those of the original are converging exponentially fast. In both cases the Euler method keeps the error under control, but in the former case the initial error must be unrealistically small if it is to remain small.

To ensure that the errors in the Euler method for (5.1) do not grow, that is the bound $M \leq 1$ in (5.2), we need to take step sizes less than 2^{-3}. The situation does not improve if we use the higher order Heun method (1.6). However, the implicit trapezoidal method (1.5) offers a substantial improvement. In this case it is

(5.3)
$$y_{n+1} = y_n + \frac{1}{2} \{-16 y_n - 16 y_{n+1}\} \Delta,$$

which we can solve explicitly to get

$$y_{n+1} = \left(\frac{1 - 8\Delta}{1 + 8\Delta} \right) y_n.$$

Here

(5.4)
$$\left| \frac{1 - 8\Delta}{1 + 8\Delta} \right| < 1$$

for any $\Delta > 0$.

Problem 3.5.2 *Solve equation (5.1) numerically on the interval $[0,1]$ with $x_0 = 1$ by the Heun method (1.6) and the implicit trapezoidal method (1.5) for $\Delta = 2^{-3}$. Plot the linearly interpolated paths in the same picture.*

Program 3.5.2 Heun and implicit trapezoidal methods

```
...
NN:=TRUNC((T-TO)/DELTA+0.1);

{ Generation of the approximations : }

YT1[0]:=X0;YT2[0]:=X0; { initial values }
ABSCISSA[0]:=TO;
I:=0;
REPEAT
I:=I+1; { time step }
ABSCISSA[I]:=ABSCISSA[I-1]+DELTA; { value of the x-axis }

{ Heun scheme : }

YT1[I]:=(1.0-16.*(1.0-8*DELTA)*DELTA)*YT1[I-1];

{ Implicit trapezoidal scheme : }
```

```
YT2[I]:=(1.0-8.*DELTA)*YT2[I-1]/(1.0+8*DELTA);

UNTIL I=NN;
    ...
```

Notes 3.5.2   The program is similar to Program 3.1.1. We list only the iterative routines for the Heun and implicit Euler methods. Note that we have rearranged the latter as an explicit method. For a nonlinear differential equation we usually have to solve an implicit algebraic equation for y_{n+1} numerically at each step.

To obtain an indication of suitable values of Δ it is useful to consider a class of test equations. These are the complex-valued linear differential equations

$$(5.5) \qquad \frac{dx}{dt} = \lambda x,$$

with $\lambda = \lambda_r + \imath\lambda_i$ where $\imath = \sqrt{-1}$, which have oscillating solutions when $\lambda_i \neq 0$. We can write (5.5) equivalently as a 2-dimensional differential equation

$$\frac{d}{dt}\begin{pmatrix} x^1 \\ x^2 \end{pmatrix} = \begin{bmatrix} \lambda_r & -\lambda_i \\ \lambda_i & \lambda_r \end{bmatrix}\begin{pmatrix} x^1 \\ x^2 \end{pmatrix}$$

where $x = x^1 + \imath x^2$. The suitable values of the step size $\Delta > 0$ are expressed in terms of the *region of absolute stability* for the method, consisting of the complex numbers $\lambda\Delta$ for which an error in y_0 at t_0 will not grow in subsequent iterations of the method applied to the differential equation (5.5). Essentially, these are the values of λ and Δ producing a bound $M \leq 1$ in (5.2). For the Euler method we thus require

$$|1 + \lambda\Delta| \leq 1,$$

so its region of absolute stability is the unit disc in the complex plane centered on $z = -1 + 0\imath$.

Another interesting class of equations consists of the 2-dimensional linear differential equations

$$(5.6) \qquad \frac{d}{dt}\begin{pmatrix} x^1 \\ x^2 \end{pmatrix} = \begin{bmatrix} -\alpha_1 & 0 \\ 0 & -\alpha_2 \end{bmatrix}\begin{pmatrix} x^1 \\ x^2 \end{pmatrix}$$

for which the two eigenvalues of the coefficient matrix are negative and very different, that is with $-\alpha_2 \ll -\alpha_1 \leq 0$. The components of (5.6) are uncoupled, so they can be solved separately to give

$$(5.7) \qquad x^1(t) = e^{-\alpha_1 t}, \qquad x^2(t) = e^{-\alpha_2 t}.$$

for the initial value $(x_0^1, x_0^2) = (1,1)$. Since $-\alpha_1$ is much larger than $-\alpha_2$ the first component shows a very slow exponential decay in comparison with the second. In other words the two components have widely differing time scales. In the literature such a system of equations is often called a *stiff system*. In the general d-dimensional case we shall say that a linear system is *stiff* if the largest and smallest real parts of the eigenvalues of the coefficient matrix differ considerably.

Now, if we apply the Euler method to (5.6), for the second component to remain within the region of absolute stability we need a step size

$$\Delta \leq \frac{2}{\alpha_2}.$$

We saw in Figure 3.1.3 that there is a lower bound on the step size for which the influence of the roundoff error remains acceptable. But the upper bound $2/\alpha_2$ might already be too small to allow for the control over the propagation of roundoff errors in the first component.

Problem 3.5.3 *Apply the Euler method with step size* $\Delta = 0.125$ *to equation (5.6) on* $[0, 2.0]$ *with* $x_0^1 = x_0^2 = 1$ *and with* $\alpha_1 = 1.0$ *and* $\alpha_2 = 15.0$. *Plot the calculated result and the exact solution for each component.*

Notes 3.5.3

(i) We omit Program 3.5.3 which is essentially the same as Program 3.1.1 applied to each component since these are uncoupled. One should run the program first with ALPHA equal to α_1 and then with it replaced by α_2.

(ii) The program also includes the implicit Euler method which we will discuss below. Press the <RETURN> key after examining the results for the explicit method to see the corresponding results for the implicit method.

A much more stable result is obtained when we apply the *implicit Euler method*

$$(5.8) \qquad\qquad y_{n+1} = y_n + a\,(t_{n+1}, y_{n+1})\,\Delta$$

to equation (5.6).

Problem 3.5.4 *Repeat Problem 3.5.3 with the implicit Euler method (5.8).*

We achieve a better result in Problem 3.5.4 than in Problem 3.5.3 due to the fact that with (5.8) for equation (5.6) we obtain

$$y_n^i = (1 + \alpha_i \Delta)^{-n} y_0^i$$

for $i = 1$ and 2. Hence for all $\Delta > 0$ we have

$$|y_n - \tilde{y}_n| \leq |y_0 - \tilde{y}_0|$$

for $n = 0, 1, \ldots$ and any two solutions y_n, $\tilde{y}_n$ of (5.8). Thus the implicit Euler method (5.8) applied to the stiff system (5.6) still behaves in a stable manner in its first component even when $\Delta > 2/\alpha_1$.

B. Stiff Stochastic Differential Equations

To characterize stiff stochastic differential equations we start with a linear d-dimensional Stratonovich equation

$$(5.9) \qquad dZ_t = AZ_t \, dt + \sum_{k=1}^{m} B^k Z_t \circ dW_t^k$$

for $t \geq 0$. $Z_0 \in R^d$. We use the d *Lyapunov exponents* $\lambda_d \leq \lambda_{d-1} \leq \ldots \leq \lambda_1$ which exist and are non random real numbers. These numbers are just the upper limits

$$(5.10) \qquad \lambda(t_0, z_0) = \limsup_{t \to \infty} \frac{1}{t - t_0} \ln |Z_t^{t_0, z_0}|,$$

when the process Z starts at t_0 in $Z_0 = z_0$ with z_0 belonging to certain random subsets $E_d(w), E_{d-1}(w), \ldots, E_1(w)$ which partition R^d.

We call (5.9) a *stiff stochastic differential equation* if its largest and smallest Lyapunov exponents differ considerably, that is if

$$(5.11) \qquad \lambda_1 \gg \lambda_d.$$

This generalizes the deterministic definition of stiffness given at the end of the preceding subsection to the stochastic context because the real parts of the eigenvalues of the coefficient matrix of a deterministic linear differential equation are its Lyapunov exponents. Thus, stochastic stiffness also refers to the presence of two or more widely differing time scales in the solutions. In particular, a stiff linear ordinary differential equation is also stiff in the stochastic sense.

More generally, we shall say that a stochastic differential equation is *stiff* if its linearization about a stationary solution is stiff in the above sense. We remark that this is just one way to characterize some cases of stiffness.

Problem 3.5.5 (PC-Exercise 9.8.2)   *Simulate the exact solution and the Euler approximation with step size $\delta = \Delta = 2^{-4}$ corresponding to the same realization of the Wiener process for the 1-dimensional stochastic differential equation*

$$dX_t = \alpha \, X_t \, dt + dW_t$$

with $X_0 = 1$ on the interval $[0, T]$, where $T = 1$ and $\alpha = 5.0$. Plot both paths on the same X versus t axes. Note that the exact solution here (see (2.2.4)) can be simulated using the correlated Gaussian random variables

$$\Delta E_n = \int_{\tau_n}^{\tau_{n+1}} \left(e^{-\alpha(s - \tau_n)} - 1 \right) dW_s \quad and \quad \Delta W_n = \int_{\tau_n}^{\tau_{n+1}} dW_s$$

for which

$$E(\Delta E_n) = 0, \quad E(\Delta W_n) = 0, \quad Var(\Delta W_n) = \Delta$$

$$Var(\Delta E_n) = \frac{3}{2\alpha} + e^{2\alpha\Delta} \left(\frac{1}{2\alpha} - \Delta \right) + 2 e^{\alpha\Delta} \left(\Delta - \frac{1}{\alpha} \right),$$

$$E(\Delta E_n \, \Delta W_n) \simeq \Delta(e^{\alpha \Delta} - 1).$$

Notes 3.5.5 Program 3.5.5 is given on the diskette. It combines Program 3.2.1 with a modification of Program 1.3.5 to generate the correlated Gaussian random variables.

From Problem 3.5.5 it is apparent that as the time interval $[t_0, T]$ becomes relatively large, the propagated error of a numerically stable scheme, which is theoretically still under control, may, in fact, become so unrealistically large as to make the approximation useless for practical purposes.

C. Asymptotic Numerical Stability

Let Y denote a discrete time approximation with maximum time step size $\delta > 0$ starting at time t_0 at Y_{t_0}, with $\bar{Y}$ denoting the corresponding approximation starting at $\bar{Y}_{t_0}$. We shall say that a discrete time approximation Y is *asymptotically numerically stable* for a given stochastic differential equation if there exists a positive constant Δ_a such that for each $\epsilon > 0$ and step size $\delta \in (0, \Delta_a)$

$$(5.12) \qquad \lim_{|Y_{t_0} - \bar{Y}_{t_0}| \to 0} \lim_{T \to \infty} P\left(\sup_{t_0 \leq t \leq T} |Y_{n_t} - \bar{Y}_{n_t}| \geq \epsilon \right) = 0.$$

Problem 3.5.6 (PC-Exercise 9.8.3) *Repeat Problem 3.5.5 for $\alpha = -15$ and step sizes $\Delta = 2^{-3}$ and 2^{-4}.*

Notes 3.5.6 See the diskette for Program 3.5.6. It is more general than Program 3.5.5 as it also includes the implicit Euler scheme needed in Exercise 3.5.1 below.

We can consider asymptotical numerical stability of a stochastic scheme with respect to an appropriately restricted class of stochastic differential equations. This would generalize the well-known concept of A-stability for deterministic differential equations. We shall choose the class of complex-valued linear test equations

$$(5.13) \qquad dX_t = \lambda \, X_t \, dt + dW_t,$$

where the parameter λ is a complex number with real part $\text{Re}(\lambda) < 0$ and W is a real-valued standard Wiener process. This stochastic generalization simply includes additive noise in the deterministic test equation (5.5).

Suppose that we can write a given scheme with equidistant step size $\Delta \equiv \delta$ applied to the test equations (5.13) with $\text{Re}(\lambda) < 0$ in the recursive form

$$(5.14) \qquad Y_{n+1} = Y_n G\left(\lambda \Delta\right) + Z_n,$$

for $n = 0, 1, \ldots$, where G is a mapping of the complex plane onto itself and the $Z_0, Z_1, \ldots$ are random variables which do not depend on λ or on the Y_0,

$Y_1, \ldots$. Then, we shall call the set of complex numbers $\lambda\Delta$ with

(5.15) $\mathrm{Re}(\lambda) < 0$ and $|G(\lambda\Delta)| < 1$

the *region of absolute stability* of the scheme. From this region we can determine appropriate step sizes such that an error in the approximation by this scheme of a particular test equation from the class (5.13) will not grow in subsequent iterations. Obviously, the scheme is asymptotically numerically stable for such a test equation if $\lambda\Delta$ belongs to the region of absolute stability.

The Euler scheme with equidistant step size $\Delta > 0$ for the stochastic differential equation (5.13) is

$$Y_{n+1} = Y_n(1 + \lambda\Delta) + \Delta W_n.$$

Thus

$$|Y_n - \bar{Y}_n| \le |1 + \lambda\Delta|^n |Y_0 - \bar{Y}_0|,$$

where $\bar{Y}_n$ is the solution starting at $\bar{Y}_0$. The additive noise terms cancel out here, so we obtain the same region of absolute stability as in the deterministic case which was a unit disc centered on $z = -1 + 0\imath$.

The *implicit Euler scheme*

(5.16) $Y_{n+1} = Y_n + a(\tau_{n+1}, Y_{n+1})\Delta + b(\tau_n, Y_n)\Delta W_n$

takes the form

$$Y_{n+1} = Y_n + \lambda Y_{n+1}\Delta + \Delta W_n$$

for the test equation (5.13), from which we obtain

$$|Y_n - \bar{Y}_n| \le |1 - \lambda\Delta|^{-n} |Y_0 - \bar{Y}_0|$$

and therefore the same region of absolute stability as for the deterministic implicit Euler scheme. Thus, for any λ with $\mathrm{Re}(\lambda) < 0$ the step size $\Delta > 0$ can be chosen quite large. Generalizing the usual deterministic definition, we shall say that a stochastic scheme is *A-stable* if its region of absolute stability is the whole of the left half of the complex plane. Hence the implicit Euler scheme (5.16) is A-stable. Note that an A-stable stochastic scheme is also A-stable in the deterministic sense.

Exercise 3.5.1   *Repeat Problem 3.5.6 using the implicit Euler scheme (5.16).*

The required program is already built into Program 3.5.6. To view the results simply press the <RETURN> key after the corresponding explicit Euler result has been examined.

Literature for Chapter 3

There is an extensive literature on numerical methods for deterministic ordinary differential equations. Standard textbooks, which cover the material to different depths, include Henrici (1962), Gear (1971), Stoer & Bulirsch (1980), Butcher (1987) and Hairer, Nørsett & Wanner (1987). The Euler scheme was very early investigated in Maruyama (1955). Most of the papers on numerical methods for stochastic differential equations refer to it. Very different strong convergence criteria are used in the literature, see Rao et. al. (1974), Milstein (1974) and Platen (1981). The above strong convergence criterion from Kloeden & Platen (1989) coincides for an ordinary deterministic differential equation with the usual deterministic criterion as e.g. in Gear (1971). Milstein (1978) introduced the above weak convergence criterion. For numerical stability in the deterministic case we refer to Dahlquist & Bjorck (1974) and the excellent book by Hairer, Nørsett & Wanner (1987) and Hairer & Wanner (1991). For the stochastic case one can find results in Milstein (1988a), Hernandez & Spigler (1991), Drummond & Mortimer (1991), Petersen (1990), Kloeden & Platen (1992a,b) and Saito & Mitsui (1992).

Chapter 4

Strong Approximations

In this chapter we shall apply explicit one-step methods such as strong Taylor schemes and explicit schemes, as well as multi-step and implicit schemes, to approximate the solutions of stochastic differential equations with respect to the strong convergence criterion. Numerical experiments will be used to investigate the convergence of these schemes.

4.1 Strong Taylor Schemes

We shall consider discrete time approximations of various strong orders that have been derived from stochastic Taylor expansions by including appropriately many terms. To allow us to describe these schemes succinctly for a general d-dimensional Ito process satisfying the stochastic differential equation

$$(1.1) \qquad X_t = X_{t_0} + \int_{t_0}^t a(s, X_s)\, ds + \sum_{j=1}^m \int_{t_0}^t b^j(s, X_s)\, dW_s^j$$

in Ito form, for $t \in [t_0, T]$, or

$$(1.2) \qquad X_t = X_{t_0} + \int_{t_0}^t \underline{a}(s, X_s)\, ds + \sum_{j=1}^m \int_{t_0}^t b^j(s, X_s) \circ dW_s^j$$

in its equivalent Stratonovich form, we shall use the following generalizations of the operators that were introduced in Section 2.3:

$$(1.3) \qquad L^0 = \frac{\partial}{\partial t} + \sum_{k=1}^d a^k \frac{\partial}{\partial x^k} + \frac{1}{2} \sum_{k,l=1}^d \sum_{j=1}^m b^{k,j} b^{l,j} \frac{\partial^2}{\partial x^k \partial x^l},$$

$$(1.4) \qquad \underline{L}^0 = \frac{\partial}{\partial t} + \sum_{k=1}^d \underline{a}^k \frac{\partial}{\partial x^k}$$

and

$$(1.5) \qquad L^j = \underline{L}^j = \sum_{k=1}^d b^{k,j} \frac{\partial}{\partial x^k}$$

for $j = 1, 2, \ldots, m$, with the corrected drift

$$(1.6) \qquad \underline{a}^k =: a^k - \frac{1}{2} \sum_{j=1}^m \underline{L}^j b^{k,j}$$

for $k = 1, 2, \ldots, d$. In addition, as in Chapter 2, we shall abbreviate multiple Ito integrals by

$$(1.7) \qquad I_{(j_1,\ldots,j_l)} = \int_{\tau_n}^{\tau_{n+1}} \cdots \int_{\tau_n}^{s_2} dW_{s_1}^{j_1} \cdots dW_{s_l}^{j_l}$$

and multiple Stratonovich integrals by

$$(1.8) \qquad J_{(j_1,\ldots,j_l)} = \int_{\tau_n}^{\tau_{n+1}} \cdots \int_{\tau_n}^{s_2} \circ dW_{s_1}^{j_1} \cdots \circ dW_{s_l}^{j_l}$$

for $j_1, \ldots, j_l \in \{0, 1, \ldots, m\}$, $l = 1, 2, \ldots$ and $n = 0, 1, \ldots$ with the convention that

$$(1.9) \qquad\qquad\qquad W_t^0 = t$$

for all $t \in \Re^+$. We shall also use the abbreviation

$$f = f(\tau_n, Y_n),$$

for $n = 0, 1, \ldots$, in the schemes for any given function f defined on $\Re^+ \times \Re^d$, and usually not explicitly mention the initial value Y_0 or the step indices $n = 0, 1, \ldots$.

A. Euler Scheme

We begin with the Euler scheme, also called the Euler–Maruyama scheme, which we have already looked at in Chapter 3. It is the simplest strong Taylor approximation, containing only the time and Wiener integrals of multiplicity one from the Ito-Taylor expansion (2.3.17), and usually attains the order of strong convergence $\gamma = 0.5$.

In the 1-dimensional case $d = m = 1$ the *Euler scheme* has the form

$$(1.10) \qquad\qquad Y_{n+1} = Y_n + a\,\Delta_n + b\,\Delta W_n,$$

where

$$(1.11) \qquad\qquad \Delta_n = \int_{\tau_n}^{\tau_{n+1}} dt = \tau_{n+1} - \tau_n$$

is the length of the time discretization subinterval $[\tau_n, \tau_{n+1}]$ and

$$(1.12) \qquad\qquad \Delta W_n = \int_{\tau_n}^{\tau_{n+1}} dW_t = W_{\tau_{n+1}} - W_{\tau_n}$$

is the $N(0; \Delta_n)$ distributed increment of the Wiener process W on $[\tau_n, \tau_{n+1}]$.

For the general multi-dimensional case with $d, m = 1, 2, \ldots$ the *Euler scheme* has the form of the kth component of the *Euler scheme* has the form

$$(1.13) \qquad\qquad Y_{n+1}^k = Y_n^k + a^k\,\Delta_n + \sum_{j=1}^{m} b^{k,j}\,\Delta W_n^j$$

where

(1.14)
$$\Delta W_n^j = \int_{\tau_n}^{\tau_{n+1}} dW_t^j = W_{\tau_{n+1}}^j - W_{\tau_n}^j$$

is the $N(0; \Delta_n)$ distributed increment of the jth component of the m-dimensional standard Wiener process W on $[\tau_n, \tau_{n+1}]$; thus $\Delta W_n^{j_1}$ and $\Delta W_n^{j_2}$ are independent for $j_1 \neq j_2$.

Problem 4.1.1 (PC-Exercise 10.2.1) *Determine explicitly the 1-dimensional Ito process X satisfying*

(1.15)
$$dX_t = -\frac{1}{2}X_t \, dt + X_t \, dW_t^1 + X_t \, dW_t^2$$

on the time interval $[0,T]$ with $T = 1$ for the initial value $X_0 = 1$, where W^1 and W^2 are two independent standard Wiener processes. Then simulate $M = 20$ batches each of $N = 100$ trajectories of X and their Euler approximations corresponding to the same sample paths of the Wiener processes with equidistant time steps of step size $\Delta = 2^{-3}$. Determine the 90%-confidence interval for the absolute error ϵ at time T. Repeat the calculations for step sizes 2^{-4}, 2^{-5} and 2^{-6}, and plot $\log_2 \epsilon$ against $\log_2 \Delta$.

Program 4.1.1   Euler scheme for two driving Wiener processes

```
...
{ Generation of the Euler approximation and its absolute error : }

  I:=0;
  TI:=T0; { initial time                    }
  YT:=X0; { initial value of the approximation }
  WHILE TI<T DO
    BEGIN
     I:=I+1;              { time step                              }
     TI:=TI+DELTA_Y;      { time                                   }
     GENERATE(G1,G2);     { uses Polar Marsaglia method            }
     DW1TI:=G1*SQDELTA_Y; { Wiener process increment  W1(t(i+1)) - W1(ti) }
     DW2TI:=G2*SQDELTA_Y; { Wiener process increment  W2(t(i+1)) - W2(ti) }
     W1T:=W1T+DW1TI;      { value of the Wiener process W1(t) at t = TI   }
     W2T:=W2T+DW2TI;      { value of the Wiener process W2(t) at t = TI   }

    { Euler scheme : }

     YT:=YT+A(TI-DELTA_Y,YT)*DELTA_Y                    { drift part      }
        +B1(TI-DELTA_Y,YT)*DW1TI+B2(TI-DELTA_Y,YT)*DW2TI; { diffusion part }

    END;{ WHILE }
    ...
```

Notes 4.1.1   The program is similar to Program 3.3.4 except that two independent Wiener processes are now involved. Their increments are generated simultaneously by the Polar Marsaglia routine GENERATE.

Under Lipschitz and linear growth conditions on the coefficients a and b, it has been proved that the Euler approximation has the order of strong convergence

$\gamma = 0.5$, as should be indicated by the slope 0.5 of the curve in the output to Program 4.1.1. In special cases the Euler scheme may actually achieve a higher order of strong convergence. For example, when the noise is *additive*, that is when the diffusion coefficient has the form

$$(1.16) \qquad\qquad b(t, x) \equiv b(t)$$

for all $(t, x) \in \Re^+ \times \Re^d$, it turns out that the Euler scheme has order $\gamma = 1.0$ of strong convergence under appropriate smoothness assumptions on a and b. We remark that additive noise is sometimes understood to have $b(t)$ as constant. Usually the Euler scheme gives good numerical results when the drift and diffusion coefficients are nearly constant. In general, however, it is not particularly satisfactory and the use of higher order schemes is recommended.

B. Milstein Scheme

If, in the 1-dimensional case with $d = m = 1$, we add to the Euler scheme (1.10) the term

$$(1.17) \qquad\qquad bb' J_{(1,1)} = \frac{1}{2} bb' \left\{ (\Delta W_n)^2 - \Delta_n \right\}$$

from the Ito-Taylor expansion (2.3.17), then we obtain the *Milstein scheme*

$$(1.18) \qquad Y_{n+1} = Y_n + a \Delta_n + b \Delta W_n + \frac{1}{2} bb' \left\{ (\Delta W_n)^2 - \Delta_n \right\}. \qquad P.78$$

We can rewrite this as

$$(1.19) \qquad\qquad Y_{n+1} = Y_n + \underline{a} \Delta_n + b \Delta W_n + \frac{1}{2} bb' (\Delta W_n)^2$$

since

$$\underline{a} = a - \frac{1}{2} bb',$$

which is what we would obtain from the Stratonovich-Taylor expansion (2.2.23) by including the term

$$(1.20) \qquad\qquad bb' J_{(1,1)} = \frac{1}{2} bb' (\Delta W_n)^2.$$

Under the assumption that a is once and b twice continuously differentiable it can be shown that the Milstein scheme has the order of strong convergence $\gamma = 1.0$. Thus, with the addition of just one more term to the Euler scheme we can increase the order of strong convergence from $\gamma = 0.5$ to $\gamma = 1.0$. The strong order $\gamma = 1.0$ of the Milstein scheme corresponds to that of the Euler scheme in the deterministic case without any noise, that is with $b \equiv 0$; see Section 3.1. The additional term in the Milstein scheme marks the point of divergence of stochastic numerical analysis from the deterministic.

Problem 4.1.2 (PC-Exercise 10.3.2)   *Consider the Ito process X satisfying the linear stochastic differential equation*

$$dX_t = a X_t \, dt + b X_t \, dW_t$$

on the time interval $[0, T]$ with $T = 1$, $X_0 = 1.0$, $a = 1.5$, $b = 0.1$; see (3.2.14). Generate $M = 20$ batches each of $N = 100$ simulations of X_T and of the values $Y(T)$ of the Milstein scheme (1.18) corresponding to the same sample path of the Wiener process for equidistant time steps with $\delta = \Delta = 2^{-3}$. Then evaluate the 90%-confidence interval for the absolute error ϵ. Repeat for step sizes 2^{-4}, 2^{-5} and 2^{-6}, and plot $\log_2 \epsilon$ against $\log_2 \Delta$.

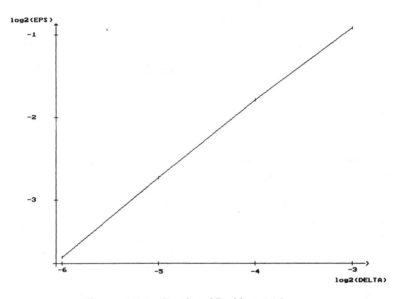

Figure 4.1.1   Results of Problem 4.1.2.

Notes 4.1.2   See Program 3.3.3 and Notes 3.3.3. The program is the same apart from the additional term in the Milstein scheme.

In the general multi-dimensional case with d, $m = 1, 2, \ldots$ the kth component of the *Milstein scheme* has the form

$$(1.21) \qquad Y_{n+1}^k = Y_n^k + a^k \, \Delta_n + \sum_{j=1}^{m} b^{k,j} \Delta W_n^j + \sum_{j_1,j_2=1}^{m} L^{j_1} b^{k,j_2} I_{(j_1,j_2)}$$

in terms of multiple Ito integrals $I_{(j_1,j_2)}$, or

$$(1.22) \qquad Y_{n+1}^k = Y_n^k + \underline{a}^k \, \Delta_n + \sum_{j=1}^{m} b^{k,j} \Delta W_n^j + \sum_{j_1,j_2=1}^{m} \underline{L}^{j_1} b^{k,j_2} J_{(j_1,j_2)}$$

if multiple Stratonovich integrals $J_{(j_1,j_2)}$ are used. Here $\underline{a}$ is the Stratonovich corrected drift, see (1.6).

As seen in Chapter 2, when $j_1 = j_2$ we have

$$(1.23) \qquad I_{(j_1,j_1)} = \frac{1}{2}\{(\Delta W_n^{j_1})^2 - \Delta_n\} \quad \text{and} \quad J_{(j_1,j_1)} = \frac{1}{2}(\Delta W_n^{j_1})^2.$$

On the other hand, when $j_1 \neq j_2$ we have

(1.24) $J_{(j_1,j_2)} = I_{(j_1,j_2)}$,

but we cannot express these just in terms of the increments $\Delta W_n^{j_1}$ and $\Delta W_n^{j_2}$ as in (2.20). We can, however, use the approximation $J_{(j_1,j_2)}^p$ introduced in (2.3.32) and a corresponding approximation $I_{(j_1,j_2)}^p$ for the multiple Ito integral, defined by

$$I_{(j_1,j_2)}^p = J_{(j_1,j_2)}^p$$

when $j_1 \neq j_2$. In view of (2.3.37) and the inequality (2.3.38) we need to pick

$$p \geq p(\Delta) = K/\Delta,$$

for some constant K, to ensure that the error introduced by the use of such approximations of the multiple integrals is less than the $\gamma = 1.0$ discretization error of the Milstein scheme.

Problem 4.1.3 (PC-Exercise 10.3.3) *Consider the scalar Ito process (1.15) in Problem 4.1.1 with $d = 1$ and $m = 2$ on the time interval $[0,T]$ with $T = 1$. Compute $M = 20$ batches each of $N = 100$ simulations of X_T and of the values $Y(T)$ of the Milstein scheme corresponding to the same sample paths of the Wiener processes for equidistant time steps with step size $\delta = \Delta = 2^{-3}$. Evaluate the 90% confidence interval for the absolute error ϵ. Repeat for step sizes 2^{-4}, 2^{-5} and 2^{-6}, and plot $\log_2 \epsilon$ against $\log_2 \Delta$. Approximate the multiple integral $J_{(1,2)}$ by $J_{(1,2)}^p$ with $p = 2$. Repeat with $p = 10$:*

Program 4.1.3   Milstein scheme with multiple stochastic integrals

```
...
PROCEDURE MULTIINTJ12(PL,PR:INTEGER;ROP,DELTA,GSI1,GSI2,MU1,MU2:REAL;
                             VAR SUMVAL,J12:REAL);
VAR
 R:INTEGER;
 ETA,ZETA:ARRAY[1..2,1..100] OF REAL;
BEGIN
 FOR R:=PL TO PR DO
   BEGIN
     GENERATE(ZETA[1,R],ZETA[2,R]);GENERATE(ETA[1,R],ETA[2,R]);
     SUMVAL:=SUMVAL+(1/R)*(ZETA[1,R]*(SQRT(2)*GSI2+ETA[2,R])
                    -ZETA[2,R]*(SQRT(2)*GSI1+ETA[1,R]));
   END;
 J12:=SUMVAL*DELTA/PI;
 J12:=J12+DELTA*(GSI1*GSI2/2+SQRT(ROP)*(MU1*GSI2-MU2*GSI1));
END;{ MULTIINTJ12 }
...
```

```
   { Approximation of the multiple Stratonovich integral J12 : }

   SUMVAL:=0.0;
   GENERATE(MU1,MU2); { Gaussian random numbers MU1 and MU2 }
   MULTIINTJ12(1,P,ROP,DELTA_Y,G1,G2,MU1,MU2,SUMVAL,J12P);
```

```
J21P:=DW1TI*DW2TI-J12P;

{ Milstein scheme(Ito version) : }

I1:=DW1TI;I11:=0.5*(DW1TI*DW1TI-DELTA_Y);I12P:=J12P; { Ito integrals }
I2:=DW2TI;I22:=0.5*(DW2TI*DW2TI-DELTA_Y);I21P:=J21P;
AA:=A(TI-DELTA_Y,YT);
BB1:=B1(TI-DELTA_Y,YT);DB1DX:=DB1X(TI-DELTA_Y,YT);
BB2:=B2(TI-DELTA_Y,YT);DB2DX:=DB2X(TI-DELTA_Y,YT);
L1B11:=BB1*DB1DX;L1B12:=BB1*DB2DX;
L2B12:=BB2*DB2DX;L2B11:=BB2*DB1DX;
YT:=YT+AA*DELTA_Y  { drift part }
+BB1*I1+BB2*I2+L1B11*I11+L1B12*I12P+L2B11*I21P+L2B12*I22;{ diffusion }
...
```

Notes 4.1.3 The multiple integral $J_{(1,2)}$ is approximated by the routine MULTIINTJ12, and $J_{(2,1)}$ is obtained from it via formula (1.2.6). A larger truncation index P will give better approximations, but will require more computational time. Apart from the generation of these additional multiple integral terms in the Milstein scheme, the program is the same as Program 4.1.1.

In many practical problems the diffusion coefficients often have special properties which allow the Milstein scheme to be simplified in a way that avoids the use of double stochastic integrals involving different components of the Wiener process.

An important special case is that of commutative noise in which the diffusion matrix satisfies the *commutativity condition*

$$(1.25) \qquad \underline{L}^{j_1} b^{k,j_2} = \underline{L}^{j_2} b^{k,j_1}$$

for all $j_1, j_2 = 1, \ldots, m$, $k = 1, \ldots, d$ and $(t,x) \in \Re^+ \times \Re^d$. Since

$$(1.26) \qquad J_{(j_1,j_2)} + J_{(j_2,j_1)} = \Delta W^{j_1} \Delta W^{j_2}$$

for $j_1, j_2 = 1, \ldots, m$ with $j_1 \neq j_2$, we see that the *Milstein scheme for commutative noise* reduces to

$$(1.27) \quad Y_{n+1}^k = Y_n^k + \underline{a}^k \Delta + \sum_{j=1}^m b^{k,j} \Delta W^j + \frac{1}{2} \sum_{j_1,j_2=1}^m \underline{L}^{j_1} b^{k,j_2} \Delta W^{j_1} \Delta W^{j_2}.$$

The scalar Ito process (1.15) satisfies the commutativity condition, so it was not necessary in Problem 4.1.3 to use approximate multiple Stratonovich integrals because the equivalent scheme (1.27) could be applied.

Exercise 4.1.1 *Repeat Problem 4.1.3 with the scheme (1.27).*

Finally, we note that in many cases we obtain almost identical results if we use the Ito or the Stratonovich version of the Milstein scheme.

Problem 4.1.4 *Repeat Problem 4.1.2 using the Stratonovich version of the Milstein scheme.*

C. Order 1.5 Strong Taylor Scheme

Multiple stochastic integrals provide additional information about the sample paths of the driving Wiener process within a discretization subinterval. The necessity of their inclusion in higher order schemes is a fundamental difference between the numerical analysis of stochastic and deterministic differential equations.

By including more terms from the Ito-Taylor expansion (2.3.17) in the autonomous 1-dimensional case $d = m = 1$, we obtain the *order 1.5 strong Taylor scheme*

$$(1.28) \quad Y_{n+1} = Y_n + a\,\Delta_n + b\,\Delta W_n + \frac{1}{2}\,bb'\left\{(\Delta W_n)^2 - \Delta_n\right\}$$

$$+ a'b\,\Delta Z_n + \frac{1}{2}\left(aa' + \frac{1}{2}b^2a''\right)\Delta_n^2$$

$$+ \left(ab' + \frac{1}{2}b^2b''\right)\left\{\Delta W_n\,\Delta_n - \Delta Z_n\right\}$$

$$+ \frac{1}{2}b\left(bb'' + (b')^2\right)\left\{\frac{1}{3}(\Delta W_n)^2 - \Delta_n\right\}\Delta W_n.$$

An additional random variable ΔZ_n defined by

$$(1.29) \qquad\qquad \Delta Z_n = \int_{\tau_n}^{\tau_{n+1}}\int_{\tau_n}^{s_2} dW_{s_1}\,ds_2$$

is required. It is normally distributed with mean $E(\Delta Z_n) = 0$, variance $E((\Delta Z_n)^2) = \frac{1}{3}\Delta_n^3$ and covariance $E(\Delta Z_n\Delta W_n) = \frac{1}{2}\Delta_n^2$, and can be generated along with ΔW_n from two independent $N(0;1)$ distributed random variables as in Problem 1.3.5. The last term in (1.28) contains the triple Ito integral

$$(1.30) \qquad\qquad I_{(1,1,1)} = \frac{1}{2}\left\{\frac{1}{3}(\Delta W_n)^2 - \Delta_n\right\}\Delta W_n,$$

see (2.3.45).

Problem 4.1.5 (PC-Exercise 10.4.2)   *Repeat Problem 4.1.2 with the order 1.5 strong Taylor approximation.*

Program 4.1.5   1.5 strong Taylor scheme

```
   . . .
   { Generation of the Taylor approximation and its absolute error : }

   I:=0;
   TI:=T0; { initial time                                          }
   YT:=X0; { initial value of the approximation }
   WHILE TI<T DO
    BEGIN
```

```
I:=I+1;              { time step                                    }
TI:=TI+DELTA_Y;      { current time                                 }
GENERATE(G1,G2);     { uses the Polar Marsaglia method              }
DWTI:=G1*SQDELTA_Y;  { Wiener process increment  W(t(i+1)) - W(ti) }
WT:=WT+DWTI;         { value of the Wiener process W(t) at t = TI   }

{ Generation of the multiple integrals : }

I1:=DWTI;I11:=0.5*(DWTI*DWTI-DELTA_Y);I00:=0.5*DELTA_Y*DELTA_Y;
I10:=0.5*SQDELTA_Y*DELTA_Y*(G1+G2/SQRT3);I01:=DWTI*DELTA_Y-I10;
I111:=0.5*(DWTI*DWTI/3-DELTA_Y)*DWTI;

{ 1.5 order strong Taylor scheme(Ito version) : }

TIOLD:=TI-DELTA_Y;
AA:=A(TIOLD,YT);DADX:=DAX(TIOLD,YT);
BB:=B(TIOLD,YT);DBDX:=DBX(TIOLD,YT);
LOA11:=DAT(TIOLD,YT)+AA*DADX+0.5*BB*BB*DDAXX(TIOLD,YT);
LOB11:=DBT(TIOLD,YT)+AA*DBDX+0.5*BB*BB*DDBXX(TIOLD,YT);
L1A11:=BB*DADX;L1B11:=BB*DBDX;
L1L1B11:=BB*(DBDX*DBDX+BB*DDBXX(TIOLD,YT));
YT:=YT+AA*DELTA_Y+BB*I1+L1B11*I11+L1A11*I10+LOB11*I01+LOA11*I00
    +L1L1B11*I111;

END;{ WHILE }
...
```

Notes 4.1.5 The program differs from Program 4.1.2 by the inclusion of the additional terms appearing in scheme (1.28) and the need to generate the random variable ΔZ_n as well as ΔW_n. This pair of correlated Gaussian random variables is obtained from the independent pair generated by the Polar Marsaglia method as in Program 1.3.5.

In the general multi-dimensional case with d, $m = 1, 2, \ldots$ the kth component of the *order* 1.5 *strong Taylor scheme* has the form

$$(1.31) \quad Y_{n+1}^k = Y_n^k + a^k \Delta_n + \frac{1}{2} L^0 a^k \Delta_n^2$$

$$+ \sum_{j=1}^{m} \left(b^{k,j} \Delta W_n^j + L^0 b^{k,j} I_{(0,j)} + L^j a^k I_{(j,0)} \right)$$

$$+ \sum_{j_1,j_2=1}^{m} L^{j_1} b^{k,j_2} I_{(j_1,j_2)}$$

$$+ \sum_{j_1,j_2,j_3=1}^{m} L^{j_1} L^{j_2} b^{k,j_3} I_{(j_1,j_2,j_3)}$$

for $k = 1, 2, \ldots, d$. To implement this scheme we use the exact values, such as (1.23) and (1.30) for multiple Ito integrals of a single Wiener process and the relations with multiple Stratonovich integrals and their approximations for the others as described in Subsection 2.3.F.

The scheme simplifies considerably if the diffusion matrix b satisfies the *second commutativity condition*

$$(1.32) \qquad L^{j_1} L^{j_2} b^{k,j_3}(t,x) = L^{j_2} L^{j_1} b^{k,j_3}(t,x)$$

for $k = 1, \ldots, d$ and $j_1, j_2, j_3 = 1, \ldots, m$ for all $(t,x) \in \Re^+ \times \Re^d$, in addition to the commutativity condition (1.25). It then involves only the noise increments ΔW_n^j and $\Delta Z_n^j = I_{(j,0)}$ for $j = 1, 2, \ldots, m$.

D. Order 2.0 Strong Taylor Scheme

We can derive higher order schemes by including even more terms from the stochastic Taylor expansions. The schemes become increasingly more complicated in the most general case, but are often reasonably simple in special cases. Relations between different multiple stochastic integrals can also be used to advantage.

For the multi-dimensional case $d = 1, 2, \ldots$ with *scalar noise*, that is with $m = 1$, the kth component of the *order 2.0 strong Taylor scheme* is given by

$$
\begin{aligned}
(1.33) \quad Y_{n+1}^k \;=\;& Y_n^k + \underline{a}^k \, \Delta_n + b^k \, \Delta W_n + \frac{1}{2!} \underline{L}^1 b^k \, (\Delta W_n)^2 + \underline{L}^1 \underline{a}^k \, \Delta Z_n \\
& + \frac{1}{2} \underline{L}^0 \underline{a}^k \, \Delta_n^2 + \underline{L}^0 b^k \, \{\Delta W_n \, \Delta_n - \Delta Z_n\} \\
& + \frac{1}{3!} \underline{L}^1 \underline{L}^1 b^k \, (\Delta W_n)^3 + \frac{1}{4!} \underline{L}^1 \underline{L}^1 \underline{L}^1 b^k \, (\Delta W_n)^4 \\
& + \underline{L}^0 \underline{L}^1 b^k \, J_{(0,1,1)} + \underline{L}^1 \underline{L}^0 b^k \, J_{(1,0,1)} \\
& + \underline{L}^1 \underline{L}^1 \underline{a}^k \, J_{(1,1,0)},
\end{aligned}
$$

Here the Gaussian random variables ΔW_n and ΔZ_n are the same as in the preceding subsection. The multiple Stratonovich integrals can be generated approximately as described in Subsection 2.3.E. There is also a version of the above scheme with multi-dimensional noise. But in many situations considerable simplification is possible because of some special structure of the equation. In the following problem involving a linear stochastic differential equation the threefold multiple Stratonovich integrals can be avoided by means of relations like

$$(1.34) \quad J_{(1,1,0)} + J_{(1,0,1)} + J_{(0,1,1)} = J_{(1,1)} \, \Delta, \qquad J_{(0,1)} + J_{(1,0)} = J_{(1)} \, \Delta;$$

see (2.3.40), (2.3.46) and (2.3.47).

Problem 4.1.6 (PC-Exercise 10.5.2) *Repeat Problem 4.1.2 with the order 2.0 strong Taylor approximation (1.33). Use the relations (1.34) to obtain a simplified scheme.*

Notes 4.1.6 The scheme simplifies considerably for the given SDE using the relations (1.34), so no multiple integrals have to be generated in Program 4.1.6. The running time is reasonable in strong contrast to that for Program 4.1.7 (on the diskette) which has a more general structure to allow easy modification to other SDEs but requires the generation of several multiple stochastic integrals.

Problem 4.1.7 (PC-Exercise 10.5.2, continuation) *Repeat Problem 4.1.6 not using the computational advantages offered by the use of (1.34). Compare with Problem 4.1.6 the computational time required.*

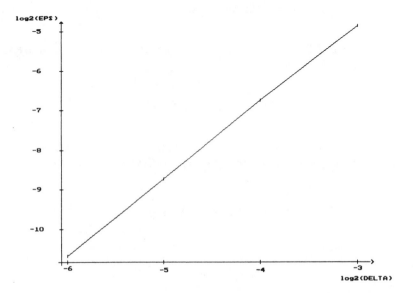

Figure 4.1.2 Results of Problem 4.1.6 and 4.1.7 without and with approximation of the multiple stochastic integrals.

We see from Figure 4.1.2 that exploiting the special structure of the underlying stochastic differential equation leads to a more accurate and less time consuming scheme. In writing down a numerical scheme for a given stochastic differential equation one should always check whether some commutativity property or symmetry allows a simplification of the Taylor scheme of desired strong order.

4.2 Explicit Strong Schemes

A disadvantage of the strong Taylor approximations is that the derivatives of various orders of the drift and diffusion coefficients must be determined and evaluated at each step in addition to the coefficients themselves. In this section we shall examine strong schemes which avoid the use of derivatives in much the same way that Runge-Kutta schemes do in the deterministic setting.

A. Explicit Order 1.0 Strong Schemes

In view of the differences between deterministic and stochastic calculi, heuristic generalizations to stochastic differential equations of the widely used deterministic numerical schemes such as the Runge-Kutta schemes have limited value. The next example illustrates this problem.

Problem 4.2.1 (PC-Exercise 11.1.1)   *Consider the Ito process X satisfying the stochastic differential equation*

$$(2.1) \qquad\qquad dX_t = a\, X_t\, dt + b\, X_t\, dW_t,$$

with $X_0 = 1.0$, $a = 1.5$ and $b = 0.1$ on the time interval $[0, T]$, where $T = 1$, and the following heuristic generalization of the Heun method (3.1.6)

$$(2.2) \quad Y_{n+1} = Y_n + \frac{1}{2}\left\{ a\left(\bar{\Upsilon}_n\right) + a\left(Y_n\right)\right\} \Delta_n + \frac{1}{2}\left\{ b\left(\bar{\Upsilon}_n\right) + b\left(Y_n\right)\right\} \Delta W_n,$$

with supporting value

$$\bar{\Upsilon}_n = Y_n + a\left(Y_n\right) \Delta_n + b\left(Y_n\right) \Delta W_n$$

and initial value $Y_0 = 1.0$; here $a(x) = ax$ and $b(x) = bx$. Generate $M = 20$ batches each of $N = 100$ simulations of X_T and of the values $Y(T)$ of (2.2) corresponding to the same sample paths of the Wiener process for equidistant time discretizations with step size $\delta = \Delta_n = 2^{-3}$, 2^{-4}, 2^{-5} and 2^{-6}. Plot $\log_2$ of the absolute error against $\log_2 \delta$.

Notes 4.2.1   The program, which is based on Program 3.3.3, is given on the diskette. The Heun method for the specified SDE only is considered. Confidence intervals can also be plotted.

While the heuristically generalized Heun method (2.2) may at first seem acceptable, the results of Problem 4.2.1 plotted in Figure 4.2.1 indicate that it does not converge. The scheme is not consistent with the Ito calculus.

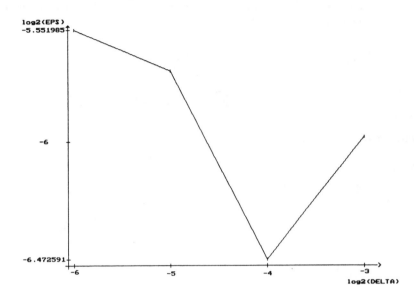

Figure 4.2.1 Results of Problem 4.2.1.

In the 1-dimensional case with $d = m = 1$ an *explicit order* 1.0 *strong scheme* due to Platen is given by

$$(2.3) \qquad Y_{n+1} = Y_n + \underline{a}\,\Delta_n + b\,\Delta W_n + \frac{1}{2\sqrt{\Delta_n}}\left\{b(\tau_n, \bar{\Upsilon}_n) - b\right\}(\Delta W_n)^2$$

with the supporting value

$$(2.4) \qquad \bar{\Upsilon}_n = Y_n + \underline{a}\,\Delta_n + b\,\sqrt{\Delta_n},$$

where $\underline{a}$ is the Stratonovich corrected drift, see (1.6). The ratio

$$\frac{1}{\sqrt{\Delta_n}}\left\{b\left(\tau_n, Y_n + \underline{a}\,\Delta_n + b\,\sqrt{\Delta_n}\right) - b\left(\tau_n, Y_n\right)\right\}$$

here is a forward difference approximation for $b\,\frac{\partial b}{\partial x}$ at (τ_n, Y_n) if we neglect higher order terms. Thus (2.3) is a derivative free counterpart of the Milstein scheme (1.19).

Problem 4.2.2 *Repeat Problem 4.2.1 using the explicit order* 1.0 *strong scheme (2.3)–(2.4).*

Notes 4.2.2 The program here is also a simplification of Program 3.3.3 to the specified scheme and SDE.

In the general multi-dimensional case $d, m = 1, 2, \ldots$ the *explicit order* 1.0 *strong scheme* has kth component

$$(2.5) \quad Y_{n+1}^k = Y_n^k + \underline{a}^k \Delta_n + \sum_{j=1}^m b^{k,j} \Delta W_n^j$$

$$+ \frac{1}{\sqrt{\Delta_n}} \sum_{j_1,j_2=1}^m \left\{ b^{k,j_2} \left(\tau_n, \bar{\Upsilon}_n^{j_1} \right) - b^{k,j_2} \right\} J_{(j_1,j_2)}$$

with vector supporting values

$$(2.6) \qquad\qquad \bar{\Upsilon}_n^j = Y_n + \underline{a} \Delta_n + b^j \sqrt{\Delta_n}$$

for $j = 1, 2, \ldots$. For *commutative* noise, see (1.25), we can also use the scheme

$$(2.7) \qquad Y_{n+1}^k = Y_n^k + \underline{a}^k \Delta_n + \frac{1}{2} \sum_{j=1}^m \left\{ b^{k,j} \left(\tau_n, \bar{\Upsilon}_n \right) + b^{k,j} \right\} \Delta W_n^j$$

with vector supporting values

$$\bar{\Upsilon}_n = Y_n + \underline{a} \Delta_n + \sum_{j=1}^m b^j \Delta W_n^j.$$

Let us remark that there are also versions of explicit order 1.0 strong schemes involving the Ito drift coefficient a. For instance, for general noise, Platen proposed the scheme

$$(2.8) \qquad Y_{n+1} = Y_n + a \Delta_n + \sum_{j=1}^m b^j \Delta W_n^j$$

$$+ \frac{1}{\sqrt{\Delta_n}} \sum_{j_1,j_2=1}^m \left\{ b^{j_2} \left(\tau_n, \bar{\Upsilon}_n^{j_1} \right) - b^{j_2} \right\} I_{(j_1,j_2)}$$

with

$$\bar{\Upsilon}_n^j = Y_n + a \Delta_n + b^j \sqrt{\Delta_n}.$$

Exercise 4.2.1 (PC-Exercise 11.1.3) *Repeat Problem 4.2.2 using the scheme (2.8).*

The above schemes all converge with strong order $\gamma = 1.0$ under conditions similar to those for the Milstein scheme.

Problem 4.2.3 (PC-Exercise 11.1.5) *Repeat Problem 4.1.1 with the scheme (2.7).*

Notes 4.2.3 The program is now based on Program 4.1.1, which has been simplified to the given scheme and SDE again. Confidence intervals can also be plotted.

The $a\,\Delta_n$ term in (2.4) can be omitted without losing the strong order of convergence $\gamma = 1.0$, as the following example shows.

Problem 4.2.4 (PC-Exercise 11.1.6) *Repeat Problem 4.2.1 for the explicit order 1.0 strong scheme (2.3) using the supporting value*

$$(2.9) \qquad \tilde{\Upsilon}_n = Y_n + b\sqrt{\Delta_n}$$

instead of (2.4). Compare the results with those of Problem 4.1.2 and Problem 4.2.2.

Notes 4.2.4 A simple change to Program 4.2.2 gives the required program. The computational results of Problem 4.2.4 should show that while the order of convergence is similar, the results are slightly worse than for the original explicit order 1.0 strong scheme. Essentially, (2.4) provides a better approximation of the higher order terms then does (2.9).

B. Explicit Order 1.5 Strong Scheme

There exist analogous derivative free schemes of order 1.5 obtained by replacing the derivatives in the order 1.5 strong Taylor scheme (1.28) by corresponding finite differences.

In the general multi-dimensional autonomous case with d, $m = 1, 2, \ldots$ the kth component of the *explicit order 1.5 strong scheme* satisfies

$$(2.10) \quad Y_{n+1}^k \;=\; Y_n^k + a^k\,\Delta_n + \sum_{j=1}^m b^{k,j}\,\Delta W_n^j$$

$$+ \frac{1}{2\sqrt{\Delta_n}} \sum_{j_2=0}^m \sum_{j_1=1}^m \left\{ b^{k,j_2}\left(\tilde{\Upsilon}_+^{j_1}\right) - b^{k,j_2}\left(\tilde{\Upsilon}_-^{j_1}\right)\right\} I_{(j_1,j_2)}$$

$$+ \frac{1}{2\Delta_n} \sum_{j_2=0}^m \sum_{j_1=1}^m \left\{ b^{k,j_2}\left(\tilde{\Upsilon}_+^{j_1}\right) - 2b^{k,j_2} + b^{k,j_2}\left(\tilde{\Upsilon}_-^{j_1}\right)\right\} I_{(0,j_2)}$$

$$+ \frac{1}{2\Delta_n} \sum_{j_1,j_2,j_3=1}^m \left[b^{k,j_3}\left(\bar{\Phi}_+^{j_1,j_2}\right) - b^{k,j_3}\left(\bar{\Phi}_-^{j_1,j_2}\right)\right.$$

$$\left. - b^{k,j_3}\left(\tilde{\Upsilon}_+^{j_1}\right) + b^{k,j_3}\left(\tilde{\Upsilon}_-^{j_1}\right)\right] I_{(j_1,j_2,j_3)}$$

with

$$(2.11) \qquad \tilde{\Upsilon}_\pm^j = Y_n + \frac{1}{m}\,a\,\Delta_n \pm b^j\sqrt{\Delta_n}$$

and

$$(2.12) \qquad \bar{\Phi}_\pm^{j_1,j_2} = \tilde{\Upsilon}_+^{j_1} \pm b^{j_2}\left(\tilde{\Upsilon}_+^{j_1}\right)\sqrt{\Delta_n},$$

where we have written $b^{k,0}$ for a^k in the summation terms. The multiple Ito integrals here can be approximated by multiple Stratonovich integrals as described in Subsections 2.3.D and E.

Problem 4.2.5 (PC-Exercise 11.2.1) *Repeat Problem 4.2.1 using the explicit order 1.5 strong scheme (2.10)–(2.12) and compare the results with those of Problem 4.1.4.*

Notes 4.2.5 The program is adapted from Program 4.2.2. Confidence intervals can also be plotted. The results obtained will be in fact better than order 1.5 for the linear SDE under consideration.

The above scheme can often be simplified by making use of any special structure of a particular stochastic differential equation under consideration, for instance, in the case of commutative noise of the second kind (1.32) or additive noise (1.16).

C. Explicit Order 2.0 Strong Scheme

We can also take advantage of some special structure of the equations under consideration to obtain a relatively simple second order explicit scheme which does not involve derivatives of the drift and diffusion cofficients. To illustrate this we restrict our attention to the case of *additive noise* (1.16), where the diffusion coefficient satisfies $b(t, x) \equiv b(t)$ for all t and x. In the *nonautonomous* multi-dimensional case $d = 1$, 2, ... with $m = 1$ the *explicit order 2.0 strong scheme for scalar additive noise* has kth component

$$(2.13) \quad Y_{n+1}^k = Y_n^k + \frac{1}{2}\left\{ \underline{a}^k \left(\tau_n + \frac{1}{2}\Delta_n, \bar{\Upsilon}_+ \right) + \underline{a}^k \left(\tau_n + \frac{1}{2}\Delta_n, \bar{\Upsilon}_- \right) \right\} \Delta_n$$

$$+ b^k \, \Delta W_n + \frac{1}{\Delta_n} \left\{ b^k \left(\tau_{n+1} \right) - b^k \right\} \left\{ \Delta W_n \Delta_n - \Delta Z_n \right\}$$

with

$$\bar{\Upsilon}_\pm = Y_n + \frac{1}{2}\underline{a}\,\Delta_n + \frac{1}{\Delta_n} b \left\{ \Delta Z_n \pm \sqrt{2 J_{(1,1,0)}\,\Delta_n - (\Delta Z_n)^2} \right\},$$

for $k = 1$, ..., d. It generalizes a scheme first proposed by Chang. The random variables ΔW_n, ΔZ_n and $J_{(1,1,0)}$ here can be generated or approximated as in (2.3.31).

We shall use the nonautonomous linear stochastic differential equation with additive noise

$$(2.14) \qquad dX_t = \left(\frac{2}{1+t} X_t + (1+t)^2 \right) dt + (1+t)^2 \, dW_t,$$

$t \geq 0$, with initial value $X_0 = 1$ and exact solution

$$X_t = (1+t)^2 \, (1 + W_t + t)$$

in the following problem.

Problem 4.2.6 (PC-Exercise 11.3.2) *Generate $M = 20$ batches each of $N = 100$ simulations of X_T for (2.14), with $T = 0.5$, and of the value $Y(T)$ of the explicit order 2.0 strong scheme (2.13) corresponding to the same sample paths of the Wiener process for step sizes $\delta = \Delta_n = 2^{-3}$, 2^{-4}, 2^{-5} and 2^{-6}. Plot $\log_2$ of the absolute error against $\log_2 \delta$.*

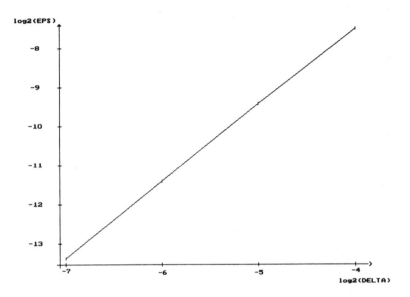

Figure 4.2.2   Result for explicit order 2.0 strong scheme.

Notes 4.2.6   Since the modified drift coefficient is linear in (2.14), the terms involving the multiple integral $J_{(1,1,0)}$ cancel so there is no need to approximate $J_{(1,1,0)}$. Program 4.2.2 has been adapted to the scheme (2.13) for the equation (2.14). Confidence intervals can also be plotted. For general SDEs the multiple integral terms will not cancel and an approximation $J^p_{(1,1,0)}$ for $J_{(1,1,0)}$ will be required.

Exercise 4.2.2   *Repeat Problem 4.2.6 without cancelling the terms involving the multiple integral $J_{(1,1,0)}$. Use $J^p_{(1,1,0)}$ given in (2.31) with $p = 15$ to approximate $J_{(1,1,0)}$.*

D. Two-Step Order 1.0 Strong Scheme

Multi-step methods are often more efficient computationally than one-step methods of the same order because they require essentially only one new evaluation of the right hand side of the differential equation for each iteration. In this subsection we shall briefly introduce a stochastic two-step scheme. As with deterministic multi-step schemes, a one-step scheme will be used to generate the initial steps needed to start the two-step scheme.

Also here one can sometimes take advantage of the structure of certain types of stochastic differential equations to obtain relatively simple multi-step schemes. Illustrative of this is the 2-dimensional Ito system

$$(2.15) \quad dX_t^1 \;=\; X_t^2 \, dt$$

$$dX_t^2 \;=\; \left\{ -a(t)\, X_t^2 + b\left(t, X_t^1\right)\right\} dt + \sum_{j=1}^{m} c^j \left(t, X_t^1\right) dW_t^j,$$

for the first component of which Lépingle and Ribémont proposed the two-step scheme

$$(2.16) \quad Y_{n+2}^1 \;=\; \left\{ 2 - a(\tau_n)\, \Delta_n \right\} Y_{n+1}^1 - \left\{ 1 - a(\tau_n)\, \Delta_n \right\} Y_n^1$$

$$+ b\left(\tau_n, Y_n^1\right) \Delta_n^2 + \sum_{j=1}^{m} c^j \left(\tau_n, Y_n^1\right) \Delta W_n^j \, \Delta_n,$$

This has strong order $\gamma = 1.0$ and can, in fact, be derived from the 2-dimensional Milstein scheme for (2.15) by eliminating the second component.

Problem 4.2.7 (PC-Exercise 11.4.1)   *Use the two-point scheme (2.16) with time step $\delta = \Delta_n = 2^{-4}$ to simulate and plot a linearly interpolated approximate trajectory of the first component of system (2.15) with*

$$a(t) \equiv 1.0, \quad b(t, x) = x^1 \left(1 - \left(x^1\right)^2\right), \quad c(t, x) \equiv 0.01$$

and $X_0^1 = -2.0$, $X_0^2 = 0$ on the interval $[0, 8]$. Use the first equation of the 2-dimensional Milstein scheme as starting routine.

Notes 4.2.7   The graphics routines for the program are based on Program 1.7.2. The suggested starting routine provides the values

$$Y_0^1 = X_0^1, \qquad\qquad Y_1^1 = X_0^1 + X_0^2 \, \Delta_n$$

for the two-step scheme (2.16). See the diskette for more details of the program.

E. Two-Step Order 1.5 Strong Scheme

A slight variation of the stochastic Ito-Taylor expansion yields in the general multi-dimensional case $d, m = 1, 2, \ldots$ to the following *two-step order 1.5 strong scheme*

$$(2.17) \quad Y_{n+1} = Y_{n-1} + 2a\, \Delta_n - \sum_{j=1}^{m} L^j a \left(\tau_{n-1}, Y_{n-1}\right) \Delta W_{n-1}^j \Delta_n + V_n + V_{n-1}$$

with

$$
V_n = \sum_{j=1}^{m} \left[b^j \, \Delta W_n^j + L^0 b^j \left\{ \Delta W_n^j \Delta_n - \Delta Z_n^j \right\} + L^j a \, \Delta Z_n^j \right]
$$

$$
+ \sum_{j_1,j_2=1}^{m} L^{j_1} b^{j_2} \, I_{(j_1,j_2),\tau_n,\tau_{n+1}}
$$

$$
+ \sum_{j_1,j_2,j_3=1}^{m} L^{j_1} L^{j_2} b^{j_3} \, I_{(j_1,j_2,j_3),\tau_n,\tau_{n+1}},
$$

where the notation of the operators L^0, L^j and the multiple Ito integrals is the same as in (1.3)–(1.6). Again the multiple Ito integrals can be approximated.

There is also a derivative free counterpart of the above scheme, which can simplify considerably for specific equations. Note that (2.17) can be interpreted as a higher strong order stochastic generalization of the deterministic midpoint method (3.1.12).

Problem 4.2.8 (PC-Exercise 11.4.2) *Repeat Problem 4.1.2 for the scheme (2.17) with the explicit order 1.5 strong scheme as starting routine.*

Notes 4.2.8 The program is adapted from Program 4.2.5, which can also be used for the starting routine. Note that Y_n, ΔW_n and V_n must be saved to become Y_{n-1}, ΔW_{n-1} and V_{n-1}, respectively, in the next iteration. As in Problem 4.2.5 an order of convergence slightly better than the theoretic $\gamma = 1.5$ is obtained for the linear equation under consideration.

4.3   Implicit Strong Approximations

In this section we shall consider implicit strong schemes. These schemes usually have a wide range of step sizes suitable for the approximation of stochastic dynamical systems, in particular those involving vastly different time scales, without the excessive accumulation of unavoidable roundoff errors. Thus implicit schemes are well suited for simulating the solutions of stiff stochastic differential equations.

In implementing an implicit scheme we need to solve an additional algebraic equation at each time step. This can usually be done with standard numerical methods such as the Newton-Raphson method.

A. Implicit Euler Scheme

The simplest implicit scheme is the implicit Euler scheme which has strong order $\gamma = 0.5$. To ensure the existence of the solution of the scheme, only the drift term in it is implicit. That is, the $a(Y_n)\Delta_n$ term in the explicit Euler scheme (3.2.10) is replaced by $a(Y_{n+1})\Delta_n$ while the diffusion term $b(Y_n)\Delta W_n$ remains unchanged.

By interpolating between the fully implicit and the explicit Euler schemes we obtain a family of implicit Euler schemes. In the general multi-dimensional case $d, m = 1, 2, \ldots$ the *family of implicit Euler schemes* has kth component

$$(3.1) \quad Y_{n+1}^k = Y_n^k + \left\{ \alpha a^k \left(\tau_{n+1}, Y_{n+1} \right) + (1 - \alpha) a^k \right\} \Delta_n + \sum_{j=1}^{m} b^{k,j} \, \Delta W_n^j$$

where $\alpha \in [0,1]$ denotes the degree of implicitness. For $\alpha = 0$ we have the explicit Euler scheme and for $\alpha = 1$ the fully implicit one. Note that we could use different values of the implicitness parameter in each component of a scheme.

We shall use the following 2-dimensional linear Ito stochastic differential equation in problems to test the above family of implicit schemes and others to be introduced later. Let $W = \{W_t, t \geq 0\}$ be a 1-dimensional Wiener process and define matrices

$$(3.2) \qquad A = \begin{bmatrix} -a & a \\ a & -a \end{bmatrix} \quad \text{and} \quad B = bI = \begin{bmatrix} b & 0 \\ 0 & b \end{bmatrix}$$

where a and b are given real numbers and I is the unit matrix. For these matrices

$$(3.3) \qquad\qquad dX_t = AX_t \, dt + BX_t \, dW_t$$

is a 2-dimensional, homogeneous linear Ito stochastic differential equation. It can be shown that it has the explicit solution

$$(3.4) \qquad\qquad X_t = \exp\left(\left(A - \frac{1}{2} B^2 \right) t + B \, W_t \right) X_0.$$

It is reasonable to say that for $a = 5$ and $b = 0.01$, for example, equation (3.3) is a stiff stochastic differential equation. We shall use it to test implicit schemes. For such a linear equation these implicit schemes can be manipulated algebraically to provide an explicit recursion formula, thus avoiding the need for a numerical procedure such as the Newton-Raphson method at each time step.

Problem 4.3.1 (PC-Exercise 12.2.3)   *Consider the 2-dimensional Ito process X satisfying the stochastic differential equation (3.3) with coefficient matrices (3.2) with $a = 5$ and $b = 0.01$ on the time interval $[0, T]$ with initial value $X_0 = (1, 0)$ and $T = 1$. Compute $M = 20$ batches each of $N = 100$ simulations of X_T and of the values $Y(T)$ of the implicit Euler approximation with $\alpha_1 = \alpha_2 = 0$, 0.5 and 1.0 corresponding to the same sample paths of the Wiener process for equidistant time steps with step size $\delta = \Delta_n = 2^{-2}$. Evaluate the 90% confidence intervals for the absolute errors ϵ. Repeat for step sizes $\delta = \Delta_n = 2^{-3}$, 2^{-4} and 2^{-5} and plot $\log_2 \epsilon$ versus $\log_2 \delta$ for the three cases $\alpha = 0$, 0.5 and 1.0, respectively.*

Program 4.3.1   Implicit Euler scheme

```
...
PROCEDURE EXPLSOL(TI,WT:REAL;VAR X1,X2:REAL);
VAR
 EROP,EROM,Q1,Q2:REAL;
BEGIN
 EROP:=EXP(BETA*(-0.5*BETA*(TI-T0)+WT));EROM:=EROP*EXP(-2.0*ALPHA*(TI-T0));
 Q1:=0.5*(EROP+EROM);Q2:=0.5*(EROP-EROM);
 X1:=Q1*X10+Q2*X20;X2:=Q2*X10+Q1*X20;
END;{ EXPLSOL }

FUNCTION SQRTERR(X1,X2,Y1,Y2:REAL):REAL;
BEGIN
 SQRTERR:=SQRT((X1-Y1)*(X1-Y1)+(X2-Y2)*(X2-Y2)); { Euclidean norm }
END;{ SQRTERR }

PROCEDURE COMPSAMPLEPARA(NN:INTEGER;X:VECTOR0;VAR SAVERAGE,SVARIANCE:REAL);

{ Main program : }
 ...
 DELTA_Y:=2.*DELTA;
 G:=0;
 REPEAT
  G:=G+1;                    { index of the time step size used   }
  DELTA_Y:=DELTA_Y/2;        { current time step size            }
  SQDELTA_Y:=SQRT(DELTA_Y); { square root of DELTA_Y            }
  KAPPA0:=ALPHA*DELTA_Y;     { initialization of scheme constants }
  KAPPA1:=KAPPA0*(2.0-ALPHAK1-ALPHAK2);KAPPA2:=1.0+KAPPA0*(ALPHAK1+ALPHAK2);

 { Generation for different batches : }

  J:=0;
  REPEAT
   J:=J+1;          { batch index                                }
   EPSYLON[J]:=0.0; { sum of the strong errors of the batch used }

  { Generation of different trajectories : }

   K:=0;
   REPEAT
    K:=K+1; { index of the trajectory used        }
    WT:=0.0; { value of the Wiener process at t=T0  }

   { Generation of the Euler approximation and its strong error : }

    I:=0;
    TI:=T0;   { initial time                                       }
    Y1T:=X10; { initial value of first component of the approximation  }
    Y2T:=X20; { initial value of second component of the approximation }
    HII:=Y2T-Y1T;
    WHILE TI<T DO
      BEGIN
       I:=I+1;          { time step index }
       TI:=TI+DELTA_Y; { current time    }
       IF I MOD 2 = 1 THEN GENERATE(G1,G2) { uses Polar Marsaglia method }
        ELSE G1:=G2;
       DWTI:=G1*SQDELTA_Y; { Wiener process increment  W(t(i+1)) - W(ti) }
```

```
    WT:=WT+DWTI;              { value of the Wiener process W(t) at t = TI  }

  { Implicit Euler scheme written in explicit form for the linear equation:}

    HH:=1.0+BETA*DWTI;
    HI:=HII;HII:=HII*(HH-KAPPA1)/KAPPA2;
    Y1T:=Y1T*HH+KAPPA0*(ALPHAK1*HII+(1.0-ALPHAK1)*HI);
    Y2T:=Y2T*HH-KAPPA0*(ALPHAK2*HII+(1.0-ALPHAK2)*HI);

    END;{ WHILE }

  { Summation of the errors : }

    EXPLSOL(TI,WT,X1T,X2T); { calculates the components of the solution }
    EPSYLON[J]:=EPSYLON[J]+SQRTERR(X1T,X2T,Y1T,Y2T);

   UNTIL K=N;{ REPEAT for different samples }
   EPSYLON[J]:=EPSYLON[J]/N; { estimate of the error of the batch }
  UNTIL J=M;{ REPEAT for different batches }

{ Calculation of the confidence interval and initialization of data : }

  DEL[G]:=DELTA_Y; { current time step size }
  COMPSAMPLEPARA(M,EPSYLON,AVERAGE,VARIANCE);
  EPS[G]:=AVERAGE;              { midpoint of the confidence interval }
  DIFFER[G]:=QUANTILE*SQRT(VARIANCE/M); { half the interval length }

 UNTIL G=NUM;{ REPEAT for different time step sizes }
    ...
```

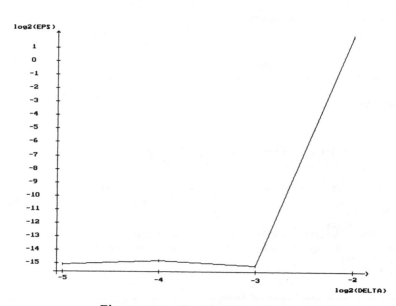

Figure 4.3.1   Explicit Euler scheme, $\alpha = 0$.

Notes 4.3.1   The program is adapted from Program 3.3.2 to the 2-dimensional linear equation and schemes under consideration. In particular, the error is defined in terms of the Euclidean norm. The program needs to be rerun for each choice of the implicitness parameter α. Note in the program that different implicitness parameters could be used in the different components.

It is apparent from Figure 4.3.1 that the explicit Euler scheme, does not give an acceptable result for the stiff system under consideration. On the other hand the generalized trapezoidal scheme, that is (3.1) with $\alpha = 0.5$, and the fully implicit Euler scheme with $\alpha = 1.0$ generally provide reasonable results and the error graphs have slope 0.5 or even better. This is the same as the theoretical strong order of the explicit Euler scheme, which is impractical here.

B. Implicit Milstein Scheme

Similar results but with slope 1.0 instead of 0.5 can be obtained with implicit Milstein schemes. These involve an additional term in their Euler counterparts. Like with the Euler schemes, we can also interpolate between the corresponding explicit and fully implicit Milstein schemes.

In the general multi-dimensional case $d, m = 1, 2, \ldots$ the *family of implicit Milstein schemes* has kth component

$$
(3.5) \qquad Y_{n+1}^k = Y_n^k + \left\{ \alpha a^k \left(\tau_{n+1}, Y_{n+1} \right) + (1 - \alpha) a^k \right\} \Delta_n
$$
$$
+ \sum_{j=1}^m b^{k,j} \, \Delta W_n^j + \sum_{j_1,j_2=1}^m L^{j_1} b^{k,j_2} I_{(j_1,j_2)}
$$

in its *Ito version* and

$$
(3.6) \qquad Y_{n+1}^k = Y_n^k + \left\{ \alpha \underline{a}^k \left(\tau_{n+1}, Y_{n+1} \right) + (1 - \alpha) \underline{a}^k \right\} \Delta_n
$$
$$
+ \sum_{j=1}^m b^{k,j} \, \Delta W_n^j + \sum_{j_1,j_2=1}^m L^{j_1} b^{k,j_2} J_{(j_1,j_2)}
$$

in its *Stratonovich version* for $k = 1, \ldots, d$, where $\alpha \in [0, 1]$. The multiple stochastic integrals $I_{(j_1,j_2)}$ and $J_{(j_1,j_2)}$ here can be approximated by the method proposed in (2.3.32).

For *commutative noise*, see (1.25), the above family of implicit Milstein schemes simplifies to

$$
(3.7) \quad Y_{n+1}^k = Y_n^k + \left\{ \alpha a^k \left(\tau_{n+1}, Y_{n+1} \right) + (1 - \alpha) a^k \right\} \Delta_n + \sum_{j=1}^m b^{k,j} \, \Delta W_n^j
$$
$$
+ \frac{1}{2} \sum_{j_1,j_2=1}^m L^{j_1} b^{k,j_2} \left\{ \Delta W_n^{j_1} \Delta W_n^{j_2} - \delta_{j_1,j_2} \Delta_n \right\}
$$

in its Ito version, where δ_{j_1,j_2} is the Kronecker delta symbol

$$\delta_{j_1,j_2} = \left\{ \begin{array}{ll} 1 & : \quad j_1 = j_2 \\ 0 & : \quad \text{otherwise,} \end{array} \right.$$

and to

$$(3.8) \quad Y_{n+1}^k = Y_n^k + \left\{ \alpha \underline{a}^k \left(\tau_{n+1}, Y_{n+1} \right) + (1-\alpha) \underline{a}^k \right\} \Delta_n$$

$$+ \sum_{j=1}^{m} b^{k,j} \, \Delta W_n^j + \frac{1}{2} \sum_{j_1,j_2=1}^{m} L^{j_1} b^{k,j_2} \Delta W_n^{j_1} \Delta W_n^{j_2}.$$

in its Stratonovich version, for $k = 1, 2, \ldots, d$.

Problem 4.3.2 (PC-Exercise 12.2.4) *Repeat Problem 4.3.1 for the Ito version of the implicit Milstein schemes with $\alpha = 0.5$ and 1.0.*

Notes 4.3.2   The program differs from Program 4.3.1 only by the inclusion of the additional multiple integral terms. We remark that the implementation allows to choose different degrees of implicitness for different components.

C. Implicit Order 1.5 and 2.0 Strong Taylor Schemes

By systematically involving higher order multiple stochastic integrals we can derive higher order implicit schemes. Here too only the coefficients of the purely deterministic multiple integrals are made implicit.

In the general multi-dimensional case $d, m = 1, 2, \ldots$ we have the *family of implicit order 1.5 strong Taylor schemes* with kth component

$$(3.9) \qquad Y_{n+1}^k = Y_n^k + \left\{ \alpha a^k \left(\tau_{n+1}, Y_{n+1} \right) + (1-\alpha) a^k \right\} \Delta_n$$

$$+ \left(\frac{1}{2} - \alpha \right) \left\{ \beta L^0 a^k \left(\tau_{n+1}, Y_{n+1} \right) + (1-\beta) L^0 a^k \right\} \Delta_n^2$$

$$+ \sum_{j=1}^{m} \left(b^{k,j} \, \Delta W_n^j + L^0 b^{k,j} I_{(0,j)} + L^j a^k \left\{ I_{(j,0)} - \alpha \Delta W_n^j \Delta_n \right\} \right)$$

$$+ \sum_{j_1,j_2=1}^{m} L^{j_1} b^{k,j_2} I_{(j_1,j_2)} + \sum_{j_1,j_2,j_3=1}^{m} L^{j_1} L^{j_2} b^{k,j_3} I_{(j_1,j_2,j_3)},$$

where $\alpha, \beta \in [0,1]$. The multiple stochastic integrals here can also be approximated as in (2.3.31)–(2.3.36).

The above scheme reduces considerably in the case of commutative noise of the second kind (1.32) and the choice of implicitness parameter $\alpha = 0.5$, giving

$$(3.10) \qquad Y_{n+1}^k = Y_n^k + \frac{1}{2} \left\{ a^k \left(\tau_{n+1}, Y_{n+1} \right) + a^k \right\} \Delta_n$$

$$+ \sum_{j=1}^{m} \left[b^{k,j} \, \Delta W_n^j + \frac{1}{2} L^j b^{k,j} \left\{ \left(\Delta W_n^j \right)^2 - \Delta_n \right\} \right.$$

$$\left. + L^0 b^{k,j} \left\{ \Delta W_n^j \, \Delta_n - \Delta Z_n^j \right\} + L^j a^k \left\{ \Delta Z_n^j - \frac{1}{2} \Delta W_n^j \, \Delta_n \right\} \right]$$

$$+ \sum_{j_1=1}^{m} \sum_{j_2=1}^{j_1-1} L^{j_1} b^{k,j_2} \, \Delta W_n^{j_1} \Delta W_n^{j_2}$$

$$+ \frac{1}{2} \sum_{\substack{j_1,j_2=1 \\ j_1 \neq j_2}}^{m} L^{j_1} L^{j_2} b^{k,j_2} \Delta W_n^{j_1} \left\{ \left(\Delta W_n^{j_2} \right)^2 - \Delta_n \right\}$$

$$+ \sum_{j_1=1}^{m} \sum_{j_2=1}^{j_1-1} \sum_{j_3=1}^{j_2-1} L^{j_1} L^{j_2} b^{k,j_3} \, \Delta W_n^{j_1} \Delta W_n^{j_2} \Delta W_n^{j_3}$$

$$+ \frac{1}{2} \sum_{j=1}^{m} L^j L^j b^{k,j} \left\{ \frac{1}{3} \left(\Delta W_n^j \right)^2 - \Delta_n \right\} \Delta W_n^j.$$

We note that this scheme, which is of strong order 1.5, involves only the correlated pairs of Gaussian random variables $(\Delta W_n^j, \Delta Z_n^j)$, $j = 1, \ldots, m$; see relation (1.29).

Problem 4.3.3 (PC-Exercise 12.2.5)   *Repeat Problem 4.3.1 for the implicit order 1.5 strong Taylor scheme (3.9) with $\alpha = 0.5$.*

Notes 4.3.3   Equation (3.3) has commutative noise of the second kind (1.32), so the scheme reduces to (3.10). In fact, since $L^0 b^{k,j} \equiv L^j a^k$ for the given equation there is no need to generate the ΔZ_n^j random variables. For greater generality, other choices of α are allowed in the program, which is similar to Program 4.3.2, but this requires retention of the other implicitness parameter β in (3.9).

There is also an implicit order 2.0 strong Taylor scheme which is not much more complicated than the order 1.5 scheme above. In the general multi-dimensional case $d, m = 1, 2, \ldots$ the kth component of the *implicit order 2.0 strong Taylor scheme* with implicitness parameter $\alpha = 0.5$ is

$$(3.11) \qquad Y_{n+1}^k = Y_n^k + \frac{1}{2} \left\{ \underline{a}^k \left(\tau_{n+1}, Y_{n+1} \right) + \underline{a}^k \right\} \Delta_n$$

$$+ \sum_{j=1}^{m} \left[b^{k,j} \, \Delta W_n^j + \underline{L}^0 b^{k,j} \, J_{(0,j)} + \underline{L}^j \underline{a}^k \left\{ J_{(j,0)} - \frac{1}{2} \Delta W_n^j \, \Delta_n \right\} \right]$$

$$+ \sum_{j_1,j_2=1}^{m} \left[\underline{L}^{j_1} b^{k,j_2} \, J_{(j_1,j_2)} + \underline{L}^0 \underline{L}^{j_1} b^{k,j_2} \, J_{(0,j_1,j_2)} \right.$$

$$+\underline{L}^{j_1}\underline{L}^0 b^{k,j_2}\, J_{(j_1,0,j_2)}$$

$$+\underline{L}^{j_1}\underline{L}^{j_2}\underline{a}^k \left\{ J_{(j_1,j_2,0)} - \frac{1}{2}\Delta_n J_{(j_1,j_2)} \right\} \Bigg]$$

$$+ \sum_{j_1,j_2,j_3=1}^{m} \underline{L}^{j_1}\underline{L}^{j_2} b^{k,j_3}\, J_{(j_1,j_2,j_3)}$$

$$+ \sum_{j_1,j_2,j_3,j_4=1}^{m} \underline{L}^{j_1}\underline{L}^{j_2}\underline{L}^{j_3} b^{k,j_4}\, J_{(j_1,j_2,j_3,j_4)},$$

where $\underline{a}$, denotes the corrected Stratonovich drift (1.6). In special cases the scheme (3.11) also simplifies substantially to ones that do not involve all of the multiple Stratonovich integrals.

Problem 4.3.4 (PC-Exercise 12.2.6)   *Repeat Problem 4.3.1 for the implicit order 2.0 strong Taylor scheme (3.11).*

Notes 4.3.4   The scheme simplifies considerably for equation (3.3) with its linear coefficients and scalar noise. The formulae (2.3.40) can be used for the surviving multiple integrals. The program, which is given on the diskette, is adapted from Program 4.3.3, except that here the implicitness parameter is fixed at $\alpha = 0.5$.

D. Implicit Strong Runge-Kutta Schemes

We shall now consider implicit schemes which avoid the use of derivatives in the terms involving non-deterministic multiple stochastic integrals. They are obtained from the corresponding implicit strong Taylor schemes by replacing the derivatives there by finite differences expressed in terms of appropriate supporting values. For this reason we shall call them implicit strong Runge-Kutta schemes, but we emphasize that they are not simply heuristic adaptations to stochastic differential equations of the deterministic Runge-Kutta schemes.

By interpolating between the fully implicit scheme and the corresponding explicit scheme, we can form a *family of implicit order 1.0 strong Runge-Kutta schemes.*   In the general multi-dimensional case d, $m = 1, 2, \dots$ these have kth component

$$(3.12)\quad Y_{n+1}^k \;=\; Y_n^k + \left\{ \alpha a\left(\tau_{n+1}, Y_{n+1}^k\right) + (1-\alpha)\, a^k \right\} \Delta_n + \sum_{j=1}^{m} b^{k,j}\, \Delta W_n^j$$

$$+ \frac{1}{\sqrt{\Delta_n}} \sum_{j_1,j_2=1}^{m} \left\{ b^{k,j_2}\left(\tau_n, \bar{\Upsilon}_n^{j_1}\right) - b^{k,j_2} \right\} I_{(j_1,j_2)}$$

with vector supporting values

$$\bar{\Upsilon}_n^j = Y_n + a\,\Delta_n + b^j\,\sqrt{\Delta_n},$$

for $j = 1, \ldots, m$, and implicitness parameter $\alpha \in [0,1]$. The multiple Ito integrals $I_{(j_1,j_2)}$ here can be approximated as in (2.3.32). There is also a *Stratonovich version* of (3.12), which for *commutative noise* reduces to

$$(3.13) \qquad Y_{n+1}^k = Y_n^k + \left\{ \alpha \underline{a}^k \left(\tau_{n+1}, Y_{n+1}^k \right) + (1-\alpha) \underline{a}^k \right\} \Delta_n$$

$$+ \frac{1}{2} \sum_{j=1}^m \left\{ b^{k,j} \left(\tau_n, \bar{\Psi}_n \right) + b^{k,j} \right\} \Delta W_n^j$$

with vector supporting values

$$\bar{\Psi}_n = Y_n + \underline{a}\,\Delta_n + \sum_{j=1}^m b^j\, \Delta W_n^j$$

and implicitness parameter $\alpha \in [0,1]$. The term $\underline{a}\,\Delta_n$ can be omitted from $\bar{\Psi}_n$ here without affecting the order of convergence.

Problem 4.3.5 (PC-Exercise 12.3.1) *Repeat Problem 4.3.1 for the scheme (3.13) with $\alpha = 1.0$. Then repeat the calculation with the supporting values*

$$\bar{\Psi}_n = Y_n + b\,\Delta W_n.$$

Notes 4.3.5 The program is adapted from Program 4.3.1 and is listed on the diskette. After the results for the first supporting values appear on the screen press the <RETURN> key and the calculations will be repeated for the other supporting values.

As above, we can also derive implicit order 1.5 strong schemes by replacing derivatives by their corresponding finite differences. For the general multidimensional case $d, m = 1, 2, \ldots$ we have the *implicit order 1.5 strong Runge-Kutta scheme* in vector form

$$(3.14) \qquad Y_{n+1} = Y_n + \frac{1}{2} \left\{ a \left(\tau_{n+1}, Y_{n+1} \right) + a \right\} \Delta_n$$

$$+ \sum_{j=1}^m \left[b^j\, \Delta W_n^j + \frac{1}{\Delta_n} \left\{ b^j \left(\tau_{n+1}, Y_n \right) - b^j \right\} I_{(0,j)} \right.$$

$$\left. + \frac{1}{2\sqrt{\Delta_n}} \left\{ a \left(\tau_n, \bar{\Upsilon}_+^j \right) - a \left(\tau_n, \bar{\Upsilon}_-^j \right) \right\} \left\{ I_{(j,0)} - \frac{1}{2} \Delta W_n^j\, \Delta_n \right\} \right]$$

$$+ \frac{1}{2\sqrt{\Delta_n}} \sum_{j_1,j_2=1}^m \left\{ b^{j_2} \left(\tau_n, \bar{\Upsilon}_+^{j_1} \right) - b^{j_2} \left(\tau_n, \bar{\Upsilon}_-^{j_1} \right) \right\} I_{(j_1,j_2)}$$

$$+ \frac{1}{2\Delta_n} \sum_{j_1,j_2=1}^m \left\{ b^{j_2} \left(\tau_n, \bar{\Upsilon}_+^{j_1} \right) - 2b^{j_2} + b^{j_2} \left(\tau_n, \bar{\Upsilon}_-^{j_1} \right) \right\} I_{(0,j_2)}$$

$$+\frac{1}{2\Delta_n} \sum_{j_1,j_2,j_3=1}^{m} \left[b^{j_3}\left(\tau_n, \bar{\Phi}_+^{j_1,j_2}\right) - b^{j_3}\left(\tau_n, \bar{\Phi}_-^{j_1,j_2}\right) \right.$$

$$\left. - b^{j_3}\left(\tau_n, \bar{\Upsilon}_+^{j_1}\right) + b^{j_3}\left(\tau_n, \bar{\Upsilon}_-^{j_1}\right) \right] I_{(j_1,j_2,j_3)}$$

with vector supporting values

$$\bar{\Upsilon}_\pm^j = Y_n + \frac{1}{m} a\,\Delta_n \pm b^j \sqrt{\Delta_n}$$

and

$$\bar{\Phi}_\pm^{j_1,j_2} = \bar{\Upsilon}_+^{j_1} \pm b^{j_1}\left(\tau_n, \bar{\Upsilon}_+^{j_2}\right) \sqrt{\Delta_n}.$$

Exercise 4.3.1 (PC-Exercise 12.3.2)   *Repeat Problem 4.3.1 with the scheme 3.14.*

Finally, we state an implicit order 2.0 strong scheme for the case of scalar additive noise, that is with $m = 1$ and time dependent diffusion coefficient only. In the multi-dimensional case $d = 1, 2, \ldots$ and $m = 1$ with *scalar additive noise* the implicit order 2.0 strong Runge-Kutta scheme has the form

$$(3.15)\quad Y_{n+1} = Y_n + \left\{ \underline{a}\left(\bar{\Upsilon}_+\right) + \underline{a}\left(\bar{\Upsilon}_-\right) - \frac{1}{2}\left(\underline{a}\left(Y_{n+1}\right) + \underline{a}\right) \right\} \Delta_n + b\,\Delta W_n$$

$$+ \frac{1}{\Delta_n}\left\{b\left(\tau_{n+1}\right) - b\right\}\left\{\Delta W_n \Delta_n - \Delta Z_n\right\}$$

with

$$\bar{\Upsilon}_\pm = Y_n + \frac{1}{2}\underline{a}\,\Delta_n + \frac{1}{\Delta_n} b\left(\Delta\tilde{Z}_n \pm \zeta\right),$$

where

$$\Delta\tilde{Z}_n = \frac{1}{2}\Delta Z_n + \frac{1}{4}\Delta W_n\,\Delta_n$$

and

$$\zeta = \sqrt{J_{(1,1,0)}\,\Delta_n - \frac{1}{2}(\Delta Z_n)^2 + \frac{1}{8}\left((\Delta W_n)^2 + \frac{1}{2}\left(2\Delta Z_n\,\Delta_n^{-1} - \Delta W_n\right)^2\right)\Delta_n^2}.$$

Problem 4.3.6 (PC-Exercise 12.3.3)   *Repeat Problem 4.2.6 with the scheme (3.15) with $\underline{a}$ determined by interpreting the time t as the first component of a 2-dimensional Ito process $\{(t, X_t), t \geq 0\}$.*

Notes 4.3.6   The program, which can be found on the diskette, is based on Program 4.2.6. For the given equation the terms involving $\tilde{\zeta}$ cancel, so there is no need to generate the multiple integrals $J_{(1,1,0)}$ here.

The above problem shows that even when the applied scheme is derivative free one can gain much by checking whether all multiple stochastic integrals are really needed for the specific equation.

E. Implicit Two-Step Strong Schemes

The number of evaluations of derivatives of the drift and diffusion coefficients can be reduced by using multi-step schemes. Here we shall describe some implicit two-step schemes which may be effective for stiff stochastic differential equations. We note that we need to start with a one-step scheme to provide the first approximation step.

An *implicit two-step order 1.0 strong scheme* in the general multi-dimensional case d, $m = 1, 2, \ldots$ is given componentwise by

$$(3.16) \qquad Y_{n+1}^k = Y_{n-1}^k + \left[a^k \left(\tau_{n+1}, Y_{n+1} \right) + a^k \right] \Delta_n + V_n^k + V_{n-1}^k$$

with

$$V_n^k = \sum_{j=1}^m b^{k,j} \Delta W_n^j + \sum_{j_1,j_2=1}^m L^{j_1} b^{k,j_2} I_{(j_1,j_2),\tau_n,\tau_{n+1}}$$

for $k = 1, 2, \ldots, d$.

Problem 4.3.7 (PC-Exercise 12.4.1) *Repeat Problem 4.3.1 for the scheme (3.16) using the implicit Milstein scheme (3.7) for $\alpha = 1$ as starting routine.*

Notes 4.3.7 The program is based on Program 4.3.2 with α fixed at the value 1.0. It also provides the starting routine. See Notes 4.2.8 about carrying values forward to the next iteration. The program is listed on the diskette.

We have in the general multi-dimensional case d, $m = 1, 2, \ldots$ the following *family of implicit two-step order 1.5 strong schemes* with kth component:

$$(3.17) \qquad Y_{n+1}^k = (1 - \gamma) Y_n^k + \gamma Y_{n-1}^k$$

$$+ \frac{1}{2} \left\{ a^k \left(\tau_{n+1}, Y_{n+1} \right) + (1 + \gamma) a^k + \gamma a^k \left(\tau_{n-1}, Y_{n-1} \right) \right\} \Delta_n$$

$$- \frac{1}{2} (1 - \gamma) \sum_{j=1}^m L^j a^k \left(\tau_{n-1}, Y_{n-1} \right) \Delta W_{n-1}^j \Delta_{n-1}$$

$$+ V_n^k + \gamma V_{n-1}^k$$

with

$$V_n^k = \sum_{j=1}^m \left[b^{k,j} \Delta W_n^j + L^0 b^{k,j} \left\{ \Delta W_n^j \Delta_n - \Delta Z_n^j \right\} \right.$$

$$\left. + L^j a^k \left\{ \Delta Z_n^j - \frac{1}{2} \Delta W_n^j \Delta_n \right\} \right]$$

$$+ \sum_{j_1,j_2=1}^m L^{j_1} b^{k,j_2} I_{(j_1,j_2),\tau_n,\tau_{n+1}}$$

$$+ \sum_{j_1,j_2,j_3=1}^{m} L^{j_1} L^{j_2} b^{k,j_3} I_{(j_1,j_2,j_3),\tau_n,\tau_{n+1}}$$

and coupling parameter $\gamma \in [0,1]$. If we set $\gamma = 0$ in (3.17) we obtain an implicit one-step scheme.

Problem 4.3.8 (PC-Exercise 12.4.2)   *Repeat Problem 4.3.1 for the scheme (3.17) with $\gamma = 1.0$ using the implicit order 1.5 strong Taylor scheme (3.9) as its starting routine.*

Notes 4.3.8   The program is based on Program 4.3.3 with the implicitness parameters frozen at the stated values. There is no need to generate the ΔZ_n^j as the terms involving them have the same coefficients here and the ΔZ_n^j cancel out.

In the multi-dimensional case with *scalar additive noise* $d = 1, 2, \ldots$ and $m = 1$ there is a *family of implicit two-step order* 2.0 *strong schemes* for which the kth component is

(3.18) $$Y_{n+1}^k = (1 - \gamma) Y_n^k + \gamma Y_{n-1}^k$$

$$+ \frac{1}{2} \left\{ \underline{a}^k \left(\tau_{n+1}, Y_{n+1} \right) + (1 + \gamma) \underline{a}^k + \gamma \underline{a}^k \left(\tau_{n-1}, Y_{n-1} \right) \right\} \Delta_n$$

$$- \frac{1}{2} (1 - \gamma) \underline{L}^1 \underline{a}^k \left(\tau_{n-1}, Y_{n-1} \right) \Delta W_{n-1} \Delta_{n-1}$$

$$- \frac{1}{4} (1 - \gamma) \underline{L}^1 \underline{L}^1 \underline{a}^k \left(\Delta W_n \right)^2 \Delta_n$$

$$+ V_n^k + \gamma V_{n-1}^k$$

with

$$V_n^k = b^k \Delta W_n + \frac{\partial b^k}{\partial t} \left\{ \Delta W_n \Delta_n - \Delta Z_n \right\}$$

$$+ \underline{L}^1 \underline{a}^k \left\{ \Delta Z_n - \frac{1}{2} \Delta W_n \Delta_n \right\}$$

$$+ \underline{L}^1 \underline{L}^1 \underline{a}^k \left\{ J_{(1,1,0),\tau_n,\tau_{n+1}} - \frac{1}{4} \left(\Delta W_n \right)^2 \Delta_n \right\},$$

where $\gamma \in [0,1]$.

Problem 4.3.9 (PC-Exercise 12.4.3)   *Repeat Problem 4.2.6 for the scheme (3.18) with $\gamma = 1.0$ using the implicit order 2.0 strong Taylor scheme (3.11) with $\alpha = 0.5$ as its starting routine.*

Notes 4.3.9   For the given equation the ΔZ_n terms have the same coefficients and also cancel here. The coefficient of the final term in V_n^k is zero, so there is also no need to generate the $J_{(1,1,0)}$ multiple stochastic integral. The program is adapted from Program 4.3.6. The stated implicitness and coupling parameters α and γ cannot be changed in the program.

4.4   Simulation Studies

This final section on strong approximations compares the efficiency and accuracy of the different explicit and implicit schemes which were described above. While most of the numerical experiments will involve linear stochastic differential equations, in concluding we shall also consider a nonlinear equation. The proposed problems will require a considerable amount of computational time.

A. Comparison of Global Errors

Let us consider again the linear stochastic differential equation

$$(4.1) \qquad dX_t = a\,X_t\,dt + b\,X_t\,dW_t$$

as in many of the problems in the preceding sections, see (3.2.14). In the following problem we simply combine the previous programs to compare the global errors for different schemes and step sizes in a single graphic.

Problem 4.4.1   *Consider the Ito process X satisfying (4.1) with $X_0 = 1.0$, $a = 1.5$ and $b = 0.1$ on the time interval $[0,T]$ with $T = 1.0$. Combine Programs 3.3.3, 4.1.2, 4.1.5 and 4.1.7 for the Euler, Milstein, strong Taylor 1.5 and strong Taylor 2.0 schemes, respectively, into one program and plot the logarithm of the absolute error against the logarithm of the step size Δ_n. Use $M = 20$ batches each of $N = 100$ simulations and step sizes $\delta = \Delta_n = 2^0, 2^{-1}, \ldots, 2^{-8}$. Do not simplify the schemes to take advantage of the specific linear structure of the given SDE.*

Notes 4.4.1

(i)   Program 4.4.1, given on the diskette, is based on Program 4.1.7 but includes the computational kernels of Programs 3.3.3, 4.1.2, 4.1.5 and 4.1.7 to simulate the Euler, Milstein, strong order 1.5 and 2.0 Taylor schemes, respectively. The computations for each of these schemes are done separately one after the other. In each case, the program indicates the currently used scheme and displays its total computational time. Finally the error graphs are plotted in a single figure, Figure 4.4.1 below. Press the <RETURN> key to view this figure.

(ii)   We have extended the routine GRAPH311 to our purpose of plotting several graphs in the same logarithmic figure and have created a new routine GRAPH441 which is stored in the new unit EXTGRAPH.

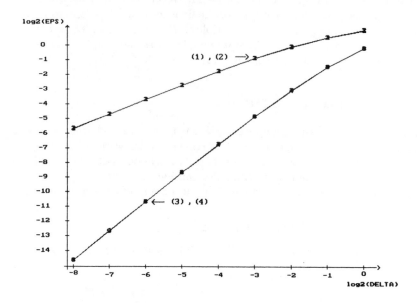

Figure 4.4.1   Logarithmic absolute errors for Euler (1), Milstein (2), Taylor order
1.5 (3) and Taylor order 2.0 (4) schemes with $b = 0.1$.

In Figure 4.4.1 we have plotted the logarithms of the absolute errors of the
schemes against the logarithms of the step sizes. Note that the graphs of the
Euler (1) and Milstein (2) schemes are quite close to each other, which is due
to the fact that we have a small noise parameter $b = 0.1$ in our example and in
this case the Euler scheme performs similarly to the next higher order scheme,
the Milstein scheme; if there were no noise, $b = 0$, then the Euler and Milstein
schemes would obviously coincide. We have a similar situation for the order
1.5 and 2.0 strong Taylor schemes in graphs (3) and (4) which are also both
quite close to each other; in the deterministic case, $b = 0$, the order 1.5 strong
Taylor scheme coincides with the order 2.0 strong Taylor scheme.

In conclusion, it seems that it may be worth choosing an order 0.5 or 1.5
strong scheme in the case of small noise, but this should be decided only after having studied the necessary computer time to achieve a desired accuracy. Before doing this we should examine the absolute error for larger noise,
which should be different for the order 0.5 and 1.0 or 1.5 and 2.0 strong Taylor
schemes.

Exercise 4.4.1   *Repeat Problem 4.4.1 with noise parameter $b = 1.0$ using
$p = 10$ in the approximations for the required multiple stochastic integrals.*

B. Efficiency of Strong Taylor Schemes

In practice what really counts is the computer time needed to obtain a desired level of accuracy. We should thus also measure the time needed by a specific scheme to generate a path solving the underlying SDE. For a comparison it is convenient to plot the logarithm of the necessary computer time versus the negative logarithm of the absolute error achieved for different step sizes. In the following we shall do this for the different strong Taylor schemes. But the necessary computer time depends very much on the implementation of the schemes, that is on how one takes advantage of the specific structure of the underlying stochastic differential equation. Our linear example (4.1) offers structural advantages which we shall also use in a second implementation of the order 2.0 strong Taylor scheme in Stratonovich form

$$Y_{i+1} = Y_i \left(1 + \underline{a}\Delta_n + b\Delta W_n + \frac{1}{2!} b^2 (\Delta W_n)^2 + \underline{a}b\Delta W_n \Delta_n + \frac{1}{2!} \underline{a}^2 \Delta_n^2 \right.$$

$$(4.2) \qquad\qquad \left. + \frac{1}{2!} \underline{a}b^2 (\Delta W_n)^2 \Delta_n + \frac{1}{3!} b^3 (\Delta W_n)^3 + \frac{1}{4!} b^4 (\Delta W_n)^4 \right),$$

which we shall call the *bilinear order 2.0 Taylor scheme*. In particular, it avoids the need to generate higher order multiple stochastic integrals.

Problem 4.4.2   *Consider the Ito process X as in Problem 4.4.1. Include the bilinear order 2.0 Taylor scheme (4.2) in addition to the Euler, Milstein, order 1.5 and 2.0 Taylor schemes. Estimate the absolute error ϵ and plot $\log_2$ of the computer time necessary to simulate one path versus $-\log_2 \epsilon$.*

Notes 4.4.2

(i)   We have adapted Program 4.4.1 to our special task by adding a routine for the bilinear order 2.0 Taylor scheme which takes considerable advantage of the linear structure of (4.1).

(ii) Here we also use the unit EXTGRAPH to plot the desired five efficiency curves in double logarithmic scaling applying the procedure GRAPH441. The necessary computer time for the different schemes together with the given parameters are shown first.

Figure 4.4.2 illustrates the relative efficiency of the Taylor schemes considered for low noise. We observe that the curves for the Euler scheme (1) and the Milstein scheme (2) have nearly the same slope of magnitude 1, whereas all three remaining schemes have a slope of magnitude 0.5 which is consistent with second order strong convergence. In this low noise case, the Euler scheme (1) appears to have strong order 1.0 and the order 1.5 strong Taylor scheme (3) gives the impression of being an order 2.0 strong scheme. Thus, for very small noise intensities, the Euler scheme or order 1.5 strong Taylor scheme may be considered preferrable as they show nearly the same performance of the schemes of next higher integer order. Finally, we note that the bilinear order 2.0 Taylor scheme takes considerable advantage of the specific structure

of the underlying SDE. Such a scheme would be preferrable if it applies to the
task at hand.

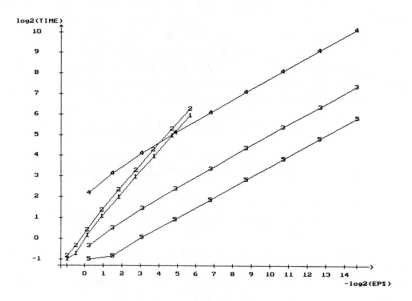

Figure 4.4.2 $\log_2$ (computer time) versus $-\log_2 \epsilon$ for the Euler (1), Milstein (2),
order 1.5 Taylor (3), order 2.0 Taylor (4) and bilinear order 2.0 Taylor (5) schemes
with $b = 0.1$.

Problem 4.4.3   *Repeat Problem 4.4.2 with larger noise parameter $b = 1.0$.*

We see clearly from Figure 4.4.3 the advantage of using a higher order strong
scheme to achieve high accuracy. The excellent performance of the bilinear order 2.0 Taylor scheme shows the enormous benefits to be won from exploiting
the specific structure of the underlying stochastic differential equation such as
the bilinearity of (4.1) here.

Exercise 4.4.2   *Repeat Problem 4.4.2 for an SDE with additive noise $b(x) \equiv$
1.5 and the same linear drift coefficient $a(x) = 1.5\,x$. A similar but more
complicated case will be treated in Subsection E.*

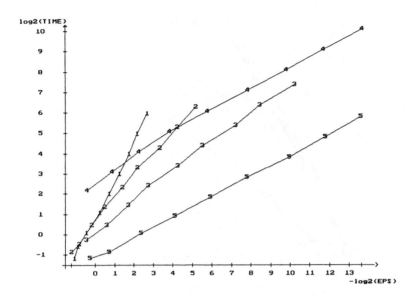

Figure 4.4.3 $\log_2$ (computer time) versus $-\log_2 \epsilon$ for Euler (1), Milstein (2),
order 1.5 Taylor (3), order 2.0 Taylor (4) and bilinear order 2.0 Taylor (5) schemes
with $b = 1.0$.

C. Efficiency of Explicit Schemes

Here we shall investigate the numerical performance of several one-step explicit
strong schemes that we first considered in Section 4.2. We shall take advantage of the specific structure of the given stochastic differential equation. For
comparison we consider the Euler scheme (1.10) (label (1)), explicit order 1.0
(2.3), (2.4) (label (2)), explicit order 1.0 (2.8) (label (3)), explicit order 1.0
(2.3), (2.9) (label (4)), explicit order 1.5 (2.10), (2.11), (2.12) (label (5)), bilinear Euler scheme (label (6)) and bilinear Taylor 2.0 (4.2) (label (7)) in the
following numerical experiment.

Problem 4.4.4   *Repeat Problem 4.4.2 with the above mentioned schemes.*

Notes 4.4.4   Apart from the change in computational kernels for the new
schemes, the program is essentially the same as Programs 4.4.1 and 4.4.2. See
Notes 4.4.1 and 4.4.2. Press the <RETURN> key after each figure has been
viewed to proceed to the next one. When finished press <ESC>.

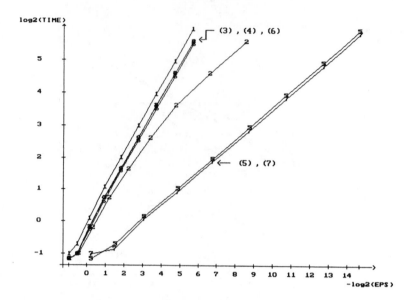

Figure 4.4.4 $\log_2$ (computer time) versus $-\log_2 \epsilon$ for Problem 4.4.4.

We observe from Figure 4.4.4 that the higher order explicit schemes perform in our example much better than the lower order ones. There is no remarkable difference between schemes of the same order.

D. Efficiency of Implicit Schemes

Let us now compare the efficiency of the different implicit strong schemes that we introduced in Section 4.3. As a test example we take once again the two-dimensional linear stochastic differential equation (3.3)

$$(4.3) \qquad\qquad dX_t = A\,X_t dt + B\,X_t dW_t$$

with coefficient matrices

$$(4.4) \qquad A = a \begin{bmatrix} -1 & 1 \\ 1 & -1 \end{bmatrix} \quad \text{and} \quad B = bI = b \begin{bmatrix} 1 & 0 \\ 0 & 1 \end{bmatrix},$$

where a and b are real numbers. Its solution is given explicitly in (3.4).

The linear structure of this equation allows us to reorganize the implicit schemes so we do not have to solve an algebraic equation numerically at each step.

In the following numerical experiment we shall compare the efficiency of the implicit Euler scheme (3.1) with $\alpha = 0.5$, the implicit Milstein scheme (3.8) with $\alpha = 0.5$, the implicit order 1.0 strong Runge-Kutta scheme (3.12) with $\alpha = 0.5$, the implicit order 1.5 Taylor scheme (3.10) and the implicit order 2.0 Taylor scheme (3.11).

Problem 4.4.5 *Consider the two-dimensional stochastic differential equation (4.3) with $X_0 = (1,0)$, $T = 1$, $a = 5$ and $b = 0.01$. Compute $M = 20$ batches each of $N = 100$ simulations of X_T and of the values $Y(T)$ of the implicit strong schemes (3.1), (3.8), (3.12), (3.10) and (3.11) for $\alpha = 0.5$ and equidistant step sizes $\delta = \Delta_n = 2^0, 2^{-1}, \ldots, 2^{-3}$. Evaluate the absolute error ϵ and plot $\log_2$ of the necessary computer time versus $-\log_2 \epsilon$.*

Notes 4.4.5 Program 4.4.5 is also a modification of Program 4.4.1. Many of the schemes here simplify considerably for equation (3.3) and it is not necessary to generate many of the multiple stochastic integrals.

The result of the above numerical experiment should show that for all step sizes the implicit schemes perform satisfactorily.

Exercise 4.4.3 *Repeat Problem 4.4.5 with the degree of implicitness $\alpha = 0.0$.*

E. Efficiency of Strong Schemes for a Nonlinear SDE

To gain additional insights into the numerical behaviour of different strong simulation algorithms we shall now test them on a nonlinear example. For this purpose it is important that the exact value $X_T(\omega)$ of the realizations of the solution at time T can be evaluated, so we can determine the absolute error of the approximations.

We consider the following nonlinear Ito stochastic differential equation

$$(4.5) \qquad dX_t = -\left(\sin 2X_t + \frac{1}{4}\sin 4X_t\right)dt + \sqrt{2}(\cos X_t)^2 dW_t,$$

$t \in [0, T]$, with initial value $X_0 = 1$. By the Ito formula (2.1.19) we can show that its solution is

$$(4.6) \qquad\qquad X_t = \arctan(V_t)$$

where

$$(4.7) \qquad\qquad V_t = V_0\, e^{-t} + \sqrt{2}\int_0^t e^{s-t}\, dW_s$$

and

$$(4.8) \qquad\qquad V_0 = \tan(1).$$

For the equidistant time discretization $\tau_i = i\Delta$, $\Delta = T/N_0$, $i = 0, 1, \ldots, N_0$ we are able to generate the exact value of the realizations of the solution at time T by applying the following facts:

$$(4.9) \qquad V_T = e^{-T}\left(V_0 + \sqrt{2}\sum_{i=1}^{N_0} e^{\tau_{i-1}}(\Delta W_{i-1} + \Delta U_{i-1} + \Delta Q_{i-1})\right)$$

with

(4.10) $$\Delta U_{i-1} = \int_{\tau_{i-1}}^{\tau_i} (s - \tau_{i-1})\, dW_s$$

and

(4.11) $$\Delta Q_{i-1} = \int_{\tau_{i-1}}^{\tau_i} \left(\sum_{k=2}^{\infty} \frac{(s - \tau_{i-1})^k}{k!} \right) dW_s,$$

where, by the Ito formula (2.1.19),

(4.12) $$\Delta U_{i-1} = \Delta W_{i-1}\,\Delta - \Delta Z_{i-1}$$

with

(4.13) $$\Delta Z_{i-1} = I_{(1,0)} = \int_{\tau_{i-1}}^{\tau_i} \int_{\tau_{i-1}}^{s} dW_r\, ds,$$

so we obtain the representation

(4.14) $$V_T = e^{-T} \left(V_0 + \sqrt{2} \sum_{i=1}^{N_0} e^{\tau_{i-1}} (\Delta W_{i-1}(1 + \Delta) + \Delta Z_{i-1} + \Delta Q_{i-1}) \right),$$

where the random vector $(\Delta W_i, \Delta Z_i, \Delta Q_i)$ is normally distributed with zero mean and correlations:

$$E\left((\Delta W_i)^2\right) = \Delta, \qquad E\left((\Delta Z_i)^2\right) = \Delta^3/3,$$

$$E\left((\Delta Q_i)^2\right) = (e^{2\Delta} - 1)/2 - 2\Delta e^{\Delta} + \Delta + \Delta^2 + \Delta^3/3,$$

$$E\left(\Delta Q_i\,\Delta W_i\right) = e^{\Delta} - (1 + \Delta + \Delta^2/2), \qquad E\left(\Delta W_i\,\Delta Z_i\right) = \Delta^2/2,$$

$$E\left(\Delta Q_i\,\Delta Z_i\right) = e^{\Delta} - (1 + \Delta + \Delta^2/2 + \Delta^3/6),$$

for $i = 1, 2, \ldots, N_0$. By generating ΔW_i, ΔZ_i, ΔQ_i at each time step we can evaluate the increment in V, and hence in X, as well as the increments of time discrete approximations up to strong order $\gamma = 1.5$.

In this numerical experiment we shall only apply explicit schemes as there is no stiffness in the problem and this will simplify the implementation of the algorithms.

Problem 4.4.6 *Consider the stochastic differential equation (4.5) with $X_0 = 1$ and $T = 1.0$. Compute $M = 20$ batches each with $N = 100$ simulations of X_T together with the values $Y(T)$ of the Euler scheme (1.10), Milstein scheme (1.18), explicit order 1.0 strong scheme (2.8), order 1.5 Taylor scheme (1.28) and the explicit order 1.5 strong scheme (2.10) for equidistant step sizes $\delta = \Delta_n = \Delta = 2^0, 2^{-1}, \ldots, 2^{-8}$. Evaluate the absolute error ϵ and plot $\log_2$ of the necessary computer time versus $-\log_2 \epsilon$.*

Notes 4.4.6 The program is adapted from earlier ones in this section and applied to the nonlinear test equation (4.5).

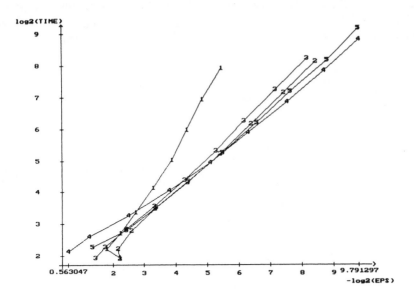

Figure 4.4.5 Results of Problem 4.4.6 for the Euler (1), Milstein (2), explicit order 1.0 (3), order 1.5 Taylor (4) and explicit order 1.5 (5) schemes.

From Figure 4.4.5 we see that the higher order explicit schemes appear more efficient. We note again that Taylor schemes provide a slight advantage over the corresponding explicit schemes.

Exercise 4.4.4 *Repeat Problem 4.4.6 for the stochastic differential equation with explicit solution*

$$X_t = \left(\frac{2W_t}{t+1} + 10 \right)^2.$$

Literature for Chapter 4

First results on strong convergence of the Euler approximation can be found in Maruyama (1955) and Gikhman & Skorokhod (1979). Higher order discrete time strong approximations have been proposed, for example, by Milstein (1974), Mc Shane (1974), Rao, Borwankar & Ramakrishna (1974), Kloeden & Pearson (1977), Wagner & Platen, (1978), Clark (1978), Platen (1980, 1981), Clark & Cameron (1980), Talay (1982), Rümelin (1982), Chang (1987), Milstein (1988a,b), Nakazawa (1990), Kloeden & Platen (1992a,b). Implicit strong approximations were considered in Talay (1982), Klauder & Petersen (1985), Milstein (1988a), Smith & Gardiner (1988), Drummond & Mortimer (1991), Hernandez & Spigler (1991), Petersen (1990), Kloeden & Platen (1992a,b) or Saito & Mitsui (1992). Simulation studies can be found, for example, in

Pardoux & Talay (1985) (see (4.5)), Liske & Platen (1987), Newton (1991), Klauder & Petersen (1985), Kloeden, Platen & Schurz (1993). More references on strong approximations are given in Kloeden & Platen (1992a).

Chapter 5

Weak Approximations

The aim of this chapter is to introduce weak approximations such as weak Taylor schemes, explicit and implicit weak schemes and their extrapolations to approximate functionals of the solutions of stochastic differential equations. Such functionals include, for example, moments of the solution at a given time instant as well as expectations of integrals of functions of such solutions over a given time interval. Simulation studies will provide an indication of the numerical efficiency of the schemes. Finally, some variance reduction methods will be discussed.

5.1 Weak Taylor Schemes

In this section we shall use truncated stochastic Taylor expansions as we did in Section 4.1, but now to derive discrete time approximations corresponding to the weak convergence criterion. We shall call the approximations so obtained *weak Taylor approximations* and shall investigate the corresponding *weak schemes*. As with strong approximations, the desired order of weak convergence also determines which truncation must be used. This will, however, be different from the truncation required for strong convergence of the same order, in general involving fewer terms.

A. The Weak Euler Scheme

We recall from relation (4.1.13) that for the general multi-dimensional case d, $m = 1, 2, \ldots$ the kth component of the *Euler scheme* has the form

$$(1.1) \qquad Y_{n+1}^k = Y_n^k + a^k \, \Delta_n + \sum_{j=1}^{m} b^{k,j} \, \Delta W_n^j,$$

with initial value $Y_0 = X_0$, where

$$\Delta_n = \tau_{n+1} - \tau_n \quad \text{and} \quad \Delta W_n^j = W_{\tau_{n+1}}^j - W_{\tau_n}^j.$$

If, amongst other assumptions, a and b are sufficiently smooth, then the Euler approximation has order of weak convergence $\beta = 1.0$. In fact, it is the order 1.0 weak Taylor scheme.

In the numerical experiment that follows we shall use the Ito process

$$(1.2) \qquad X_t = 0.1 + \int_0^t \frac{3}{2} X_s \, ds + \int_0^t \frac{1}{10} X_s \, dW_s^1 + \int_0^t \frac{1}{10} X_s \, dW_s^2$$

which is driven by the two independent standard Wiener processes $W^1 = \{W_t^1, t \geq 0\}$ and $W^2 = \{W_t^2, t \geq 0\}$. It has expectation

$$(1.3) \qquad\qquad E(X_T) = 0.1 \exp\left(\frac{3}{2}T\right).$$

Problem 5.1.1 (PC-Exercise 14.1.2)   *For the Ito process X in (1.2) simulate $M = 20$ batches each of $N = 100$ trajectories of the Euler approximation Y using (1.1) with equidistant time steps of step size $\delta = \Delta_n = 2^{-2}$. Determine the 90%-confidence interval for the mean error*

$$(1.4) \qquad\qquad \mu = E(Y(T)) - E(X_T)$$

at time $T = 1$. Repeat the calculations for step sizes $\delta = 2^{-3}$, 2^{-4} and 2^{-5}, and plot $\log_2 |\mu|$ versus $\log_2 \delta$.

Notes 5.1.1
(i)   The program is similar to Program 4.1.1 with the appropriate change in the error criterion.
(ii) Other stochastic differential equations can be easily considered by changing the declared drift and diffusion coefficients, as well as the exact expected value.

With the weak convergence criterion (3.4.16) we have much freedom to choose other, simpler random variables instead of using the Gaussian increments ΔW_n^j in (1.1). Roughly speaking, such random variables have only to coincide in their lower order moments with those of ΔW_n^1 and ΔW_n^2 to provide a sufficient accurate approximation of the probability law of the Ito diffusion. For instance, in (1.1) we could use two-point distributed random variables $\Delta \hat{W}_n^j$ with

$$(1.5) \qquad\qquad P\left(\Delta \hat{W}_n^j = \pm\sqrt{\Delta_n}\right) = \frac{1}{2},$$

in which case (1.1) becomes the *simplified Euler scheme*

$$(1.6) \qquad\qquad Y_{n+1}^k = Y_n^k + a^k \Delta_n + \sum_{j=1}^m b^{k,j} \Delta \hat{W}_n^j.$$

Exercise 5.1.1 (PC-Exercise 14.1.4)   *Repeat Problem 5.1.1 using the simplified Euler scheme (1.6) with the noise increments generated by two-point distributed random variables.*

B. Order 2.0 Weak Taylor Scheme

More accurate weak Taylor schemes can be derived by including further multiple stochastic integrals from the stochastic Taylor expansion. Since the objective is to obtain more information about the probability measure of the underlying Ito process rather than about its sample paths, we also have the freedom to replace the multiple stochastic integrals by much simpler random variables as in the simplified Euler scheme (1.6).

We shall consider the weak Taylor scheme obtained by adding all of the double stochastic integrals from the Ito-Taylor expansion to the Euler scheme. In the autonomous 1-dimensional case $d = m = 1$ we obtain the *order* 2.0 *weak Taylor scheme*

$$(1.7) \qquad Y_{n+1} = Y_n + a\,\Delta_n + b\,\Delta W_n + \frac{1}{2}\,bb'\left\{(\Delta W_n)^2 - \Delta_n\right\}$$

$$+ a'b\,\Delta Z_n + \frac{1}{2}\left(aa' + \frac{1}{2}a''b^2\right)\Delta_n^2$$

$$+ \left(ab' + \frac{1}{2}b''b^2\right)\left\{\Delta W_n\,\Delta_n - \Delta Z_n\right\},$$

where ΔZ_n represents the double Ito integral

$$(1.8) \qquad\qquad \Delta Z_n = \int_{\tau_n}^{\tau_{n+1}}\int_{\tau_n}^{s_2} dW_{s_1}\,ds_2.$$

The pair of correlated Gaussian random variables $(\Delta W_n,\ \Delta Z_n)$ can be generated from a pair of independent standard Gaussian random variables as in Problem 1.3.5. The scheme (1.7) was first mentioned by Milstein and has weak order $\beta = 2.0$ under sufficient smoothness assumptions on the drift and diffusion coefficients a and b.

Problem 5.1.2 (PC-Exercise 14.2.2) *Consider the Ito process X satisfying the linear stochastic differential equation*

$$dX_t = a\,X_t\,dt + b\,X_t\,dW_t,$$

see (3.4.2), on the time interval $[0, T]$, where $T = 1$, with initial value $X_0 = 0.1$, $a = 1.5$ and $b = 0.01$. Generate $M = 20$ batches each of $N = 100$ trajectories of the order 2.0 weak Taylor scheme (1.7) with equidistant time steps $\delta = \Delta_n = 2^{-2}$. Determine the 90%-confidence interval for the mean error

$$\mu = E\left(Y(T)\right) - E\left(X_T\right).$$

Repeat the calculations for step sizes $\delta = \Delta_n = 2^{-3}$, 2^{-4} and 2^{-5}, and plot $\log_2 |\mu|$ against $\log_2 \delta$.

Notes 5.1.2

(i) The program is based on Program 3.4.2. Since the coefficients of the ΔZ_n random variables cancel for a linear stochastic differential equation, there is no need to generate these random variables here.

(ii) Note that many more trajectories may be required to give good results when the step size is very small.

(iii) We have considered the case of a small noise parameter b which is similar to the deterministic case. For stronger noise it may be difficult on a PC to obtain sufficiently small confidence intervals in an acceptable time.

As we have already mentioned, we have much more freedom in generating the noise increments for the weak convergence criterion than for the strong one. For example, in the scheme (1.7) it is possible to use only a three-point distributed random variable $\Delta \hat{W}_n$ with

$$(1.9) \qquad P\left(\Delta \hat{W}_n = \pm\sqrt{3\Delta_n}\right) = \frac{1}{6}, \qquad P\left(\Delta \hat{W}_n = 0\right) = \frac{2}{3},$$

replacing ΔW_n by $\Delta \hat{W}_n$ and ΔZ_n by $\frac{1}{2}\Delta \hat{W}_n \Delta_n$. The resulting scheme converges still with weak order $\beta = 2.0$, but is computationally simpler. Note that the two-point distributed random variable (1.5) is not sufficient for order 2.0 weak convergence.

In the general multi-dimensional case $d, m = 1, 2, \ldots$ the *simplified order 2.0 weak Taylor scheme* has kth component

$$(1.10) \qquad Y_{n+1}^k = Y_n^k + a^k \Delta_n + \frac{1}{2} L^0 a^k \Delta_n^2$$

$$+ \sum_{j=1}^{m} \left\{ b^{k,j} + \frac{1}{2}\Delta_n \left(L^0 b^{k,j} + L^j a^k\right) \right\} \Delta \hat{W}_n^j$$

$$+ \frac{1}{2} \sum_{j_1, j_2 = 1}^{m} L^{j_1} b^{k,j_2} \left(\Delta \hat{W}_n^{j_1} \Delta \hat{W}_n^{j_2} + V_{j_1, j_2}\right).$$

where the $\Delta \hat{W}_n^j$ for $j = 1, 2, \ldots, m$ and $n = 0, 1, \ldots$ are independent random variables satisfying (1.9) and the V_{j_1, j_2} are independent two-point distributed random variables with

$$(1.11) \qquad\qquad P\left(V_{j_1, j_2} = \pm\Delta_n\right) = \frac{1}{2}$$

for $j_2 = 1, \ldots, j_1 - 1$,

$$(1.12) \qquad\qquad V_{j_1, j_1} = -\Delta_n$$

and

$$(1.13) \qquad\qquad V_{j_1, j_2} = -V_{j_2, j_1}$$

for $j_2 = j_1 + 1, \ldots, m$ and $j_1 = 1, \ldots, m$. The operators L^j here were defined in (4.1.3) and (4.1.5). This scheme was proposed by Talay.

Problem 5.1.3 (PC-Exercise 14.2.3) *Repeat Problem 5.1.2 for the simplified order 2.0 weak Taylor scheme (1.10) with noise increments generated by the three-point distributed random variable (1.9).*

Notes 5.1.3

(i) The program is based on Program 5.1.2 with the main change being in the generation of the required random variables. The linearity of the test equation considerably simplifies the coefficients appearing in the scheme.

(ii) It is recommended that the reader experiments with even smaller step sizes and with different samples.

Exercise 5.1.2 *Repeat Problem 5.1.3 with step sizes $\delta = \Delta_n = 2^{-5}, 2^{-6}, 2^{-7}$.*

C. Order 3.0 Weak Taylor Scheme

To obtain the third order weak Taylor scheme we need to include all of the multiple Ito integrals of multiplicity three from the Ito-Taylor expansion. This scheme, which was proposed by Platen, involves multiple Ito integrals of higher multiplicity which are difficult to generate. But again these multiple integrals can be replaced by simpler random variables. If the underlying stochastic differential equation has some specific structure, then we often obtain a quite efficient third order weak scheme.

As an example we consider in the scalar case, $d = 1$, with *scalar noise*, $m = 1$, the *simplified order* 3.0 *weak Taylor scheme*

$$
(1.14) \quad Y_{n+1} = Y_n + a\,\Delta_n + b\Delta\tilde{W}_n + \frac{1}{2}L^1 b \left\{ \left(\Delta\tilde{W}_n\right)^2 - \Delta_n \right\}
$$

$$
+ L^1 a\,\Delta\tilde{Z}_n + \frac{1}{2}L^0 a\,\Delta_n^2 + L^0 b \left\{ \Delta\tilde{W}_n\,\Delta_n - \Delta\tilde{Z}_n \right\}
$$

$$
+ \frac{1}{6}\left(L^0 L^0 b + L^0 L^1 a + L^1 L^0 a \right)\Delta\tilde{W}_n\,\Delta_n^2
$$

$$
+ \frac{1}{6}\left(L^1 L^1 a + L^1 L^0 b + L^0 L^1 b \right)\left\{ \left(\Delta\tilde{W}_n\right)^2 - \Delta_n \right\}\Delta_n
$$

$$
+ \frac{1}{6}L^0 L^0 a\,\Delta_n^3 + \frac{1}{6}L^1 L^1 b \left\{ \left(\Delta\tilde{W}_n\right)^2 - 3\Delta_n \right\}\Delta\tilde{W}_n.
$$

Here the multiple integrals have been represented exactly or approximately in terms of the correlated Gaussian random variables

$$
(1.15) \qquad \Delta\tilde{W}_n \sim N(0; \Delta_n), \qquad \Delta\tilde{Z}_n \sim N\left(0; \frac{1}{3}\Delta_n^3\right)
$$

with covariance

$$
(1.16) \qquad E\left(\Delta\tilde{W}_n\,\Delta\tilde{Z}_n\right) = \frac{1}{2}\Delta_n^2.
$$

In particular, the mixed threefold multiple integrals have been approximated here.

Problem 5.1.4 (PC-Exercise 14.3.1) *Repeat Problem 5.1.2 with $a = -5.0$ and $b = 0.1$ for the simplified order 3.0 weak Taylor scheme (1.14).*

Notes 5.1.4
(i) The program is based on Program 5.1.2, but is specialized to the equation under consideration for which the scheme has a particularly simple structure.

(ii) The results fluctuate for different computer runs, but stabilize when larger sample sizes are used which may create problems on smaller PCs.

Relatively simple third order weak schemes similar to (1.14) are also available for multi-dimensional stochastic differential equations with multiplicative and additive noise.

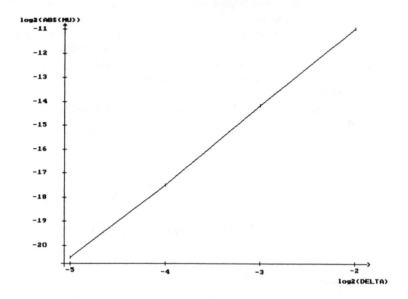

Figure 5.1.1 Simplified third order weak Taylor scheme.

D. Order 4.0 Weak Taylor Scheme

To construct the order 4.0 weak Taylor scheme we also need to include all of the multiple Ito integrals of multiplicity four from the Ito-Taylor expansion.

For the 1-dimensional case, $d = m = 1$, with *additive noise* many of the complicated integrals have zero coefficients and we obtain the *simplified order 4.0 weak Taylor scheme*

$$
(1.17) \qquad Y_{n+1} = Y_n + a\,\Delta_n + b\,\Delta\tilde{W}_n + \frac{1}{2}\,L^0 a\,\Delta_n^2 + L^1 a\,\Delta\tilde{Z}_n
$$

$$
+ L^0 b\left\{\Delta\tilde{W}_n\,\Delta_n - \Delta\tilde{Z}_n\right\}
$$

$$
+ \frac{1}{3!}\left\{L^0 L^0 b + L^0 L^1 a\right\}\Delta\tilde{W}_n\,\Delta_n^2 + L^1 L^0 a\left\{\Delta\tilde{Z}_n\,\Delta_n - \frac{1}{6}\,\Delta\tilde{W}_n\,\Delta_n^3\right\}
$$

$$
+ L^1 L^1 a\left\{2\,\Delta\tilde{W}_n\,\Delta\tilde{Z}_n - \frac{5}{6}\left(\Delta\tilde{W}_n\right)^2\Delta_n - \frac{1}{6}\,\Delta_n^2\right\}
$$

$$+ \frac{1}{3!} L^0 L^0 a \, \Delta_n^3 + \frac{1}{4!} L^0 L^0 L^0 a \, \Delta_n^4$$

$$+ \frac{1}{4!} \left\{ L^1 L^0 L^0 a + L^0 L^1 L^0 a + L^0 L^0 L^1 a + L^0 L^0 L^0 b \right\} \Delta \tilde{W}_n \, \Delta_n^3$$

$$+ \frac{1}{4!} \left\{ L^1 L^1 L^0 a + L^0 L^1 L^1 a + L^1 L^0 L^1 a \right\} \left\{ \left(\Delta \tilde{W}_n \right)^2 - \Delta_n \right\} \Delta_n^2$$

$$+ \frac{1}{4!} L^1 L^1 L^1 a \, \Delta \tilde{W}_n \left\{ \left(\Delta \tilde{W}_n \right)^2 - 3 \Delta_n \right\} \Delta_n$$

where $\Delta \tilde{W}_n$ and $\Delta \tilde{Z}_n$ are correlated Gaussian random variables as in (1.15)-(1.16), which are sufficient to approximate the surviving mixed multiple integrals.

Problem 5.1.5 (PC-Exercise 14.4.1) *Consider the Ornstein-Uhlenbeck process* X *satisfying the linear stochastic differential equation with additive noise*

$$dX_t = a \, X_t \, dt + b \, dW_t$$

on the time interval $[0, T]$, *where* $T = 1.0$, *with initial value* $X_0 = 0.1$, $a = 2.0$ *and* $b = 0.01$. *Simulate* $M = 20$ *batches each with* $N = 100$ *trajectories of the order 4.0 weak Taylor scheme (1.17) for equidistant step size* $\delta = \Delta_n = 2^0$ *and evaluate the 90%-confidence interval for the mean error* μ *at time* T. *Repeat the calculations for the step sizes* 2^{-1} *and* 2^{-2} *and plot* $\log_2 |\mu|$ *versus* $\log_2 \delta$.

Notes 5.1.5

(i)   The program is adapted from Program 5.1.4, again simplifying considerably for the given linear additive noise equation with vanishing coefficients for many of the higher order integrals. Here $E(X_T) = E(X_0) \, e^{aT}$.

(ii) To avoid large fluctuations in the results for different computer runs, larger batch sizes are recommended.

We observe that there is a general problem in the weak approximation of functionals due to the fact that we have a fixed variance for the random variable which we generate. To approximate it sufficiently accurately we may need a very large sample size to obtain sufficiently small confidence intervals. However as already mentioned in (3.4.8) the length of the confidence interval can be reduced only proportional to the reciprocal square root of the sample size. Later we will see that variance reduction methods can substantially decrease the necessary computing time.

5.2   Explicit Weak Schemes and Extrapolation Methods

Higher order weak Taylor schemes require the determination and evaluation of derivatives of various orders of the drift and diffusion coefficients. These derivatives can be avoided in Runge-Kutta type weak schemes which we shall call *explicit weak schemes*. Note that these are not heuristic adaptations of the well-known Runge-Kutta schemes for deterministic equations. We shall see that the order of convergence of weak schemes can be increased significantly by the application of extrapolation methods.

A. Explicit Order 2.0 Weak Schemes

By using additional supporting points to evaluate the drift and diffusion coefficients it is possible to obtain schemes which do not involve derivatives of these coefficients, but have similar properties to the simplified order 2.0 weak Taylor scheme (1.10).

For example, in the general autonomous case with $d, m = 1, 2, \ldots$ Platen obtained the following *explicit order 2.0 weak scheme* in vector form

$$
(2.1) \qquad Y_{n+1} = Y_n + \frac{1}{2} \left(a \left(\bar{\Upsilon} \right) + a \right) \Delta_n
$$

$$
+ \frac{1}{4} \sum_{j=1}^{m} \left[\left(b^j \left(\bar{R}_+^j \right) + b^j \left(\bar{R}_-^j \right) + 2 b^j \right) \Delta \hat{W}_n^j \right.
$$

$$
+ \sum_{\substack{r=1 \\ r \neq j}}^{m} \left(b^j \left(\bar{U}_+^r \right) + b^j \left(\bar{U}_-^r \right) - 2 b^j \right) \Delta \hat{W}_n^j \Delta_n^{-1/2} \right]
$$

$$
+ \frac{1}{4} \sum_{j=1}^{m} \left[\left(b^j \left(\bar{R}_+^j \right) - b^j \left(\bar{R}_-^j \right) \right) \left\{ \left(\Delta \hat{W}_n^j \right)^2 - \Delta_n \right\} \right.
$$

$$
+ \sum_{\substack{r=1 \\ r \neq j}}^{m} \left(b^j \left(\bar{U}_+^r \right) - b^j \left(\bar{U}_-^r \right) \right) \left\{ \Delta \hat{W}_n^j \Delta \hat{W}_n^r + V_{r,j} \right\} \right] \Delta_n^{-1/2}
$$

with supporting values

$$
\bar{\Upsilon} = Y_n + a \, \Delta_n + \sum_{j=1}^{m} b^j \, \Delta \hat{W}_n^j, \qquad \bar{R}_\pm^j = Y_n + a \, \Delta_n \pm b^j \sqrt{\Delta_n}
$$

and

$$
\bar{U}_\pm^j = Y_n \pm b^j \sqrt{\Delta_n},
$$

where the random variables $\Delta \hat{W}_n^j$ and $V_{r,j}$ are defined in (1.9), (1.11)–(1.13).

Problem 5.2.1 (PC-Exercise 15.1.1)   *Repeat Problem 5.1.2 with $a = 1.5$ and $b = 0.01$ using the explicit order 2.0 weak scheme (2.1) with $\delta = \Delta_n = 2^{-3}$, $\ldots, 2^{-6}$.*

Notes 5.2.1 The scheme simplifies considerably for the given bilinear equation. In particular, the $V_{r,j}$ random variables are not required in this 1-dimensional example. The quality of the results improves with increasing batch sizes.

Other schemes which are not completely derivative-free have also been suggested in the literature. For example, for the non-autonomous 1-dimensional case with scalar noise, $d = m = 1$, Milstein proposed the scheme

$$(2.2) \quad Y_{n+1} = Y_n + \frac{1}{2} b \, \Delta \hat{W}_n + \frac{1}{2} (a - bb') \, \Delta_n + \frac{1}{2} bb' \left(\Delta \hat{W}_n \right)^2$$

$$+ \frac{1}{2} a \left(\tau_{n+1}, Y_n + a \Delta + b \, \Delta \hat{W}_n \right) \Delta_n$$

$$+ \frac{1}{4} b \left(\tau_{n+1}, Y_n + a \Delta_n + \frac{1}{\sqrt{3}} b \, \Delta \hat{W}_n \right) \Delta \hat{W}_n$$

$$+ \frac{1}{4} b \left(\tau_{n+1}, Y_n + a \Delta_n - \frac{1}{\sqrt{3}} b \, \Delta \hat{W}_n \right) \Delta \hat{W}_n,$$

where the random variable $\Delta \hat{W}_n$ is three-point distributed as in (1.9).

Another scheme which also retains the derivative b' was proposed by Talay. In the autonomous 1-dimensional case $d = m = 1$ it has the form

$$(2.3) \quad Y_{n+1} = Y_n + \sqrt{2} b \left(\bar{\Upsilon} \right) \Delta \tilde{W}_n + \frac{1}{\sqrt{2}} b \left\{ \Delta \hat{W}_n - \Delta \tilde{W}_n \right\}$$

$$+ \left(a \left(\bar{\Upsilon} \right) - \frac{1}{2} b \left(\bar{\Upsilon} \right) b' \left(\bar{\Upsilon} \right) \right) \Delta_n$$

$$+ \frac{1}{2} bb' \left(\Delta \hat{W}_n \right)^2 + \frac{1}{2} b \left(\bar{\Upsilon} \right) b' \left(\bar{\Upsilon} \right) \left(\Delta \tilde{W}_n \right)^2$$

$$- \frac{1}{2} bb' \left(\Delta \hat{W}_n + \Delta \tilde{W}_n \right)^2$$

with the supporting value

$$\bar{\Upsilon} = Y_n + \frac{1}{\sqrt{2}} b \, \Delta \hat{W}_n + \frac{1}{2} \left(a - \frac{1}{2} bb' \right) \Delta_n + \frac{1}{4} bb' \left(\Delta \hat{W}_n \right)^2,$$

where $\Delta \hat{W}_n$ and $\Delta \tilde{W}_n$ are independent three-point distributed random variables as defined in (1.9).

Problem 5.2.2 (PC-Exercise 15.1.3) *Repeat Problem 5.1.2 for the schemes (2.2) and (2.3) with Gaussian $\Delta \tilde{W}_n$ and $\Delta \hat{W}_n$.*

Notes 5.2.2 For the given equation the scheme (2.2) is essentially the same as that in Program 5.1.2 apart from new expressions for some coefficients. The scheme (2.3) also simplifies considerably here and is also included in Program 5.2.2.

B. An Explicit Order 3.0 Weak Scheme

For important classes of stochastic differential equations special structural features such as additive or scalar noise can be exploited to construct efficient
derivative free higher order schemes.

For example, for scalar additive noise, that is $m = 1$ and $b \equiv const.$, in the
autonomous case $d = 1, 2, \ldots$ we have the *explicit order 3.0 weak scheme* in
vector form

$$(2.4) \quad Y_{n+1} = Y_n + a\,\Delta_n + b\,\Delta\hat{W}_n$$

$$+ \frac{1}{2}\left(a_\zeta^+ + a_\zeta^- - \frac{3}{2}a - \frac{1}{4}\left(\tilde{a}_\zeta^+ + \tilde{a}_\zeta^-\right)\right)\Delta_n$$

$$+ \sqrt{\frac{2}{\Delta_n}}\left(\frac{1}{\sqrt{2}}\left(a_\zeta^+ - a_\zeta^-\right) - \frac{1}{4}\left(\tilde{a}_\zeta^+ - \tilde{a}_\zeta^-\right)\right)\zeta\,\Delta\hat{Z}_n$$

$$+ \frac{1}{6}\left[a\left(Y_n + \left(a + a_\zeta^+\right)\Delta_n + (\zeta + \rho)\,b\,\sqrt{\Delta_n}\right) - a_\zeta^+ - a_\rho^+ + a\right]$$

$$\times\left[(\zeta + \rho)\,\Delta\hat{W}_n\,\sqrt{\Delta_n} + \Delta_n + \zeta\rho\left\{\left(\Delta\hat{W}_n\right)^2 - \Delta_n\right\}\right]$$

with supporting values

$$a_\phi^\pm = a\left(Y_n + a\,\Delta_n \pm b\,\sqrt{\Delta_n}\,\phi\right)$$

and

$$\tilde{a}_\phi^\pm = a\left(Y_n + 2a\,\Delta_n \pm b\,\sqrt{2\Delta_n}\,\phi\right),$$

where ϕ is either ζ or ρ, which are independent two-point distributed random
variables with

$$P(\zeta = \pm 1) = P(\rho = \pm 1) = \frac{1}{2}.$$

In addition, $\Delta\hat{W}_n$ and $\Delta\hat{Z}_n$ are correlated Gaussian random variables satisfying (1.15) and (1.16).

Problem 5.2.3 (PC-Exercise 15.2.1)   *Consider the scalar linear stochastic
differential equation with additive noise*

$$dX_t = a\,X_t\,dt + b\,dW_t$$

*on the time interval $[0,T]$ with $T = 1$, $X_0 = 10.0$ at time $t_0 = 0$, $a = 1.0$ and
$b = 0.01$. Simulate $M = 120$ batches each of $N = 1000$ values $Y(T)$ of the
explicit order 3.0 weak scheme (2.4) with step size $\delta = \Delta_n = 2^0$ and determine
the 90%-confidence interval for the mean error*

$$\mu = E\left(Y(T)\right) - E\left(X_T\right).$$

Repeat the calculations for equidistant step sizes $\delta = \Delta_n = 2^{-1}$ *and* 2^{-2} *and plot* $\log_2 |\mu|$ *against* $\log_2 \delta$.

Notes 5.2.3 The program is adapted from Program 5.2.2, with the inclusion of a two-point distributed random number generator. The results are not particularly reliable for the given batch sizes and parameters. The situation improves if a larger batch size is used.

There are also explicit order 3.0 weak schemes for general noise, but we do not state them here.

Exercise 5.2.1 (PC-Exercise 15.2.2) *Repeat Problem 5.1.2 using the explicit order 3.0 weak scheme (15.2.2) in* Kloeden & Platen (1992) *for scalar noise with* $M = 200$ *and* $N = 1000$.

C. Romberg-Richardson Extrapolation

We turn now to extrapolation methods for the simulation of functionals of Ito diffusions based on discrete time weak approximations, assuming in what follows that the function g is given, smooth and at most of polynomial growth together with all its derivatives, for example, a polynomial.

The simplest stochastic extrapolation method is of second weak order and uses the Euler approximation Y^δ generated by (1.1) with step size δ. One first simulates the functional

$$(2.5) \qquad V_{1,1}^\Delta = E\left(g(Y^\Delta(T))\right)$$

for the step size $\Delta = \delta$, then for the double step size $\Delta = 2\delta$, and finally the two results are combined to yield the second order extrapolation

$$(2.6) \qquad V_{1,2}^\delta = 2V_{1,1}^\delta - V_{1,1}^{2\delta}.$$

This is obviously a stochastic generalization of the well-known Romberg or Richardson extrapolation (3.1.8) method. It was proposed by Talay and Tubaro.

Let us check experimentally whether we obtain from the first weak order $\beta = 1.0$ Euler scheme a method of weak order $\beta = 2.0$ by using the extrapolation method (2.6).

Problem 5.2.4 (PC-Exercise 15.3.1) *Consider the Ito process X satisfying the linear stochastic differential equation*

$$(2.7) \qquad dX_t = a\,X_t\,dt + b\,X_t\,dW_t$$

with $X_0 = 0.1$, $a = 1.5$ and $b = 0.01$ on the time interval $[0,T]$ where $T = 1$. Use the Euler scheme (1.1) to simulate the order 2.0 weak extrapolation $V_{1,2}^\delta$ in (2.6) for $g(x) = x$ and $\delta = 2^{-3}$. Generate $M = 20$ batches of $N = 100$ trajectories each and determine the 90% confidence interval for

$$\mu_2 = V_{1,2}^\delta - E\left(g(X_T)\right).$$

Repeat the calculations for step sizes $\delta = 2^{-4}, 2^{-5}$ and 2^{-6} and plot the results on $\log_2 |\mu_2|$ versus $\log_2 \delta$ axes.

Program 5.2.4   Extrapolation

```
   ...
FUNCTION A(TI,XI:REAL):REAL;
FUNCTION B(TI,XI:REAL):REAL;
FUNCTION GXT(X:REAL):REAL;
FUNCTION MEANGXT(TI:REAL):REAL;
PROCEDURE COMPSAMPLEPARA(NN:INTEGER;X:VECTORO;VAR SAVERAGE,SVARIANCE:REAL);

{ Main program : }
   ...
{ Estimation of the confidence intervals for the mean errors MU }
{ for different time step sizes :                               }

  EGXT:=MEANGXT(T); { calculates E g(X(T)) }
  DELTA_Y:=2.*DELTA;
  G:=0;
  REPEAT
   G:=G+1;                    { index of the time step size used }
   DELTA_Y:=DELTA_Y/2;        { current time step size           }
   SQDELTA_Y:=SQRT(DELTA_Y);  { square root of DELTA_Y           }

  { Generation for different batches : }

   J:=0;
   REPEAT
    J:=J+1;         { batch index                               }
    EGYT[J]:=0.0;   { estimate of E g(Y(T)) for the batch used  }
    EGY2T[J]:=0.0;  { estimate of E g(Y2(T)) for the batch used }

   { Generation of different trajectories : }

    K:=0;
    REPEAT
     K:=K+1; { index of the trajectory used }

    { Generation of the weak approximation Y(T) and         }
    { Estimation of the mean error MU of the extrapolation : }

     I:=0;
     TI:=T0;  { initial time                                            }
     YT:=X0;  { initial value of the Euler approximation for DELTA_Y    }
     Y2T:=X0; { initial value of the Euler approximation for 2 DELTA_Y }
     WHILE TI<T DO
       BEGIN
        I:=I+1;             { time step index }
        TI:=TI+DELTA_Y; { current time    }
        IF I MOD 2 = 1 THEN GENERATE(G1,G2) { uses Polar Marsaglia method }
         ELSE BEGIN DW2TI:=DWTI;G1:=G2; END;
        DWTI:=G1*SQDELTA_Y; { Wiener process increment  W(t(i+1)) - W(ti) }

       { Euler scheme for DELTA_Y : }

        YT:=YT+A(TI-DELTA_Y,YT)*DELTA_Y+B(TI-DELTA_Y,YT)*DWTI;
```

```
{ Euler scheme for 2 * DELTA_Y : }

  IF I MOD 2 = 0 THEN
    BEGIN
      DW2TI:=DW2TI+DWTI; { corresponding Wiener process increment }
      Y2T:=Y2T+2.*A(TI-2.*DELTA_Y,Y2T)*DELTA_Y+B(TI-2.*DELTA_Y,Y2T)*DW2TI;
    END;

  END;{ WHILE }

{ Summation of the functional values g(Y(T)) : }

  EGYT[J]:=EGYT[J]+GXT(YT);
  EGY2T[J]:=EGY2T[J]+GXT(Y2T);

UNTIL K=N;{ REPEAT for different samples }
EGYT[J]:=EGYT[J]/N; { estimate of E g(Y(T)) for the batch }
EGY2T[J]:=EGY2T[J]/N; { estimate of E g(Y2(T)) for the batch }
ERRV[J]:=2.*EGYT[J]-EGY2T[J]-EGXT; { error of the extrapolation }
UNTIL J=M;{ REPEAT for different batches }

{ Calculation of the confidence intervals and initialization of data : }

DEL[G]:=DELTA_Y; { current time step size }
COMPSAMPLEPARA(M,ERRV,AVERAGE,VARIANCE);
MU[G]:=AVERAGE;            { midpoint of the confidence interval }
DIFFER[G]:=QUANTILE*SQRT(VARIANCE/M); { half the interval length }

UNTIL G=NUM;{ REPEAT for different time step sizes }
  ...
```

Notes 5.2.4

(i) The program is based on Program 3.4.2. Wiener increments are generated for time step δ and added to give increments for time step 2δ. The Euler paths for the two time steps are calculated at the same time. The mean error for each batch is stored as a component of the vector ERRV.

(ii) The drift and diffusion coefficients can be changed, but the known MEANGXT, i.e. $E(g(X(T)))$, should be changed too.

One realizes that extrapolation is an elegant and simple way to obtain weak higher order methods. The construction of such methods is based on the existence of an asymptotic expansion of the error with respect to powers of the time step size, which is possible for most explicit one-step schemes. The choice of weights in the linear combination (2.6) of the outcomes from the Euler approximations with step sizes δ and 2δ causes the error term of order δ in the leading error expansion to cancel out asymptotically. The remaining error terms involve only the powers $\delta^2, \delta^3, \ldots$ indicating that we have a second order weak scheme.

D. Higher Order Extrapolations

There are many ways to construct higher order weak methods by extrapolation. For instance, if we start from the second order weak Taylor approximation Y^Δ given in (1.7) or (1.10), we could use the functionals

$$(2.8) \qquad\qquad V_{2,2}^\Delta = E(g(Y^\Delta(T)))$$

for the step sizes $\Delta = \delta$ and $\Delta = 2\delta$ to obtain the third order extrapolation

$$(2.9) \qquad\qquad V_{2,3}^\delta = \frac{1}{3}\left(4V_{2,2}^\delta - V_{2,2}^{2\delta}\right).$$

The third order extrapolation (2.9) could also use most other second order approximations and still remain a third order weak method. For instance, we could apply the second order extrapolation (2.6) in (2.9) using the Euler scheme for step sizes $\delta, 2\delta$ and 4δ to obtain the third order method

$$(2.10) \qquad\qquad V_{1,3}^\delta = \frac{1}{3}\left(8V_{1,1}^\delta - 6V_{1,1}^{2\delta} + V_{1,1}^{4\delta}\right).$$

Similarly, if we extrapolate a third order discrete time approximation Y^Δ used to evaluate the functional

$$(2.11) \qquad\qquad V_{3,3}^\Delta = E(g(Y^\Delta(T))),$$

to obtain a fourth order method such as

$$(2.12) \qquad\qquad V_{3,4}^\delta = \frac{1}{7}\left(8V_{3,3}^\delta - V_{3,3}^{2\delta}\right).$$

Alternatively, we could use other third order approximations such as (2.9) in (2.12) to obtain the fourth order extrapolation

$$(2.13) \qquad\qquad V_{2,4}^\delta = \frac{1}{21}\left(32V_{2,2}^\delta - 12V_{2,2}^{2\delta} + V_{2,2}^{4\delta}\right)$$

which is based on second order approximations. In this we could then use second order approximations like (2.6) to obtain the extrapolation

$$(2.14) \qquad\qquad V_{1,4}^\delta = \frac{1}{21}\left(64V_{1,1}^\delta - 56V_{1,1}^{2\delta} + 14V_{1,1}^{4\delta} - V_{1,1}^{8\delta}\right)$$

which involves only functionals $V_{1,1}$ evaluated with the Euler scheme.

In the same way we can start from a fourth order approximation Y^Δ to compute the functional

$$(2.15) \qquad\qquad V_{4,4}^\Delta = E(g(Y^\Delta(T)))$$

and extrapolate to obtain a fifth order method

$$(2.16) \qquad\qquad V_{4,5}^\delta = \frac{1}{15}\left(16V_{4,4}^\delta - V_{4,4}^{2\delta}\right).$$

Applying the fourth order extrapolations (2.12), (2.13) and (2.14), respectively, we obtain the following fifth order methods based on numerical schemes of successively lower order

$$(2.17) \qquad V_{3,5}^{\delta} = \frac{1}{105} \left(128 V_{3,3}^{\delta} - 24 V_{3,3}^{2\delta} + V_{3,3}^{4\delta} \right),$$

$$(2.18) \qquad V_{2,5}^{\delta} = \frac{1}{315} \left(512 V_{2,2}^{\delta} - 224 V_{2,2}^{2\delta} + 28 V_{2,2}^{4\delta} - V_{2,2}^{8\delta} \right),$$

$$(2.19) \qquad V_{1,5}^{\delta} = \frac{1}{315} \left(1024 V_{1,1}^{\delta} - 960 V_{1,1}^{2\delta} + 280 V_{1,1}^{4\delta} - 30 V_{1,1}^{8\delta} + V_{1,1}^{16\delta} \right).$$

Using a Romberg sequence of step sizes $\delta_j = 2^0 \delta, 2^1 \delta, 2^2 \delta, 2^3 \delta, \ldots$, we can formulate a general relation

$$(2.20) \qquad V_{\alpha,\varrho}^{\delta} = \frac{1}{2^{\varrho-1} - 1} \left(2^{\varrho-1} V_{\alpha,\varrho-1}^{\delta} - V_{\alpha,\varrho-1}^{2\delta} \right)$$

which allows us to construct extrapolations $V_{\alpha,\varrho}^{\delta}$ of given order $\varrho \in \{1, 2, \ldots\}$ recursively from approximations $V_{\alpha,\alpha}^{\delta_j}$ of order $\alpha \in \{1, 2, \ldots, \varrho - 1\}$. All of the above extrapolations follow from this relation which is a generalization of corresponding extrapolation formulae in the non-random case. Also, in the stochastic case we can choose the sequence of step sizes more freely, such as $\delta_l = d_l \delta$, $l = 1, 2, \ldots$ with $0 < d_1 < d_2 < \cdots$, and obtain quite general extrapolations analogous to the deterministic case.

For instance. it is straightforward to derive the following order 6.0 weak extrapolation

$$(2.21) \qquad V_{3,6}^{\delta} = \frac{1}{2905} \left[4032 V_{3,3}^{\delta} - 1512 V_{3,3}^{2\delta} + 448 V_{3,3}^{3\delta} - 63 V_{3,3}^{4\delta} \right]$$

from third order weak approximations such as (1.14), (2.9) or (2.10). Here we double the weak order of convergence through extrapolation.

As a numerical example let us apply a fourth order extrapolation method to the same equation as in Problem 5.2.4.

Problem 5.2.5 (PC-Exercise 15.3.2) *Repeat Problem 5.2.4 with $a = -5.0$ and $b = 2.0$ using the order 4.0 weak extrapolation (2.13) and the order 2.0 weak Taylor scheme (1.7) for $\delta = 2^{-2}, 2^{-3}$ and 2^{-4}, with*

$$\mu_4 = V_{2,4}^{\delta} - E(g(X_T))$$

instead of μ_2.

Notes 5.2.5 Program 5.2.5 is built up from Program 5.2.4 in that the paths for different time steps are now calculated separately. The values $V_{2,2}^{k\delta}$, $k = 1, 2, 4$ are stored in the 3×3 matrix EGYT.

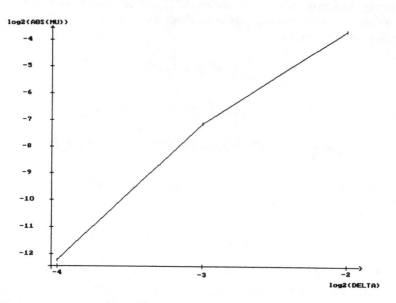

Figure 5.2.1   A fourth order extrapolation.

We have seen that, in principle, by extrapolation we can obtain high order weak methods which are easy to implement and run. They offer an efficient means of computing highly accurate approximations of functionals of diffusions, but they will work only if the underlying schemes are numerically stable for the step sizes involved. Some extrapolation results may still be disappointing because the cancellation effect in the error expansion of these schemes may not be achieved in practice for the step size and sample size used. It is recommended to extrapolate higher order one step schemes in their stability region.

5.3   Implicit Weak Approximations

In this section we shall introduce some implicit and predictor-corrector schemes which are often numerically more stable than many explicit schemes and are thus suitable for stiff stochastic differential equations. However, implicit schemes generally require an algebraic equation to be solved at each time step, which imposes an additional computational burden.

A. Implicit Euler Schemes

The simplest implicit weak scheme is the *implicit Euler scheme,* which in the general multi-dimensional case d, $m = 1, 2, \ldots$ has the general form

$$(3.1) \qquad Y_{n+1} = Y_n + \{(1 - \alpha)a\left(\tau_n, Y_n\right) + \alpha a\left(\tau_{n+1}, Y_{n+1}\right)\} \, \Delta_n$$

$$+ \sum_{j=1}^{m} b^j \left(\tau_n, Y_n \right) \Delta \hat{W}_n^j$$

where the parameter $\alpha \in [0,1]$ indicates the degree of implicitness and the $\Delta \hat{W}_n^j$ for $j = 1, \ldots, m$ and $n = 1, 2, \ldots$ are independent two-point distributed random variables with

$$(3.2) \qquad P \left(\Delta \hat{W}_n^j = \pm \sqrt{\Delta_n} \right) = \frac{1}{2}.$$

For $\alpha = 0.0$ the scheme (3.1) reduces to the explicit simplified Euler scheme (1.6), while with $\alpha = 0.5$ it is a stochastic generalization of the trapezoidal method (3.1.5). Under assumptions of sufficient regularity, the scheme (3.1) converges weakly with order $\beta = 1.0$. The implicit Euler scheme (3.1) is A-stable for $\alpha \in [0.5, 1]$, but for $\alpha \in [0, 0.5)$ its region of absolute stability is the interior of the circle of radius $r = (1 - 2\alpha)^{-1}$ centered at $-r + 0\imath$ in the complex plane.

As a test equation we use the two-dimensional linear stochastic differential equation (4.3.3) again, namely

$$(3.3) \qquad dX_t = A X_t dt + B X_t dW_t$$

with matrices

$$(3.4) \qquad A = a \begin{bmatrix} -1 & 1 \\ 1 & -1 \end{bmatrix} \quad \text{and} \quad B = bI = b \begin{bmatrix} 1 & 0 \\ 0 & 1 \end{bmatrix}$$

where a and b are given real numbers. We note from (2.2.20) that the expectation $E(X_t)$ satisfies the two-dimensional linear ordinary differential equation

$$\frac{dm}{dt} = A m,$$

so

$$
\begin{aligned}
E(X_t) &= E(X_0) \exp(A t) \\
&= E(X_0) \begin{bmatrix} 1 + 2e^{-at} & 1 - e^{-at} \\ 1 - e^{-at} & 1 + e^{-at} \end{bmatrix}.
\end{aligned}
$$

For PCs we shall also encounter the problem of obtaining sufficiently small confidence intervals for our estimates of the functional $E(X_T)$. We shall thus choose the noise parameter b small enough to ensure reasonable results.

Problem 5.3.1 (PC-Exercise 15.4.1) *Consider the 2-dimensional Ito process X satisfying the stochastic differential equation (3.3) with coefficient matrices given by (3.4) with $a = 5.0$ and $b = 0.001$ and initial value $X_0 = (1.0, 0)$ on the interval $[0, T]$ where $T = 1$. Compute $M = 20$ batches each of $N = 100$ values $Y(T)$ of the implicit Euler scheme (3.1) with $\alpha = 0.0$, 0.5 and 1.0 for* *step size $\delta = \Delta_n = 2^{-3}$. Evaluate the 90% confidence intervals for the mean error*

$$\mu = E\left(Y(T)\right) - E\left(X_T\right)$$

at time $T = 1$. Repeat the calculations for step sizes $\delta = \Delta_n = 2^{-4}$, 2^{-5} and 2^{-6} and plot the results on separate μ versus δ axes for the three cases $\alpha = 0.0, 0.5$ and 1.0. Finally, replot the results on $\log_2 |\mu|$ versus $\log_2 \delta$ axes.

Notes 5.3.1

(i) The program is based on Program 5.1.1 with adjustable implicitness parameters ALPHA and ETA and function NORM to evaluate the Euclidean norm of a vector. Note that the program applies only to the specific stochastic differential equation (3.3)–(3.4).

(ii) The expectations and confidence intervals of each component at $T = 1$ are first displayed on the screen. On pressing any key other than the <ESC> key, the norm of the expectation vector is calculated and plotted against the stepsize DELTA in logarithmic coordinates.

We cannot construct a meaningful implicit Euler scheme by just making also the diffusion coefficient in the Euler scheme implicit in the same way as the drift coefficient. In view of the definition of stochastic integrals, the solution of such a scheme will not, in general, converge to that of the given Ito equation. To obtain a weakly converging implicit approximation we need to modify appropriately the drift term.

In the general multi-dimensional case d, $m = 1, 2, \ldots$ and vector notation, we have a *family of implicit Euler schemes*

$$(3.5) \quad Y_{n+1} = Y_n + \left\{\alpha \bar{a}_\eta \left(\tau_{n+1}, Y_{n+1}\right) + (1-\alpha)\bar{a}_\eta \left(\tau_n, Y_n\right)\right\} \Delta_n$$

$$+ \sum_{j=1}^{m} \left\{\eta b^j \left(\tau_{n+1}, Y_{n+1}\right) + (1-\eta)b^j \left(\tau_n, Y_n\right)\right\} \Delta \hat{W}_n^j$$

with implicitness parameters $\alpha, \eta \in [0,1]$, where the $\Delta \hat{W}_n^j$ are as described in (3.2) and the corrected drift coefficient $\bar{a}_\eta$ is defined by

$$(3.6) \qquad\qquad \bar{a}_\eta = a - \eta \sum_{j_1, j_2 = 1}^{m} \sum_{k=1}^{d} b^{k, j_1} \frac{\partial b^{j_2}}{\partial x^k}.$$

The choice $\alpha = \eta = 1$ in (3.5) gives us the fully implicit Euler scheme. For $\eta = 0.5$ the corrected drift $\bar{a}_\eta = \underline{a}$ is the corrected drift of the corresponding Stratonovich equation, and for $\alpha = 0.5$ the scheme (3.5) yields additional stochastic generalizations of the deterministic trapezoidal method (3.1.5).

Exercise 5.3.1 (PC-Exercise 15.4.5)   *Repeat Problem 5.3.1 for the implicit Euler scheme (3.5) with $\alpha = \eta = 1$.*

B. Implicit Order 2.0 Weak Schemes

Corresponding to the order 2.0 weak Taylor scheme (1.10) one finds for the degree of implicitness $\alpha = 0.5$ the *implicit order 2.0 weak Taylor scheme*

$$
(3.7) \qquad Y_{n+1} = Y_n + \frac{1}{2} \left\{ a\left(\tau_{n+1}, Y_{n+1}\right) + a \right\} \Delta_n
$$

$$
+ \sum_{j=1}^{m} b^j \, \Delta \hat{W}_n^j + \frac{1}{2} \sum_{j=1}^{m} L^0 b^j \, \Delta \hat{W}_n^j \, \Delta_n
$$

$$
+ \frac{1}{2} \sum_{j_1, j_2 = 1}^{m} L^{j_1} b^{j_2} \left(\Delta \hat{W}_n^{j_1} \Delta \hat{W}_n^{j_2} + V_{j_1, j_2} \right),
$$

where the random variables $\Delta \hat{W}_n^j$ and V_{j_1, j_2} can be chosen as in (1.11), (1.12) and (1.13). This scheme is due to Milstein. Note that the last two terms in (3.7) vanish for additive noise. The scheme (3.7) is A-stable.

Problem 5.3.2 (PC-Exercise 15.4.6) *Repeat Problem 5.3.1 with $a = 5.0$, $b = 10^{-6}$ and $\delta = \Delta_n = 2^{-5}, \ldots, 2^{-8}$ for the implicit order 2.0 weak Taylor scheme (3.7).*

We can also avoid derivatives in weak implicit schemes by using similar approximations to those in the explicit order 2.0 weak scheme (2.1).

In the autonomous case, $d = 1, 2, \ldots$, with scalar noise $m = 1$, Platen proposed the *implicit order 2.0 weak scheme*

$$
(3.8) \qquad Y_{n+1} = Y_n + \frac{1}{2} \left(a + a\left(Y_{n+1}\right) \right) \Delta_n
$$

$$
+ \frac{1}{4} \left(b\left(\bar{\Upsilon}^+\right) + b\left(\bar{\Upsilon}^-\right) + 2b \right) \Delta \hat{W}_n
$$

$$
+ \frac{1}{4} \left(b\left(\bar{\Upsilon}^+\right) - b\left(\bar{\Upsilon}^-\right) \right) \left\{ \left(\Delta \hat{W}_n\right)^2 - \Delta_n \right\} \Delta_n^{-1/2}
$$

with supporting values

$$
\bar{\Upsilon}^\pm = Y_n + a \, \Delta_n \pm b \sqrt{\Delta_n},
$$

and $\Delta \hat{W}_n$ as in (3.7). There is also a generalization of (3.8) for multi-dimensional noise. The scheme (3.8) is A-stable.

Exercise 5.3.2 (PC-Exercise 15.4.7) *Repeat Problem 5.3.2 for the implicit order 2.0 weak scheme (3.8). Note, that for our equation the scheme (3.8) coincides with (3.7).*

C. Predictor-Corrector Method

Deterministic predictor-corrector methods, for instance (3.1.6), are used mainly because of their numerical stability, which they inherit from the implicit counterparts of their corrector schemes. In addition, the difference between the predicted and the corrected values at each time step provides an indication of the local error. In principle, these advantages carry over to the stochastic case. Here we shall describe a few predictor-corrector methods for stochastic differential equations proposed by Platen. They use as their predictors weak Taylor or explicit schemes and as their correctors the corresponding implicit schemes which are made explicit by using the predicted value $\bar{Y}_{n+1}$ instead of Y_{n+1} on the right hand side of the implicit scheme.

Using the family of implicit Euler schemes (3.5) as corrector, in the general multi-dimensional case d, $m = 1, 2, \ldots$ we can form the following *family of order* 1.0 *weak predictor-corrector methods* with corrector

$$(3.9) \quad Y_{n+1} \; = \; Y_n + \left\{ \alpha \bar{a}_\eta \left(\tau_{n+1}, \bar{Y}_{n+1} \right) + (1 - \alpha) \bar{a}_\eta \left(\tau_n, Y_n \right) \right\} \Delta_n$$

$$+ \sum_{j=1}^{m} \left\{ \eta b^j \left(\tau_{n+1}, \bar{Y}_{n+1} \right) + (1 - \eta) b^j \left(\tau_n, Y_n \right) \right\} \Delta \hat{W}_n^j$$

for $\alpha, \eta \in [0, 1]$. and predictor

$$(3.10) \qquad\qquad \bar{Y}_{n+1} = Y_n + a \Delta_n + \sum_{j=1}^{m} b^j \Delta \hat{W}_n^j,$$

where $\bar{a}_\eta$ is defined in (3.6) and the $\Delta \hat{W}_n^j$ are as in (3.2).

We note that the corrector (3.9) for $\eta > 0$ includes some degree of implicitness in the diffusion term too. In the following problem some of the parameters have been changed from those in the original PC-Exercise to enhance the quality of the results.

Problem 5.3.3 (PC-Exercise 15.5.1)   *Repeat Problem 5.3.1 with $a = 5.0$ and $b = 0.01$ for the predictor-corrector method (3.9)–(3.10) with $\alpha = 1.0$ and $\eta = 0$ using $M = 100$ batches of $N = 100$ samples each.*

From Figure 5.3.1 we can expect that the predictor-corrector method (3.9)–(3.10) with $\alpha = 1.0$ and $\eta = 0$ is numerically stable and, in fact, of weak order 2.0 for the example under consideration. The higher than expected order here is a consequence of the special structure of both the scheme and the particular example.

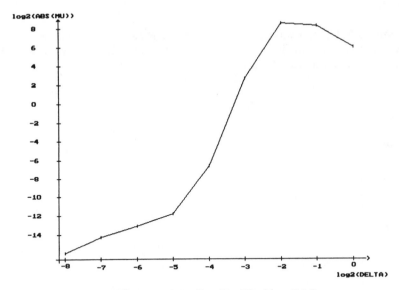

Figure 5.3.1   Result of Problem 5.3.3.

D. Order 2.0 Predictor-Corrector Methods

Let us now combine the order 2.0 weak Taylor scheme (1.10) and its implicit counterpart (3.7) to form an order 2.0 predictor-corrector method.

In the autonomous multi-dimensional scalar noise case, d, $m = 1, 2, \ldots$, a possible *order 2.0 weak predictor-corrector method* has corrector

$$(3.11) \qquad Y_{n+1} = Y_n + \frac{1}{2} \left\{ a \left(\bar{Y}_{n+1} \right) + a \right\} \Delta_n + \Psi_n$$

with

$$\Psi_n = \sum_{j=1}^{m} \left\{ b^j + \frac{1}{2} L^0 b^j \Delta_n \right\} \Delta \hat{W}_n^j + \frac{1}{2} \sum_{j_1, j_2 = 1}^{m} L^{j_1} b^{j_2} \left(\Delta \hat{W}_n^{j_1} \Delta \hat{W}_n^{j_2} + V_{j_1, j_2} \right),$$

(3.12)

and predictor

$$(3.13) \quad \bar{Y}_{n+1} = Y_n + a \Delta_n + \Psi_n + \frac{1}{2} L^0 a \Delta_n^2 + \frac{1}{2} \sum_{j=1}^{m} L^j a \Delta W_n^j \Delta_n,$$

where the independent random variables $\Delta \hat{W}_n^j$ and V_{j_1, j_2} can be chosen as in (1.11), (1.12) and (1.13).

Note that even for nonlinear equations we do not have to solve an algebraic equation at each iteration as for implicit schemes. The following numerical experiment illustrates the numerical performance of the above scheme in a linear situation.

Problem 5.3.4 (PC-Exercise 15.5.2)   *Repeat Problem 5.3.1 with a = 5.0 and b = 0.01 for the predictor-corrector method (3.11)–(3.13) using M = 100 batches with N = 100 samples each.*

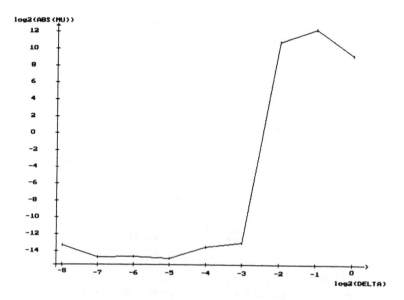

Figure 5.3.2   Results of Problem 5.3.4.

Figure 5.3.2 shows that the predictor-corrector method (3.11)–(3.13) is numerically stable and, experimentally at least, of weak order greater than three for this particular linear example. This result suggests that predictor-corrector schemes of the above type can be advantageous in many practical situations.

The following second order weak predictor-corrector method avoids also the computation of derivatives of the drift and diffusion coefficients. In the autonomous case $d = 1, 2, \ldots$ with scalar noise $m = 1$, a *derivative free order 2.0 weak predictor-corrector method* has corrector

$$(3.14) \qquad Y_{n+1} = Y_n + \frac{1}{2} \left\{ a \left(\bar{Y}_{n+1} \right) + a \right\} \Delta_n + \phi_n,$$

where

$$\phi_n = \frac{1}{4} \left(b \left(\bar{\Upsilon}^+ \right) + b \left(\bar{\Upsilon}^- \right) + 2b \right) \Delta \hat{W}_n$$

$$+ \frac{1}{4} \left(b \left(\bar{\Upsilon}^+ \right) - b \left(\bar{\Upsilon}^- \right) \right) \left\{ \left(\Delta \hat{W}_n \right)^2 - \Delta_n \right\} \Delta_n^{-1/2}$$

with supporting values

$$\bar{\Upsilon}^{\pm} = Y_n + a\,\Delta_n \pm b\,\sqrt{\Delta_n},$$

and with predictor

(3.15) $$\bar{Y}_{n+1} = Y_n + \frac{1}{2}\left\{a\left(\bar{\Upsilon}\right) + a\right\}\Delta_n + \phi_n$$

with the supporting value

$$\bar{\Upsilon} = Y_n + a\,\Delta_n + b\,\Delta\hat{W}_n.$$

Here the $\Delta\hat{W}_n$ can be chosen as (1.9). Essentially, $\bar{\Upsilon}$ is used as an initial predictor which is corrected in (3.15), with the output $\bar{Y}_{n+1}$ being itself corrected in (3.14). A general multi-dimensional generalization of this method has also been proposed by Platen; see equations (15.5.14)–(15.5.15) in Kloeden and Platen (1992).

As in the deterministic case, predictor-corrector methods provide possibilities for step size control which is very important in practice. At each step in the above schemes we first compute the predicted approximate value $\bar{Y}_{n+1}$ and then the corrected value Y_{n+1}. Their difference $Z_{n+1} = \bar{Y}_{n+1} - Y_{n+1}$ provides us with information about the local error at each step, which we could use on-line to improve the simulations. For instance, if the mean of Z_{n+1} is too large, we should change to a finer time step and repeat the calculation.

Exercise 5.3.3 (PC-Exercise 15.5.3)   *Repeat Problem 5.3.4 with $a = 5.0$ and $b = 0.01$ using the multi-dimensional version of the predictor-corrector method (3.14)–(3.15).*

5.4   Simulation Studies

As in Chapter 4 on strong approximations, we also include some simulation studies for weak approximations. Here hands-on computational experience is perhaps even more important because of the need to keep the confidence intervals of the approximated functional sufficiently small to obtain meaningful results. To remain within the capacity of most PCs we have had to restrict our numerical examples to quite small noise intensities in order to indicate the principal effects. Readers with access to workstations will find it informative to repeat the experiments with larger noise intensities.

A. Efficiency of Weak Taylor Schemes

To illustrate the effect of using higher order weak Taylor schemes we shall study numerically the simple one-dimensional, $d = m = 1$, linear stochastic differential equation

(4.1) $$dX_t = a\,X_t dt + b\,dW_t$$

with drift coefficient $a(x) = ax$ and diffusion coefficient $b(x) \equiv b$ on an interval $[0, T]$ for an initial value X_0. Since our example has additive noise we can easily implement all of the higher order discrete time Taylor approximations mentioned in Section 5.1. Our aim is to simulate a functional $E(g(X_T))$, where for simplicity we have chosen $g(x) = x$ to obtain the first moment of X_T for which we know the exact value

$$(4.2) \qquad\qquad E\left(X_T\right) = E\left(X_0\right) \exp\{aT\}$$

Problem 5.4.1   *Apply to (4.1) with $a = -2$, $b = 0.01$, $X_0 = 0.1$ and $T = 1.0$ the Euler scheme (1.1), the simplified Euler scheme (1.6), the order 2.0 simplified weak Taylor scheme (1.7), the simplified order 2.0 weak Taylor scheme (1.10), the simplified order 3.0 weak Taylor scheme (1.14) and the simplified order 4.0 weak Taylor scheme (1.17) to estimate the functional (4.2) from samples with $M = 20$ batches each of $N = 100$ trajectories for step sizes $\delta = \Delta_n = 2^0, 2^{-1}, \ldots, 2^{-8}$. From these simulations estimate the 90% confidence intervals for the mean error*

$$\mu = E(g(Y(T))) - E(g(X_T))$$

and plot the $\log_2$ of the necessary computer time against $-\log_2 |\mu|$.

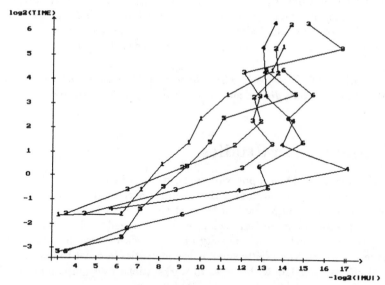

Figure 5.4.1 $\log_2$ (computer time) versus $-\log_2 |\mu|$ for Euler (1), simplified Euler (2), order 2.0 weak Taylor (3), simplified order 2.0 weak Taylor (4), simplified order 3.0 weak Taylor (5) and simplified order 4.0 weak Taylor (6) schemes.

Notes 5.4.1

(i) Program 4.4.2 is adapted to the weak approximation context under consideration. The parameter G and NUM here refer to the number of schemes and time steps, respectively, that are used.

(ii) Confidence intervals at different time steps for each scheme first appear on the screen. Press any key other than the <ESC> key to obtain the plot of computer time against mean error in logarithmic coordinates.

The results of the above numerical experiment for different step sizes are interpolated linearly for each scheme and shown together in Figure 5.4.1. The curves (1) – (6) correspond to the Euler, simplified Euler, order 2.0, simplified order 2.0, simplified order 3.0 and simplified order 4.0 weak Taylor schemes, respectively. In view of the small sample sizes and special form of the stochastic differential equation, some caution is required in making general conclusions from these results. Nevertheless, it seems that higher order schemes are more efficient and that simplified schemes are more efficient than their nonsimplified counterparts. To get an impression of how the results depend on the noise parameter b, the following exercise should be attempted if a sufficiently fast PC is available. In principle the variance in these examples is still rather large. It is recommended to plot also the corresponding confidence intervals.

Exercise 5.4.1 *Repeat Problem 5.4.1 with noise parameter $b = 0.1$ and batch size of $N = 1000$.*

B. Efficiency of Second Order Weak Schemes

Different second order weak approximations are compared in the following numerical experiment. These are the order 2.0 weak Taylor scheme (1.7), the explicit order 2.0 weak scheme (2.1), the second order extrapolation (2.6), the implicit order 2.0 weak scheme (3.8) and the order 2.0 weak predictor-corrector method (3.14), (3.15).

Problem 5.4.2 *Repeat Problem 5.4.1 for the second order weak schemes (1.7), (2.1), (2.6), (3.8), (3.14) and (3.15).*

Notes 5.4.2 The program is based on Program 5.4.1 and Notes 5.4.1 apply here too. The extrapolation scheme has parameters KAPPA and variable Y2T here.

The different second order weak schemes perform differently for the given test equation here, with the implicit order 2.0 weak scheme apparently more efficient than the others. Also here we note that for high accuracies the variance in our experiments was too large.

Exercise 5.4.2 *Repeat Problem 5.4.2 with diffusion coefficient $b = 0.1$ and batch size of $N = 1000$ trajectories.*

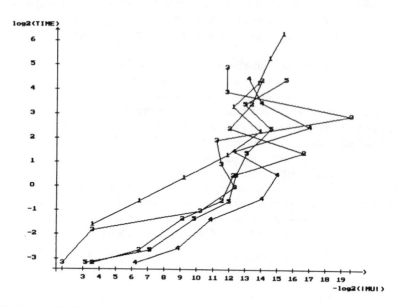

Figure 5.4.2 $\log_2$ (computer time) versus $-\log_2 |\mu|$ for order 2.0 weak Taylor (1),
explicit order 2.0 weak (2), second order extrapolation (3), implicit order 2.0 weak
(4) and order 2.0 weak predictor-corrector (5) methods.

C. Third Order Weak Methods

Let us now look at several third order weak methods, namely the simplified
order 3.0 weak Taylor scheme (1.14), the explicit order 3.0 weak scheme (2.4),
the third order extrapolation (2.9) based on the order 2.0 weak Taylor scheme
(1.7) and the third order extrapolation (2.10) based on the Euler scheme (1.1).

Problem 5.4.3 *Repeat Problem 5.4.1 for the third order weak methods
(1.14), (2.4), (2.9) and (2.10).*

Notes 5.4.3 See Note 5.4.2. Extrapolation parameters are now KAPPA,
ETA with variables Y2T, Y4T.

Before any definite conclusions are made, the results should be checked for
different runs, parameters and sample sizes. But the extrapolation based on
the order 2.0 weak Taylor scheme performed quite well and better than that
based on the Euler approximation.

Exercise 5.4.3 *Repeat Problem 5.4.3 with $b = 0.1$ and batch size $N = 1000$.*

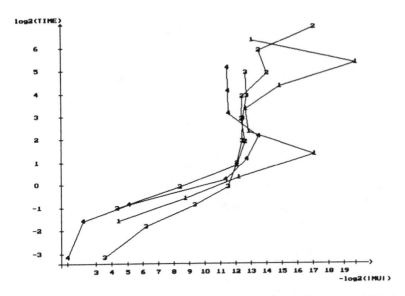

Figure 5.4.3 $\log_2$ (computer time) versus $-\log_2 |\mu|$ for order 3.0 weak Taylor (1), explicit order 3.0 weak (2), extrapolations $V^\delta_{2,3}$ (3) and $V^\delta_{1,3}$ (4).

D. Some Higher Order Weak Methods

We shall now compare fourth order approximations, namely the simplified order 4.0 weak Taylor scheme (1.17), and the fourth order extrapolation (2.12) (based on the simplified order 3.0 weak Taylor scheme (1.14)), the extrapolation (2.13) (based on the order 2.0 weak Taylor scheme (1.7)) and (2.14) with the Euler scheme.

Problem 5.4.4 *Repeat Problem 5.4.1 for the fourth order weak methods (1.17), (2.12), (2.13) and (2.14).*

Notes 5.4.4 As above with the additional extrapolation parameter ZETA and variable Y8T. Program 5.4.5 below is similar with the appropriate new parameters and variables.

To also provide a comparison between different fifth order extrapolation methods, we include the following problem which studies the numerical efficiency for the extrapolations (2.16), (2.17), (2.18) and (2.19) which are based on corresponding weak Taylor schemes (1.17), (1.14), (1.7) and (1.1), respectively.

Problem 5.4.5 *Repeat Problem 5.4.1 for the fifth order extrapolations (2.16), (2.17), (2.18) and (2.19).*

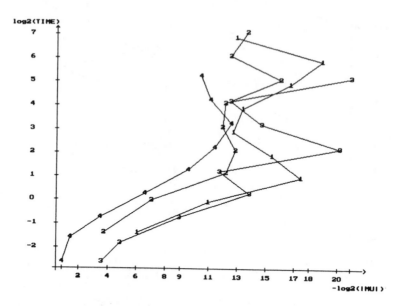

Figure 5.4.4 $\log_2$ (computer time) versus $-\log_2 |\mu|$ for fifth order extrapolations $V_{1,5}^{\delta}$ (1), $V_{2,5}^{\delta}$ (2), $V_{3,5}^{\delta}$ (3) and $V_{4,5}^{\delta}$ (4).

The extrapolation $V_{1,5}^{\delta}$ based on the Euler scheme is obviously less efficient than the other extrapolations based on higher order schemes in this simulation study.

To summarize, we see from the above results that for a given stochastic differential equation extrapolations based on higher order discrete time approximations can be more efficient and more stable numerically than those based on the Euler scheme. This seems to be due to the dominant effect of the numerical round-off error in low order schemes for small step sizes. The higher order schemes and extrapolations based on them do not require such small step sizes to achieve a comparable accuracy, whereas a wide range of step sizes is required to extrapolate a low order scheme to give a high order result. Moreover, an extrapolated low order scheme for some equations may not be stable numerically for a given step size.

It must be realized that numerical efficiency depends strongly on the leading error coefficients in the global error expansion of the method for the given stochastic differential equation. Further, the statistical error for the estimators of the approximated functionals may overshadow the whole computation if insufficiently many sample paths are used. We shall see in Section 5.5 that it is sometimes possible to reduce the variance of estimators. This seems to be a recommendable way of constructing efficient weak schemes.

5.5   Variance Reducing Approximations

In the preceding sections we used weak approximations of an Ito process to evaluate a given functional. Until now we have only considered direct discrete time approximations in the sense that the Ito solution is discretized and approximated directly in time. In this section we shall introduce additional methods which allow a reduction in the variance of an estimator of the functional.

A. Measure Transformation Method

Suppose we are given a d-dimensional Ito process $X^{s,x} = \{X_t^{s,x}, s \le t \le T\}$, starting at $x \in \Re^d$ at time $s \in [0, T)$, in terms of the stochastic equation

$$(5.1) \qquad X_t^{s,x} = x + \int_s^t a\left(z, X_z^{s,x}\right) dz + \sum_{j=1}^{m} \int_s^t b^j\left(z, X_z^{s,x}\right) dW_z^j.$$

Our aim is to approximate the functional

$$(5.2) \qquad u(s, x) = E\left(g\left(X_T^{s,x}\right)\right)$$

for a given real-valued function g.

When applying a discrete time weak approximation Y previously, we evaluated (5.2) with the functional

$$(5.3) \qquad E\left(g\left(Y_{n_T}\right)\right)$$

using the estimator

$$(5.4) \qquad \eta_N = \frac{1}{N} \sum_{r=1}^{N} g\left(Y_{n_T}\left(\omega_r\right)\right),$$

which is just the arithmetic mean of N independent simulations of the random variable $g\left(Y_{n_T}\right)$. Here $Y_{n_T}\left(\omega_r\right)$ denotes the rth simulation of Y at time $\tau_{n_T} = T$.

We have already seen in Section 3.4, see (3.4.12), that the error

$$(5.5) \qquad \hat{\mu} = \eta_N - E\left(g\left(X_T\right)\right)$$

is asymptotically Gaussian distributed with mean zero and variance

$$(5.6) \qquad \text{Var}\left(\hat{\mu}\right) = \frac{1}{N} \text{Var}\left(g\left(Y_{n_T}\right)\right).$$

In particular, we noticed from (3.4.7)–(3.4.8) that the length of the confidence interval decreases only with order $N^{-1/2}$ as $N \to \infty$. Hence, in order to obtain sufficiently small confidence intervals it is important to begin with a small variance in the random variable $g\left(Y_{n_T}\right)$. With a direct simulation method one fixes the variance of $g\left(Y_{n_r}\right)$ to a value which is close to that of the variance of $g\left(X_T\right)$. However, this variance, which depends totally on g and the given stochastic differential equation, may sometimes be extremely large. This problem leads to the question of whether it is possible to construct other estimators which have nearly the same expectation, but smaller variance.

Now, let us consider the system of stochastic equations for the process $(\tilde{X}, \Theta)$ given by

$$(5.7) \quad \tilde{X}_t^{0,x} = x + \int_0^t \left[a\left(z, \tilde{X}_z^{0,x}\right) - \sum_{j=1}^m b^j\left(z, \tilde{X}_z^{0,x}\right) d^j\left(z, \tilde{X}_z^{0,x}\right) \right] dz$$

$$+ \sum_{j=1}^m \int_0^t b^j\left(z, \tilde{X}_z^{0,x}\right) dW_z^j$$

$$\Theta_t = \Theta_0 + \sum_{j=1}^m \int_0^t d^j\left(z, \tilde{X}_z^{0,x}\right) \Theta_z \, dW_z^j$$

with $\Theta_0 \neq 0$, where the d^j denote given real-valued functions for $j = 1, \ldots, m$. Note that $\tilde{X}^{0,x}$ is d-dimensional, whereas the correction process Θ is only 1-dimensional.

By a measure transformation, the Girsanov transformation, it follows that the following expectations are equal

$$(5.8) \quad E\left(g\left(X_T^{0,x}\right)\right) = E\left(g\left(\tilde{X}_T^{0,x}\right)\Theta_T/\Theta_0\right).$$

Hence, we can use an estimate of the expectation of the random variable

$$(5.9) \quad g\left(\tilde{X}_T^{0,x}\right)\Theta_T/\Theta_0$$

to evaluate the functional (5.2) at time T. As this result does not depend on the choice of the functions d^j, $j = 1, \ldots, m$, we can use them as adjustable parameters to reduce the variance of the random variable (5.9).

Suppose that $u(t, x) > 0$ everywhere and choose the parameter functions d^j as

$$(5.10) \quad d^j(t, x) = -\frac{1}{u(t, x)} \sum_{k=1}^d b^{k,j}(t, x) \frac{\partial u}{\partial x^k}(t, x)$$

for all $(t, x) \in [0, T] \times \Re^d$ and $j = 1, \ldots, m$. Using the Ito formula (2.2.15) one can show that the expression

$$(5.11) \quad u\left(T, \tilde{X}_T^{0,x}\right)\Theta_T/\Theta_0 = g\left(\tilde{X}_T^{0,x}\right)\Theta_T/\Theta_0 = u(0, x)$$

is nonrandom, so its variance is zero. If we could simulate the expression (5.11) approximately using our proposed weak approximations for the stochastic equations (5.7), we would expect, in general, to obtain a small variance for the estimator. Unfortunately, for the construction of the parameter functions in (5.10) we need to know the function u, but this is exactly what we are trying to determine by means of the simulation.

It is often possible to find or guess a function $\bar{u}$ which is similar to u and can be used instead of u in (5.10) to define the parameter functions

$$(5.12) \qquad d^j(t,x) = -\frac{1}{\bar{u}(t,x)} \sum_{k=1}^{d} b^{k,j}(t,x)\frac{\partial \bar{u}}{\partial x^k}(t,x)$$

for all $(t,x) \in [0,T] \times \Re^d$ and $j = 1, \ldots, m$. Then the expression

$$g\left(\tilde{X}_T^{0,x}\right)\Theta_T/\Theta_0$$

will still be random, but with small variance if $\bar{u}$ is chosen sufficiently close to u.

The weak schemes, especially the higher order ones, that we studied in the preceding chapters can be readily applied to provide weak approximations of the process $(\tilde{X}^{0,x}, \Theta)$ to estimate the functional (5.8).

To illustrate the above variance reduction technique, which we call the measure transformation method, we shall now carry out a numerical experiment.

Problem 5.5.1 (PC-Exercise 16.2.3) *Evaluate* $E\left((X_T)^2\right)$ *using the measure transformation method for the linear SDE*

$$X_t = x + \int_0^t a\,X_z\,dz + \int_0^t b\,X_z\,dW_z$$

with $a = 1.5$, $b = 1.0$, $x = 0.1$, $T = 1$ and d^1 defined by (5.10). Apply the Euler scheme (1.1) with equidistant step size $\delta = \Delta_n = 2^{-4}$ to generate $M = 20$ batches each with $N = 100$ trajectories. Determine the 90% confidence intervals for the mean error μ. Repeat the calculations for step sizes $\delta = \Delta_n = 2^{-5}$, 2^{-6} and 2^{-7}. Plot the confidence intervals against Δ_n.

Notes 5.5.1   The program is based on Program 3.4.2. Since $d^1(t,x) \equiv -b$ here, the two SDEs in (5.7) are linear and decoupled. While coefficient values can be changed, the linear structure of the equations cannot. Here $\Theta_0 = 0.1$ is used.

Exercise 5.5.1   *Repeat Problem 5.5.1 with the predictor-corrector method (3.11)-(3.13).*

B. Variance Reducing Estimators

There are several other methods developed mainly in Monte-Carlo theory which allow variance reduced estimators for the functional (5.2) to be constructed. Here we discuss one such method which was proposed by Wagner.

For simplicity, we restrict attention to the scalar noise case, $m = 1$, and denote by $p(s,y;t,x)$ the transition density of the Ito process (5.1) for the transition from the point y at time s to the point x at time t, where $t \geq s$. In addition, we denote the probability measure of the initial value X_0 by $\rho(dx)$.

The functional (5.2) can be represented in the form

$$(5.13) \qquad E\left(g\left(X_T\right)\right) = \int_\Gamma F(\zeta)\, d\mu(\zeta),$$

where

$$(5.14) \qquad d\mu(\zeta) = \rho\left(dx_0\right) dx_1 \cdots dx_N$$

and

$$(5.15) \qquad F(\zeta) = \prod_{i=1}^{N} p\left(\tau_{i-1}, x_{i-1}; \tau_i, x_i\right) g\left(x_N\right),$$

for $\zeta = (x_0, \ldots, x_N) \in \Gamma := \left(\Re^d\right)^{N+1}$. A simple Monte-Carlo estimator for this integral is the one-point estimator

$$(5.16) \qquad \eta_1 = \frac{F(\zeta)}{D(\zeta)},$$

where ζ is a random variable with density $D > 0$ with respect to the measure $d\mu$. It follows from (5.13) and (5.16) that

$$(5.17) \qquad E\left(\eta_1\right) = \int_\Gamma \frac{F(\zeta)}{D(\zeta)} D(\zeta)\, d\mu(\zeta) = E\left(g\left(X_T\right)\right),$$

so the estimator (5.16) is unbiased.

Even if the function F here is not known explicitly, the estimator (5.16) is still useful if we approximate the finite-dimensional density

$$(5.18) \qquad Q\left((x_0, \ldots, x_N)\right) = \prod_{i=1}^{N} p\left(\tau_{i-1}, x_{i-1}; \tau_i, x_i\right)$$

of the Ito process by that of a discrete time weak approximation such as the Euler approximation.

In what follows we shall assume that the symmetric matrix

$$(5.19) \qquad \sigma(s, y) = b(s, y) b(s, y)^\top$$

is strictly positive definite for all $(s, y) \in [0, T] \times \Re^d$ and use the standard notation for the determinant det, the scalar product $(\cdot, \cdot)$, the matrix inverse σ^{-1}, and the vector and matrix transpose $b^\top$.

We can then represent the density of the Euler approximation in the form

$$(5.20) \qquad \tilde{Q}\left(\zeta\right) = \prod_{i:=1}^{N} \tilde{p}_{a,\sigma}\left(\tau_{i-1}, x_{i-1}; \tau_i, x_i\right)$$

for all $\zeta = (x_0, \ldots, x_N) \in \Gamma$ with the density for Gaussian increments

$$(5.21) \qquad \tilde{p}_{a,\sigma}(s, y; t, x) = \frac{1}{\sqrt{\det \sigma(s, y) (2\pi(t - s))^d}}$$

$$\times \exp\left(-\frac{1}{2(t - s)} \left(\sigma^{-1}(s, y)A, A\right)\right),$$

where

$$A = x - y - a(s, y)(t - s).$$

Now, we define the *variance reducing Euler estimator* as

$$(5.22) \qquad \tilde{\eta}_E = \frac{\tilde{Q}(\zeta)}{D(\zeta)} g(x_N),$$

where $\zeta = (x_0, \ldots, x_N) \in \Gamma$ is a random variable with density D with respect to $d\mu$. The systematic error of $\tilde{\eta}_E$ does not depend on D and thus coincides with that of the *direct Euler estimator* obtained from (5.22) by replacing D with $\tilde{Q}$ defined in (5.20), which. in turn, is equivalent to the systematic error of the weak Euler scheme (1.1) discussed in Section 5.1.

In order to determine the optimal version D_{opt} of D for the variance reduction we need to know the transition density $p(s, y; t, x)$ of the Ito process. For the one-point estimator (5.16) it is possible to show that the random variable

$$(5.23) \qquad D_{opt}(\zeta) = |F(\zeta)| \left/ \int_\Gamma |F(z)| \, d\mu(z), \right.$$

where $\zeta = (x_0, \ldots, x_N) \in \Gamma$, has variance

$$(5.24) \qquad \mathrm{Var}\left(\frac{F(\zeta)}{D_{opt}(\zeta)}\right) = \left(\int_\Gamma |F(\zeta)| \, d\mu(\zeta)\right)^2 - \left(\int_\Gamma F(\zeta) \, d\mu(\zeta)\right)^2.$$

This is the minimum possible variance, which is zero when g is only positive or only negative. We call D_{opt} in (5.23) the *optimal density* and note that it depends on the shape of the given function g. It is useful to find good approximations of D_{opt} for special classes of functions g.

So far we have considered only functions g which are sufficiently smooth, but shall now investigate what happens in the 1-dimensional case $d = m = 1$ if we use the indicator function

$$(5.25) \qquad g(y) = I_{[\underline{c}, \bar{c}]}(y) = \begin{cases} 1 & : \quad y \in [\underline{c}, \bar{c}] \\ 0 & : \quad \text{otherwise} \end{cases}$$

for all $y \in \Re^1$, where $-\infty < \underline{c} < \bar{c} < \infty$. Obviously,

$$(5.26) \qquad E(g(X_T)) = P(X_T \in [\underline{c}, \bar{c}]),$$

so with this functional we are estimating the probability that X_T lies in the interval $[\underline{c}, \bar{c}]$.

A possible approximation for the corresponding optimal density is

$$(5.27) \qquad \tilde{D}_{a,\sigma}(\zeta) = \frac{1}{\bar{c} - \underline{c}} I_{[\underline{c},\bar{c}]}(x_N) \prod_{i=1}^{N-1} D_{i,x_{i-1},x_N}(x_i),$$

where

$$(5.28) \qquad D_{i,x_{i-1},x_N}(x_i) = \frac{\tilde{p}_{a,\sigma}\left(\tau_{i-1}, x_{i-1}; \tau_i, x_i\right) \tilde{p}_{a,\sigma}\left(\tau_i, x_i; T, x_N\right)}{\tilde{p}_{a,\sigma}\left(\tau_{i-1}, x_{i-1}; T, x_N\right)}$$

for all $\zeta = (x_0, \ldots, x_N) \in \Gamma$, with $\tilde{p}_{a,\sigma}$ defined in (5.21). $D_{i,x_{i-1},x_N}(x_i)$ is the density of a Gaussian random variable x_i with vector mean

$$(5.29) \qquad x_{i-1} \frac{T - \tau_i}{T - \tau_{i-1}} + x_N \frac{\tau_i - \tau_{i-1}}{T - \tau_{i-1}}$$

and covariance matrix

$$(5.30) \qquad \sigma \frac{(T - \tau_i)(\tau_i - \tau_{i-1})}{T - \tau_{i-1}}$$

for given i, x_{i-1} and x_N.

We shall use the following example to illustrate the possibilities of the variance reduction technique. Let X be the 1-dimensional Ito process with initial value $X_0 = 0$, driven by additive noise with drift and diffusion coefficients

$$(5.31) \qquad a(s,y) = \frac{1}{2} - \frac{1}{2}\sin y \qquad \text{and} \qquad b(s,y) \equiv 1,$$

and consider the functional (5.2) with the indicator function

$$(5.32) \qquad g(y) = I_{[0.3,0.4]}(y)$$

for $y \in \Re^1$ and terminal time $T = 1$.

First we evaluate the functional without variance reduction and then with the variance reducing Euler estimator.

Problem 5.5.2 (PC-Exercise 16.3.4)   *For the above example (5.31), (5.32) use the direct Euler estimator to simulate $M = 20$ batches each of $N = 200$ trajectories and estimate the probability $P(X_T \in [0.3, 0.4])$ for equidistant step sizes $\delta = \Delta_n = \tau_{i+1} - \tau_i = 2^{-3}, \ldots, 2^{-6}$. Plot the corresponding 90% confidence intervals against Δ_n.*

Notes 5.5.2   The program is also adapted from Program 3.4.2, but here MU and MUYLON are estimates of probabilities rather than of mean errors. The results of the direct Euler estimator should be particularly satisfactory here. More batches and samples may reduce the fluctuations to some extent, but a variance reduction method is here preferrable as one can check with the next problem.

Problem 5.5.3 (PC-Exercise 16.3.5)   *Repeat Problem 5.5.2 using the vari-*
ance reducing Euler estimator (5.22), choosing as density D the expression
$\tilde{D}_{a,\sigma}$ *in (5.27) with a = 0.5, σ = 1.0.*

Program 5.5.3   Variance reduction

```
   ...
{ Generates the density of the Euler transition }

FUNCTION PE(S1,Y,S2,X:REAL):REAL;
VAR QQ:REAL;
BEGIN
 QQ:=X-Y-0.5*(1.-SIN(Y))*(S2-S1);
 PE:=(1./SQRT(2.*PI*(S2-S1)))*EXP(-QQ*QQ/(2.*(S2-S1)));
END;{ PE }

{ Generates the optimal density }

FUNCTION POD(S1,Y,S2,X:REAL):REAL;
VAR QQ:REAL;
BEGIN
 QQ:=X-Y-A*(S2-S1);
 POD:=(1./SQRT(2.*PI*SIGMA*(S2-S1)))*EXP(-QQ*QQ/(2.*SIGMA*(S2-S1)));
END;{ POD }

{ Generates the functional form of g(.) }

FUNCTION GXT(X:REAL):REAL;
BEGIN
 IF ((X<C1) OR (X>C2)) THEN GXT:=0.0
   ELSE GXT:=1.0; { provides the indicator function of the interval [C1,C2] }
END;{ GXT }

PROCEDURE COMPSAMPLEPARA(NN:INTEGER;X:VECTORO;VAR SAVERAGE,SVARIANCE:REAL);

{ Main program : }
   ...
   { Generation of Gaussian increments ZETA = ( X[I] ) and the Euler }
   { and optimal densities along ZETA :                             }

   TAOI:=T0;X[0]:=X0;
   X[NUMSTEPS]:=C1+(C2-C1)*RANDOM;
   QTILDE:=1.0; { initial value of the Euler density }
   OD:=1.0;   { initial value of the optimal density }
   FOR I:=1 TO NUMSTEPS-1 DO
     BEGIN
       TAOOLD:=TAOI;TAOI:=TAOI+DELTA_Y; { subinterval end points }
       IF I MOD 2=1 THEN GENERATE(G1,G2) ELSE G1:=G2; { uses Polar Marsaglia }
       X[I]:=SQRT(SIGMA*(T-TAOI)*DELTA_Y/(T-TAOOLD))*G1;
       X[I]:=X[I]+(X[I-1]*(T-TAOI)+X[NUMSTEPS]*DELTA_Y)/(T-TAOOLD);

     { Generation of the Euler density along ZETA = ( X[I] ) : }

       QTILDE:=QTILDE*PE(TAOOLD,X[I-1],TAOI,X[I]);

     { Generation of the optimal density along ZETA = ( X[I] ) : }
```

```
OD:=OD*POD(TAOOLD,X[I-1],TAOI,X[I])*POD(TAOI,X[I],T,X[NUMSTEPS]);
OD:=OD/POD(TAOOLD,X[I-1],T,X[NUMSTEPS]);

END;
QTILDE:=QTILDE*PE(TAOI,X[I],T,X[NUMSTEPS]);
OD:=GXT(X[NUMSTEPS])*OD/(C2-C1); { GXT(X[NUMSTEPS])=1 trivially satisfied }

{ Variance reducing Euler estimate : }

ETAEUL:=QTILDE*GXT(X[NUMSTEPS])/OD;

{ Summation of the estimates within the batch : }

MUYLON[J]:=MUYLON[J]+ETAEUL;
...
```

Notes 5.5.3

(i) The program requires quite a few changes to Program 5.5.2 on which it is based. It is specific to the particular SDE with coefficients (5.31). The declaration of densities and the computational kernel are listed above.

(ii) While the running time is roughly twice that in Problem 5.5.2, the results obtained here show a significant improvement. Arithmetic overflow may occur if the time steps are too small.

A comparison of the above results should show that a considerable reduction in the variance of the estimation of the desired functional can be achieved with the variance reducing Euler estimator. To obtain a comparable variance with the direct Euler estimator would require considerably more computer time. We remark that we assumed for this variance reduction method $bb^\top$ to be positive definite, see (5.19).

C. Unbiased Estimators

The weak approximations that we have considered so far all produce some systematic error. Wagner has applied a Monte-Carlo method due to von Neumann and Ulam to obtain unbiased estimators. With such an estimator it is possible, in principle at least, to concentrate only on the statistical error, which can be controlled by variance reduction.

The transition densities $p(\tau_{i-1}, x_{i-1}; \tau_i, x_i)$ in the estimator (5.15)–(5.17) will be replaced by corresponding unbiased Monte-Carlo estimators $q(\tau_{i-1}, x_{i-1}; \tau_i, x_i)$ to introduce the *variance reducing unbiased estimator*

$$(5.33) \qquad \eta_u = \frac{\prod\limits_{i=1}^{N} q\left(\tau_{i-1}, x_{i-1}; \tau_i, x_i\right)}{D\left((x_0, \ldots, x_N)\right)} \, g\left(x_N\right).$$

It has the property

$$E\left(\eta_u\right) = E\left(g(X_T)\right),$$

where $\zeta = (x_0, \ldots, x_N) \in \Gamma$ is a random variable with density D with respect to the measure $d\mu$, and $q(\tau_{i-1}, x_{i-1}; \tau_i, x_i)$ for given i, τ_{i-1}, x_{i-1}, τ_i, x_i denotes a random variable with the property

$$(5.34) \qquad E\left(q\left(\tau_{i-1}, x_{i-1}; \tau_i, x_i\right)\right) = p\left(\tau_{i-1}, x_{i-1}; \tau_i, x_i\right).$$

If we choose the density D in the estimator (5.33) to be the Euler density $\tilde{Q}$ from (5.20), we call η_u the *direct unbiased Euler estimator*. We can also reduce the variance by choosing D close to the optimal density D_{opt}, in which case we shall call η_u the *unbiased variance reducing Euler estimator*.

We shall now describe the unbiased estimator q for the transition density p in the simple case of additive noise of the kind $b(s,y)b(s,y)^\top \equiv \sigma\, I$, where $\sigma > 0$ is a constant and I is the $d \times d$ identity matrix. In addition, we assume that the drift $a(s,y)$ is assumed to be continuous and bounded. To simulate the unbiased estimator q, we need to generate a number l_i of auxiliary random time instants $\tau_{i,j}$ in the interval $[\tau_{i-1}, \tau_i]$ for $j = l_i, l_i - 1, \ldots, 1$ with $\tau_{i-1} = \tau_{i,l_i} < \tau_{i,l_i-1} < \cdots < \tau_{i,1} < \tau_{i,0} = \tau_i$ for each $i \in \{1, \ldots, N\}$. The number l_i, which is random, is obtained automatically by the following procedure: starting with $\tau_{i,0} = \tau_i$, for $j = 1, 2, \ldots$ choose $\tau_{i,j} = \tau_{i-1}$ with probability

$$(5.35) \qquad P\left(\tau_{i,j} = \tau_{i-1}\right) = \exp\left(-\left(\tau_{i,j-1} - \tau_{i-1}\right)\right),$$

otherwise choose $\tau_{i,j}$ as a random variable taking values in the interval $[\tau_{i-1}, \tau_{i,j-1}]$ with probability density

$$(5.36) \qquad e_{\tau_{i-1}, \tau_{i,j-1}}\left(\tau_{i,j}\right) = \frac{\exp(\tau_{i,j} - \tau_{i-1})}{\exp(\tau_{i,j-1} - \tau_{i-1}) - 1};$$

repeat this procedure until τ_{i-1} is chosen as the value for $\tau_{i,j}$ and take this j as the number l_i.

At the auxiliary times $\tau_{i,j}$ we then need to generate auxiliary points $x_{i,j} \in \Re^d$ of the trajectory for each $j = 1, \ldots, l_i - 1$ and $i = 1, 2, \ldots, N$. For convenience we shall also write $x_{i,0} = x_{i-1}$ and $x_{i,l_i} = x_i$. Analogously with (5.29)–(5.30), we generate the auxiliary points $x_{i,j}$ for $j = 1, \ldots, l_i - 1$ and $i = 1, 2, \ldots, N$ as Gaussian random variables with vector mean

$$(5.37) \qquad x_{i,j-1} \frac{\tau_i - \tau_{i,l_i-j}}{\tau_i - \tau_{i,l_i-j+1}} + x_i \frac{\tau_{i,l_i-j} - \tau_{i,l_i-j+1}}{\tau_i - \tau_{i,l_i-j+1}}$$

and covariance matrix

$$(5.38) \qquad \sigma\, I\, \frac{\left(\tau_i - \tau_{i,l_i-j}\right)\left(\tau_{i,l_i-j} - \tau_{i,l_i-j+1}\right)}{\tau_i - \tau_{i,l_i-j+1}}.$$

The unbiased estimator q for the transition density p is then given by

$$(5.39) \qquad q\left(\tau_{i-1}, x_{i-1}; \tau_i, x_i\right) = \Phi\left(\tau_{i-1}, x_{i-1}; \tau_i, x_i\right)$$

$$+ \tilde{p}_{0,\sigma I} \left(\tau_{i-1}, x_{i-1}; \tau_i, x_i \right) \exp \left(\tau_i - \tau_{i-1} \right)$$

$$\times \sum_{k=1}^{l_i-1} \left[\prod_{j=1}^{k} \frac{K \left(\tau_{i,j}, x_{i,l_i-j}; \tau_{i,j-1}, x_{i,l_i-j+1} \right)}{\tilde{p}_{0,\sigma I} \left(\tau_{i,j}, x_{i,l_i-j}; \tau_{i,j-1}, x_{i,l_i-j+1} \right)} \right]$$

$$\times \frac{\Phi \left(\tau_{i-1}, x_{i-1}; \tau_{i,k}, x_{i,l_i-k} \right)}{\tilde{p}_{0,\sigma I} \left(\tau_{i-1}, x_{i-1}; \tau_{i,k}, x_{i,l_i-k} \right) \exp \left(\tau_{i,k} - \tau_{i-1} \right)},$$

where

$$\Phi(s, y; t, x) = \frac{1}{\sqrt{(2\pi\sigma(t-s))^d}} \exp \left(-\frac{|x - y - a(t,x)(t-s)|^2}{2\sigma(t-s)} \right)$$

and

$$K(s, y; t, x) = \Phi(s, y; t, x) \frac{(a(s,y) - a(t,x), x - y - a(t,x)(t-s))}{\sigma(t-s)}.$$

Let us now compute a direct unbiased estimator for the example in the preceding subsection.

Problem 5.5.4 (PC-Exercise 16.4.2) *Repeat Problem 5.5.2 with the direct unbiased Euler estimator obtained from (5.33) by setting $D = \tilde{Q}$ as in (5.20).*

Notes 5.5.4   Program 5.5.3 has been adapted to the current problem. The particular SDE is fixed, but parameters in the density can be changed. The results should fluctuate considerably from run to run, so will be not particularly reliable. Arithmetic overflow is also possible.

Finally we repeat the calculations with a variance reducing unbiased estimator.

Problem 5.5.5 (PC-Exercise 16.4.3) *Repeat Problem 5.5.2 with the variance reducing unbiased Euler estimator (5.33) using $D = \tilde{D}_{a,\sigma}$ given in (5.27).*

Notes 5.5.5   A significant reduction in the variance will be achieved here. The program is assembled from Program 5.5.4.

A comparison of the above computations should show that the second one has smaller confidence intervals.

Literature for Chapter 5

The weak Euler scheme and an order 2.0 weak Taylor scheme appeared in Milstein (1978). Order 2.0 weak convergence is shown for a class of schemes in Talay (1984). General proofs on the convergence of weak Taylor schemes are given in Platen (1984), Milstein (1985, 1988a) and Kloeden & Platen (1992a), and for Ito processes with jump component in Mikulevicius & Platen (1991). Weak convergence for the Euler scheme under Hölder continuous coefficients follows from a result in Mikulevicius & Platen (1991).

Completely derivative free explicit order 2.0 weak schemes were first proposed in Platen (1984). Other Runge-Kutta type methods can also be found in Talay (1984), Milstein (1978, 1988a) and Kloeden & Platen (1992a). Recent results on extrapolation methods for deterministic ODEs are summarized in Deuflhard (1985). Extrapolations for stochastic differential equations were first developed by Talay & Tubaro (1990). Further results on extrapolation methods can be found in Kloeden & Platen (1992a) or Kloeden, Platen & Hofmann (1993). Milstein proposed some implicit weak schemes for additive noise in (1985, 1988a). Further general implicit and also predictor-corrector methods are given in Kloeden & Platen (1992a).

There are only a few simulation studies for weak approximations, for example, in Pardoux & Talay (1985), Klauder & Petersen (1985) and Liske & Platen (1987). The measure transformation method to reduce the variance is due to Milstein (1988a). A recent paper by Newton (1992) discusses also the control variate method for variance reduction. Other Monte-Carlo variance reduction methods can be found in Rubinstein (1981), Kalos & Whitlock (1986) and Mikhailov (1992). The results on unbiased and variance reducing estimators are due to Wagner (1988a,b, 1989a,b). Variance reduction using Hermite polynomials was proposed in Chang (1987). Further references on weak convergence can be found in Kloeden & Platen (1992a).

Chapter 6

Applications

In this final chapter we apply the strong and weak schemes introduced in the preceding chapters to typical, illustrative situations for which the numerical solution of a stochastic differential equation can provide useful information and insights. The direct simulation of trajectories by a strong scheme allows the behaviour of a stochastic dynamical system to be visualized. Theoretically derived parametric estimators and finite-state Markov chain filters can be tested using simulated data. In addition, weak schemes will be used to calculate frequency histograms of invariant measures, moments and functional integrals, such as Lyapunov exponents, of solutions of stochastic differential equations. Both strong and weak schemes are applied to investigate different aspects of a common problem in the stability and bifurcation of stochastic systems. Finally, a finance model involving the computation of option prices and hedging strategies is simulated. In all cases, the computations will be left as exercises for the reader with only the results given here.

6.1 Visualization of Stochastic Dynamics

The solution paths of a stochastic differential equation can be interpreted as the trajectories of a dynamical system governed by the stochastic differential equation. Computer plots of numerical approximations of different sample paths for the same initial value or for the same sample path for different initial values provide an effective means of visualizing the dynamical behaviour of such a stochastic system. Strong schemes are required for this direct simulation of trajectories, while weak schemes can be used to calculate frequency histograms for the invariant measure of a limiting stationary solution.

A. Bonhoeffer-Van der Pol Oscillator

The Bonhoeffer-Van der Pol equations are a 2-dimensional simplification of the famous 4-dimensional system of ordinary differential equations proposed by Hodgkin and Huxley to model the firing of a single neuron. They have the form

$$(1.1) \qquad \frac{dx_1}{dt} = c\left(x_1 + x_2 - \frac{1}{3}x_1^3 + z\right)$$

$$\frac{dx_2}{dt} = -\frac{1}{c}(x_1 + bx_2 - a)$$

where x_1 is the negative of the membrane voltage, x_2 is the membrane permeability and a, b, c, z are external parameters. When $a = 0.7$, $b = 0.8$, $c = 3.0$ and $z = -0.34$, system (1.1) has a limit cycle which is consistent in shape with the periodic slow charging and rapid discharging of a neuron that is observed in experiments. See Figure 6.1.1.

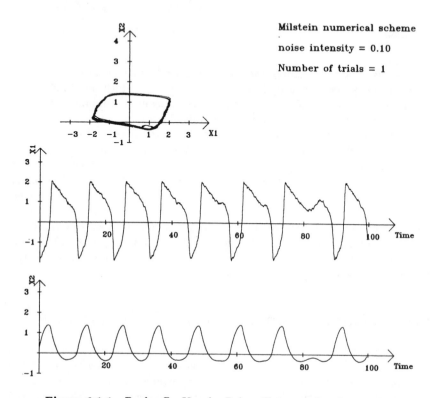

Milstein numerical scheme

noise intensity = 0.10

Number of trials = 1

Figure 6.1.1   Bonhoeffer-Van der Pol oscillator, weak noise $\sigma = 0.1$

The effects of membrane imperfections and of the firing of nearby neurons can be simulated by the inclusion of low intensity, additive noise in the first equation of system (1.2). This leads to a 2-dimensional system of Ito stochastic differential equations

$$(1.2) \qquad dX_t^1 = c\left\{X_t^1 + X_t^2 - \frac{1}{3}(X_t^1)^3 + z\right\}dt + \sigma dW_t$$

$$dX_t^2 = -\frac{1}{c}\left\{X_t^1 + bX_t^2 - a\right\}dt,$$

where $\{W_t, t \geq 0\}$ is a 1-dimensional standard Wiener process and $\sigma > 0$ is the intensity of the noise. Since real biological systems need to function effectively under variable conditions, the behaviour of the stochastic system (1.2) for small

σ should be similar to that of the deterministic system (1.1) if the latter is to be a realistic model of neuronal firing. That is, system (1.2) should have a *noisy limit cycle*. The direct simulation of solution paths of (1.2) by a strong numerical scheme provides us with a means of investigating the validity of this conjecture. To draw a reliable conclusion, however, we need to examine a large number of sample paths from a variety of representative initial values.

Exercise 6.1.1   *Apply the Milstein scheme (4.1.21) with equidistant step size $\Delta = 2^{-9}$ to the noisy Bonhoeffer-Van der Pol equations (1.2) with parameters $a = 0.7$, $b = 0.8$, $c = 3.0$, $z = -0.34$ over the time interval $0 \le t \le T = 1$ for the initial value $X_0^1 = -1.9$, $X_0^2 = 1.2$ and noise intensity $\sigma = 0$ (i.e. the noise-free case). Plot linearly interpolated paths of each component against time t on the (X^1, X^2)-plane. Repeat for 4 other sample paths. Then repeat for increasing noise intensities $\sigma = 0.05, 0.10, 0.15, 0.20, \dots, 1.00$.*

From Figure 6.1.1 it appears that the noisy Bonhoeffer-Van der Pol equations do have a noisy limit cycle. Increasing noise intensity destroys the phase relationship between the two components of the system. To exclude the possibility that the results observed are just some peculiar effect of the particular numerical scheme used, the calculations should be repeated with another scheme to see if similar behaviour is obtained.

Exercise 6.1.2   *Repeat Exercise 6.1.1 with the order 1.5 strong Taylor scheme (4.1.31).*

The output of a weak numerical scheme can also provide useful visual information about the behaviour of a stochastic dynamical system. For example, a frequency histogram can indicate the shape and support of the density of an invariant measure associated with an asymptotically stable stationary solution, such as the noisy limit cycle of the noisy Bonhoeffer-Van der Pol system (1.2). More will be said about these matters in Section 6.5. A simple box counting procedure can be used to illustrate the possibilities. Peaked regions of the histogram indicate slower passage time in these parts of the noisy limit cycle.

Exercise 6.1.3   *Apply the simplified order 2.0 weak Taylor scheme (5.1.10) with equidistant step size $\Delta = 2^{-9}$ to the noisy Bonhoeffer-Van der Pol equations (1.2) with parameters $a = 0.7$, $b = 0.8$, $c = 3.0$, $z = -0.34$ and noise intensity $\sigma = 0.10$. Evaluate 6 sample paths starting at $X_0^1 = -1.9$, $X_0^2 = 1.2$ over the time interval $0 \le t \le T = 10$. Partition the solution field into a uniform square grid and count the number of sample path iterates that fall into each grid cell, finally dividing each cell count by the total number of iterates. Repeat for noise intensities $\sigma = 0.2, 0.5$ and 1.0.*

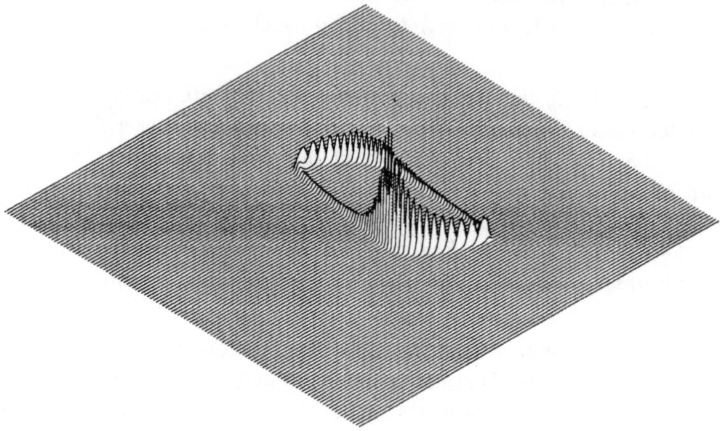

Milstein numerical scheme

noise intensity = 0.10

Number of trials = 1000

Figure 6.1.2   Frequency histogram for noise intensity $\sigma = 0.1$

A 160×160 grid is used here, with the $(1,1)$ cell located on the lower left of the solution field. Cells are located by multiplying both components by preassigned scaling factors and then rounding off and translating. The scaling factors may have to be readjusted if different parameters are used in the equations.

B. The Duffing-Van der Pol Oscillator

The solution $x(t; x)$ of an ordinary differential equation with initial value $x(0; x) = x$ depends smoothly on x under typical regularity conditions. In fact, for each fixed time instant t the mapping T_t of the state space into itself defined by $T_t(x) = x(t; x)$ is a diffeomorphism. Similarly, the mapping $T_{t,\omega}$ defined by $T_{t,\omega}(x) = X_t^x(\omega)$ for the solution X_t^x of a stochastic differential equation with deterministic initial value $X_0^x(\omega) = x$ is also a diffeomorphism, even though the sample paths themselves are only continuous in time t. This property underlies a method of visualization of global dynamical behaviour in which a large number of solutions corresponding to a grid of initial values is followed simultaneously; in the stochastic case, the same noise sample path is used for solutions starting in a grid of deterministic initial values. In view of the diffeomorphism property none of these paths can intersect each other. The ability of a strong numerical scheme to preserve this property is an indication of its accuracy.

As an example we consider a simplified version of a *Duffing-Van der Pol oscillator*

(1.3) $$\ddot{x} + \dot{x} - (\alpha - x^2)\, x = \sigma\, x\, \xi$$

driven by multiplicative white noise ξ, where α is a real-valued parameter. The corresponding Ito stochastic differential equation is 2-dimensional, with components X^1 and X^2 representing the displacement x and speed $\dot{x}$, respectively, namely

$$(1.4) \qquad dX_t^1 = X_t^2 \, dt$$

$$dX_t^2 = \left\{ X_t^1 \left(\alpha - (X_t^1)^2 \right) - X_t^2 \right\} dt + \sigma X_t^1 \, dW_t$$

where $W = \{W_t, \ t \geq 0\}$ is a 1-dimensional standard Wiener process and $\sigma \geq 0$ controls the intensity of the multiplicative noise.

We begin with the phase plane of the deterministic version of (1.4) with $\sigma \equiv 0$. It has the steady states

$$(1.5) \qquad X^1 = 0, \qquad X^2 = 0 \quad \text{for all} \quad \alpha$$

and

$$(1.6) \qquad X^1 = \pm\sqrt{\alpha}, \qquad X^2 = 0 \quad \text{for} \quad \alpha \geq 0,$$

the first of which is also a degenerate stationary state of the stochastic differential equation (1.4).

Exercise 6.1.4 (PC-Exercise 13.1.1) *Use the Milstein scheme (4.1.27) with equidistant step size $\Delta = 2^{-7}$ to simulate linearly interpolated trajectories of the Duffing-Van der Pol oscillator (1.4) with $\alpha = 1.0$ and $\sigma = 0.0$ over the time interval $[0, T]$ with $T = 8$, starting at $(X_0^1, X_0^2) = (-k\epsilon, 0)$ for $k = 11, 12, \ldots, 20$, where $\epsilon = 0.2$. Plot the results on the (X^1, X^2) phase plane.*

In the deterministic case the typical trajectory starting with nonzero displacement and zero speed is oscillatory and is attracted to one or the other of the nontrivial steady states $(\pm 1, 0)$. The regions of attraction of these two steady states could be determined by appropriately marking each initial value on the phase plane according to the steady solution which attracts the trajectory starting there.

For weak noise we might expect similar behaviour to the deterministic case, but as the noise is multiplicative here stronger noise may lead to substantial changes, particularly over a long period of time.

Exercise 6.1.5 (PC-Exercise 13.1.2) *Repeat Exercise 6.1.4 with $\alpha = 1.0$ and $\sigma = 0.2$ using the same driving sample path of the Wiener process for each trajectory starting at the different initial values.*

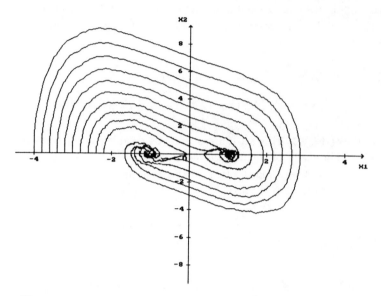

Figure 6.1.3   The Duffing–Van der Pol oscillator with weak noise.

The paths in Figure 6.1.3 are random in appearance and remain near to each other until they come close to the origin $(0,0)$, after which they separate and are attracted into the neighbourhood of either $(-1,0)$ or $(1,0)$.

Finally, we examine the effect of a stronger multiplicative noise in system (1.4) over a long period of time.

Exercise 6.1.6 (PC-Exercise 13.1.3)   *Repeat Exercise 6.1.5 with $\alpha = 1.0$ and $\sigma = 0.5$ using the same sample path of the Wiener process for each of the initial values, but now plotting the displacement component X_t^1 against time t.*

While the noisy trajectories are initially attracted by one or the other of the points $(\pm 1, 0)$, not all of them remain indefinitely in the vicinity of the same point when the noise intensity is high. Instead, after spending a period of time near one of the points, the trajectories may switch over to the other point. This might be interpreted as a tunneling phenomenon.

To convince ourselves of the reliability of the above results, we could repeat the calculations using a smaller step size or some other strong scheme. While the quantitative details may then differ, the qualitative picture should be much the same. With this check we may be able to avoid results that are only an artifice of a particular numerical scheme. It turns out that higher order schemes can, for instance, ensure the preservation over a long period of time of the important diffeomorphism property with paths starting from neighbouring initial values remaining neighbours.

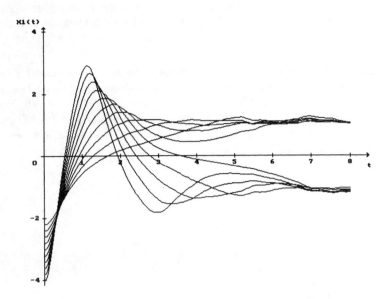

Figure 6.1.4   Displacement X_t^1 versus t for the Duffing-Van der Pol equation.

Exercise 6.1.7   *Repeat Exercise 6.1.4–6.1.5 using the order 1.5 weak Taylor scheme and look for tunneling.*

A valuable improvement in preserving the diffeomorphism property can be often achieved by using implicit schemes.

C. Stochastic Flow on a Circle

The collection of mappings $T_{t,w}$ of an initial value x to the value $X_t^x(w)$ on the sample path at time t, introduced in the previous subsection, is known as a *stochastic flow*. An example of such a flow is illustrated in Figure 6.1.3 for the noisy Duffing-Van der Pol equations (1.4) in the (X^1, X^2) phase plane. In stability and bifurcation analysis stochastic flows on manifolds such as the unit circle S^1 or the unit torus T^2 are of special interest.

The *gradient stochastic flow* on S^1 described by the Stratonovich stochastic differential equation

$$(1.7) \qquad dX_t^{0,x} = \sin\left(X_t^{0,x}\right) \circ dW_t^1 + \cos\left(X_t^{0,x}\right) \circ dW_t^2,$$

where W^1 and W'^2 are two independent standard Wiener processes, is spatially periodic with period 2π. It can thus be interpreted modulo 2π, with initial values $x \in [0, 2\pi)$, which gives the standard embedding of S^1 in $\Re^1$. We shall apply the Milstein scheme (4.1.27) to (1.7) for different initial values in $[0, 2\pi)$, but the same realization of the driving Wiener process (W^1, W^2), taking values

modulo 2π. The multiple Ito integral $I_{(1,2)}$ has to be approximated here. These
simulations enable us to visualize the gradient stochastic flow on the unit circle
S^1 identified here with the interval $[0, 2\pi)$.

Exercise 6.1.8 (PC-Exercise 13.1.4) *Simulate the gradient stochastic flow of*
$N = 10$ *particles on the unit circle* S^1 *over the time inteval* $[0, T]$ *with* $T = 5$ *by*
applying the Milstein scheme (4.1.27) with step size $\Delta = 2^{-7}$ *to the stochastic*
differential equation (1.7), modulo 2π, *for the initial values* $x = 2\pi k/N$ *where*
$k = 1, \ldots, N$. *Plot the linearly interpolated trajectories on the same* x *versus* t
axes for $0 \leq x < 2\pi$ *and* $0 \leq t \leq T$.

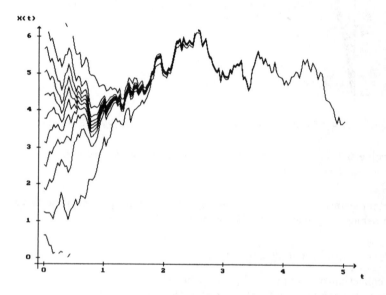

Figure 6.1.5   Stochastic flow on the unit circle.

In Figure 6.1.5 we see that the paths move closer to each other, eventually
clustering and moving together like a single path of a Wiener process. Repeating the calculations with a different realization of the driving Wiener process
produces another cluster, which behaves similarly.

Exercise 6.1.9 (PC-Exercise 13.1.5)   *Repeat Exercise 6.1.8 using a different*
seed for the random number generator.

The dynamical behaviour observed here is a consequence of the flow's having
a negative Lyapunov exponent. Such exponents are evaluated numerically in
Section 6.5.

D. Stochastic Flow on a Torus

The dynamical, and consequent visual, possibilities are much richer for a stochastic flow on a 2-dimensional torus than on a 1-dimensional circle. Recall that a torus is a surface of revolution formed by revolving a circle about a non-intersecting axis in $\Re^3$. The unit torus T^2 can be identified with the rectangle $[0, 2\pi)^2$ in $\Re^2$, with each point characterized by two angular coordinates, modulo 2π.

A parametized stochastic flow on T^2 proposed by Baxendale is based on the 2-dimensional Stratonovich stochastic differential equation

$$(1.8) \qquad dX_t^{0,x} = \sum_{j=1}^{4} b^j \left(X_t^{0,x} \right) \circ dW_t^j$$

with diffusion coefficients

$$(1.9) \qquad b^1 (x) \;=\; b^1 \left(x^1, x^2\right) = \left(\begin{array}{c} \cos \alpha \\ \sin \alpha \end{array} \right) \sin \left(x^1\right),$$

$$b^2 (x) \;=\; b^2 \left(x^1, x^2\right) = \left(\begin{array}{c} \cos \alpha \\ \sin \alpha \end{array} \right) \cos \left(x^1\right),$$

$$b^3 (x) \;=\; b^3 \left(x^1, x^2\right) = \left(\begin{array}{c} -\sin \alpha \\ \cos \alpha \end{array} \right) \sin \left(x^2\right),$$

$$b^4 (x) \;=\; b^4 \left(x^1, x^2\right) = \left(\begin{array}{c} -\sin \alpha \\ \cos \alpha \end{array} \right) \cos \left(x^2\right),$$

where W^1, W^2, W^3 and W^4 are independent, 1-dimensional standard Wiener processes and $\alpha \in [0, \pi/2]$ is a coupling parameter. Note that $\alpha = 0$ corresponds to two uncoupled 1-dimensional stochastic flows on circles. Baxendale's stochastic flow on T^2 is obtained from (1.8) by interpreting the solution $X_t^{0,x}$ modulo 2π in both components. It can thus be plotted on the rectangle $[0, 2\pi)^2$ with the points $(x, 2\pi)$ and $(2\pi, y)$ being identified with $(x, 0)$ and $(0, y)$, respectively, for all x, y in $[0, 2\pi]$. We can visualize it by applying a numerical scheme to (1.8) with the same realization of the driving Wiener process $W = (W^1, W^2, W^3, W^4)$ for a grid of initial values in $[0, 2\pi]^2$ and plotting the grid of calculated values at selected future instants of times. The inclusion of line segments joining grid points evolving from neighbouring initial values is particularly effective visually and clearly shows the increasing geometric distortion induced by the evolving stochastic flow. See Figure 6.1.6 below for the results of the weakly coupled case $\alpha = 0.1$ calculated in the following exercise.

Exercise 6.1.10 (PC-Exercise 13.1.6)   *Use the Milstein scheme (4.1.22) with step size $\Delta = 0.01$ and the same realization of the driving Wiener process W to simulate 225 trajectories of (1.8) with parameter $\alpha = 0.1$ starting at the points of a uniform 15×15 grid in $[0, 2\pi]^2$. Plot the grid of calculated values at time $T = 0.5$, using line segments to join those corresponding to adjacent initial values. Continue the calculations for larger values of T.*

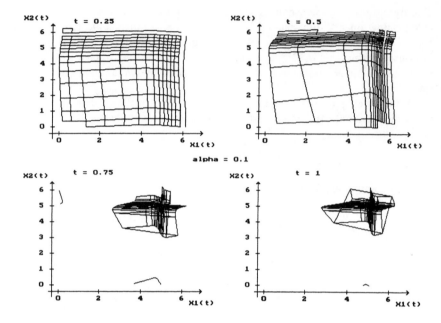

Figure 6.1.6   Stochastic flow on a torus plotted as a moving grid with $\alpha = 0.1$.

Statistical mechanics offers an alternative way of visualizing the evolution of a stochastic flow in terms of the dynamics of an ensemble of particles. While the geometric distortion induced by the flow is no longer so apparent, the clustering and meandering of compact clusters of particles now is. The dynamics here can be displayed in stroboscopic snapshots.

Exercise 6.1.11 (PC-Exercise 13.1.7)   *Repeat Exercise 6.1.10 for parameter $\alpha = 1.0$ for the times $T = 0.5$, 1.0, 2.5 and 3.0. Plot the points at each of these times in $[0, 2\pi]^2$.*

We see from Figure 6.1.7 that the particles eventually cluster into string-like forms which move randomly about the torus. It can be shown that the flow here has a negative Lyapunov exponent, which accounts for the contraction to form the strings, and a positive Lyapunov exponent which accounts for the elongation and the random motion on the strings. If we repeat the calculations for the uncoupled case $\alpha = 0.0$, we would obtain a single randomly moving cluster of particles since both of the Lyapunov exponents are now negative.

Exercise 6.1.12   *Repeat Exercise 6.1.11 with $\alpha = 0.0$.*

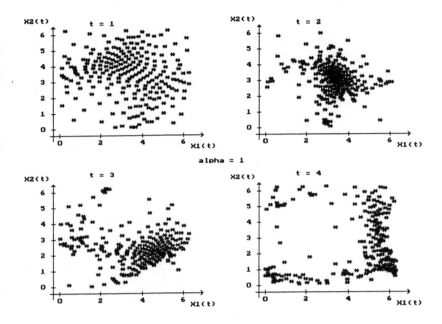

Figure 6.1.7 Stochastic flow on the torus with $\alpha = 1.0$ as an ensemble of particles.

6.2 Testing Parametric Estimators

In this section statistical properties of estimators of drift parameters for diffusion processes will be investigated using strong numerical methods for stochastic differential equations. This is particularly useful when discrete time rather than continuous time samples are used.

A. Parametric Diffusion Processes

The problem of estimating parameters in the drift coefficient when a diffusion process is observed continuously, that is at all time points in an interval, has been studied extensively. A particularly nice theory is available if the drift coefficient depends linearly on the parameters, which is common in practice.

An explicit likelihood function is, however, rarely available in the more realistic situation where a diffusion process has been observed at discrete time instants only. For this type of data one method of obtaining an estimator that is often used in practice is to construct from the data available an approximation to the estimator found in the theory for continuous observations. In general, it is difficult to study the quality of such estimators analytically. The availability of fast computers makes it feasible to use simulation methods to study the behaviour of estimators in diffusion process models.

To illustrate this procedure we shall consider parametric statistical models for d-dimensional diffusion processes defined by a class of stochastic differential equations of the form

$$(2.1) \qquad dX_t = (A_t(X_t) + B_t(X_t)\,\theta)\,dt + D_t(X_t)\,dW_t,$$

assuming that this equation has a unique strong solution for each θ from a given parameter set Θ. In (2.1) A_t, B_t and D_t are functions depending only on X_t and the time t, which for simplicity we assume are continuous in both variables. The statistical parameter θ is k-dimensional and W is a d_1-dimensional Wiener process, so B is a $d \times k$-matrix, D is a $d \times d_1$-matrix and A is a d-dimensional vector. The functions A, B and D are known, while the parameter θ is to be estimated from an observed sample path $\{X_t : t \in [0, T]\}$. Finally, the columns of the matrix $B_t(X_t)$ are assumed to be linearly independent functions of t (for otherwise the model could be parametrized by a parameter of dimension smaller than k).

The statistically important aspect of (2.1) is that the drift depends linearly on the parameter θ, while the diffusion coefficient does not depend on θ at all. For such models an explicit expression for the maximum likelihood estimator exists uniquely and has nice asymptotic properties.

B. Maximum Likelihood Estimation

Suppose we have observed the process X during the time interval $[0, T]$. Let P_θ^T denote the probability measure on the set of continuous funtions from $[0, T]$ into $\Re^d$ corresponding to the solution of (2.1) for the parameter value $\theta \in \Theta$. The likelihood function for our observation $\{X_t : t \in [0, T]\}$ is the Radon-Nikodym derivative

$$(2.2) \qquad L_T(\theta) = \frac{dP_\theta^T}{dP_0^T},$$

provided P_θ^T is dominated by P_0^T for all $\theta \in \Theta$. This is the case if the $d \times d$-matrix

$$(2.3) \qquad C_t(X_t) = D_t(X_t)\,D_t(X_t)^\top$$

is non-singular for almost all $t \in [0, T]$, and

$$(2.4) \qquad P_\theta^T(S_T < \infty) = 1$$

for all $\theta \in \Theta$, where S_T is the $k \times k$-matrix

$$(2.5) \qquad S_T = \int_0^T B_t(X_t)^\top C_t(X_t)^{-1} B_t(X_t)dt$$

which is called the information matrix.

Defining the k-dimensional random vector

$$(2.6) \qquad H_T = \int_0^T B_t(X_t)^\top C_t(X_t)^{-1} d\tilde{X}_t,$$

where

$$(2.7) \qquad \tilde{X}_t = X_t - \int_0^t A_s(X_s)ds,$$

the likelihood function is then given by

$$(2.8) \qquad L_T(\theta) = \exp\left(\theta^\top H_T - \frac{1}{2}\theta^\top S_T \theta\right).$$

The model is an exponential family of processes and is thus well-behaved statistically.

In case (2.4) is not satisfied, we cannot be sure that P_0^T dominates P_θ^T for all $\theta \in \Theta$ and hence that the likelihood function (2.2) exists. Also, when D_t depends on θ the likelihood function may not exist. However, for observations with $S_T < \infty$ we can still interprete (2.8) as a quasi-likelihood function. Estimators obtained by maximizing a quasi-likelihood function have many of the optimality properties enjoyed by the maximum likelihood estimator.

The estimator obtained by maximizing (2.8) is

$$(2.9) \qquad \hat{\theta}_T = S_T^{-1} H_T,$$

which exists because S_T is non-singular under the assumed conditions. Of course, (2.9) is only the maximum likelihood estimator if $\hat{\theta}_T \in \Theta$. The consistency and asymptotic normality of $\hat{\theta}_T$ are discussed in the literature. Note that it was not necessary to assume that X is ergodic in the derivation of the estimator (2.9).

In the following subsection we shall discretize the estimator (2.9) for some typical stochastic differential equations.

C. Ornstein-Uhlenbeck Process

Let us consider the Ornstein-Uhlenbeck process (2.2.2)

$$(2.10) \qquad X_t = X_0 + \alpha \int_0^t X_s\, ds + W_t,$$

supposing that the true value of α is -1 and that the process starts at $X_0 = 0$. From (2.9) the maximum likelihood estimator for α is

$$(2.11) \qquad \hat{\alpha}(T) = \int_0^T X_s\, dX_s \bigg/ \int_0^T (X_s)^2\, ds.$$

In practice the diffusion process is observed at a finite number of times $\tau_0 < \tau_1 < \dots < \tau_{n_T}$ rather than continuously. The exact likelihood function for such data is then the product of transition densities which can rarely be found explicitly. A simple estimation procedure for such data is to use the continuous time maximum likelihood estimator (2.9) with $\tau_0 = 0$ and $\tau_{n_T} = T$ and with suitable approximations to the integrals in H_T and S_T. If the spacing between consecutive observation times are not too large, some of the good properties of (2.9) will probably be preserved, but not without bias.

The Riemann integrals in S_T and H_T can be approximated by Riemann sums or quadrature formulae like the trapezoidal formula. The stochastic integral in H_T can also be approximated by a finite sum, but is better first transformed to a Riemann or Stieltjes integral where possible. For most one-dimensional diffusion process this can be done directly using Ito's formula, but only in special cases when X is multi-dimensional.

The simplest approximation is

$$(2.12) \qquad \hat{\alpha}_1(T) = \sum_{n=0}^{n_T-1} X_{\tau_n} \left(X_{\tau_{n+1}} - X_{\tau_n} \right) \Big/ \sum_{n=0}^{n_T-1} \left(X_{\tau_n} \right)^2 \Delta_n$$

where

$$\Delta_n = \tau_{n+1} - \tau_n.$$

A variation of this approximate estimator results if we write the integral in the numerator of (2.11) using the Ito formula (2.1.19) as

$$\int_{\tau_n}^{\tau_{n+1}} X_s \, dX_s = \frac{1}{2} \left\{ \left(X_{\tau_{n+1}} \right)^2 - \left(X_{\tau_n} \right)^2 - \Delta_n \right\}$$

and approximate the integral in the denominator of (2.11) by

$$\int_{\tau_n}^{\tau_{n+1}} \left(X_s \right)^2 \, ds \approx \frac{1}{2} \left\{ \left(X_{\tau_{n+1}} \right)^2 + \left(X_{\tau_n} \right)^2 \right\} \Delta_n.$$

Inserting these into (2.11), we obtain a second approximate estimator

$$(2.13) \quad \hat{\alpha}_2(T) = \left(\left(X_T \right)^2 - \left(X_0 \right)^2 - T \right) \Big/ \sum_{n=0}^{n_T-1} \left\{ \left(X_{\tau_{n+1}} \right)^2 + \left(X_{\tau_n} \right)^2 \right\} \Delta_n.$$

In the following exercise a simulated trajectory of the Ornstein-Uhlenbeck process (2.10) will be used to compare the two estimators $\hat{\alpha}_1(T)$ and $\hat{\alpha}_2(T)$ for sufficiently large times T.

Exercise 6.2.1 (PC-Exercise 13.2.1) *Simulate a single trajectory of the process (2.10) with $\alpha = -1$ and $X_0 = 0$ using the order 2.0 strong Taylor scheme (4.1.33) with equidistant step size $\Delta_n = \Delta = 2^{-3}$ and evaluate the estimators $\hat{\alpha}_1(T)$ and $\hat{\alpha}_2(T)$ with the same step size Δ. Plot linearly interpolated values for the same step sizes of the estimators against T for $T \in [0, 6400]$.*

We should observe that both estimators converge asymptotically to a value close to $\alpha = -1$, with $\hat{\alpha}_2(T)$ better than $\hat{\alpha}_1(T)$ which usually should show a larger bias.

D. A Two-Parameter Linear Equation

Quite often more than one parameter has to be estimated in a model. To see the difference in the maximum likelihood estimators and their discretized versions, we consider the slightly more general stochastic differential equation

$$(2.14) \qquad dX_t = (\theta_1 X_t + \theta_2)dt + c\,dW_t,$$

where θ_1 and θ_2 are unknown parameters and c is known. For $\theta_1 < 0$ the process X is stable with asymptotic mean level $-\theta_2/\theta_1$, and for $\theta_1 > 0$ the process is non ergodic and grows exponentially fast.

For the model (2.14) we find the maximum likelihood estimators

$$(2.15) \qquad \hat{\theta}_{1,T} = \left\{ T \int_0^T X_t\,dX_t - (X_T - X_0)\int_0^T X_t\,dt \right\}/N_T$$

$$(2.16) \quad \hat{\theta}_{2,T} = \left\{ (X_T - X_0)\int_0^T (X_t)^2 dt - \int_0^T X_t dX_t \int_0^T X_t\,dt \right\}/N_T$$

where

$$(2.17) \qquad N_T = T \int_0^T (X_t)^2 dt - \left(\int_0^T X_t\,dt \right)^2.$$

Comparing the estimators (2.11) and (2.15) shows that the two-parameter case is not just a direct generalization of the one-parameter situation.

In fact, we have

$$(2.18) \quad \hat{\theta}_{1,T} = \left\{ \frac{1}{2} T \left((X_T)^2 - (X_0)^2 - c^2 T\right) - (X_T - X_0)\int_0^T X_t\,dt \right\}/N_T$$

and

$$\hat{\theta}_{2,T} = \left\{ (X_T - X_0)\int_0^T (X_t)^2 dt - \frac{1}{2}\left((X_T)^2 - (X_0)^2 - c^2 T\right)\int_0^T X_t\,dt \right\}/N_T,$$

(2.19)
with N_T given by (2.17), since

$$(2.20) \qquad \int_0^T X_t\,dX_t = \frac{1}{2}\left((X_T)^2 - (X_0)^2 - c^2 T\right)$$

for equation (2.14).

We then approximate the remaining integrals in (2.18) and (2.19) by the trapezoidal formula and apply the approximate estimators

$$\hat{\theta}_{1,T}^\Delta = \frac{1}{2N_T^\Delta}\left\{ T\left((X_T)^2 - (X_0)^2 - c^2 T\right) - (X_T - X_0)\sum_{n=0}^{n_T-1}(X_{t_{n+1}} + X_{t_n})\Delta \right\}$$

(2.21)

and

$$(2.22) \quad \hat{\theta}_{2,T}^{\Delta} = \frac{1}{2N_T^{\Delta}} \left\{ (X_T - X_0) \sum_{n=0}^{n_T-1} \left((X_{t_{n+1}})^2 + (X_{t_n})^2 \right) \Delta \right.$$

$$\left. - \frac{1}{2} \left((X_T)^2 - (X_0)^2 - c^2 T \right) \sum_{n=0}^{n_T-1} (X_{t_{n+1}} + X_{t_n}) \Delta \right\},$$

with

$$(2.23) \quad N_T^{\Delta} = \frac{T}{2} \sum_{n=0}^{n_T-1} \left((X_{t_{n+1}})^2 + (X_{t_n})^2 \right) \Delta - \left(\frac{1}{2} \sum_{n=0}^{n_T-1} (X_{t_{n+1}} + X_{t_n}) \Delta \right)^2,$$

to estimate the parameters θ_1 and θ_2.

Exercise 6.2.2   *For the equation (2.14) simulate in $[0,T]$ with $T = 2000$ the corresponding process X with the parameters $\theta_1 = -1.0$, $\theta_2 = 1.0$ and $c = 1.0$ starting at $X_0 = 0$ using the order 2.0 strong Taylor scheme (4.1.33) with step size $\Delta_n = \Delta = 2$. Plot linearly interpolated values for the same step sizes of the estimators against $t \in [0,T]$.*

E. A Population Growth Model

There are many models of population growth in the literature. Those proposed by Shiga are similar to the Ito equation

$$(2.24) \qquad X_t = X_0 + \beta \int_0^t X_s (1 - X_s)\, ds + \int_0^t \sqrt{X_s}\, dW_s$$

where β is the population growth rate which must often be estimated from experimental data. We shall consider this equation with the initial value $X_0 = 1.0$.

From (2.9) the estimator for β has the form

$$\hat{\beta}(T) = \int_0^T (1 - X_s)\, dX_s \Big/ \int_0^T (1 - X_s)^2 X_s\, ds.$$

By the Ito formula (2.1.19) we can write

$$\int_{\tau_n}^{\tau_{n+1}} X_s\, dX_s = \frac{1}{2} \left\{ (X_{\tau_{n+1}})^2 - (X_{\tau_n})^2 - \int_{\tau_n}^{\tau_{n+1}} X_s\, ds \right\}$$

and thus with analogous approximations to those used above obtain the approximate estimator

$$\hat{\beta}_1(T) = \left\{ X_T - X_0 - \frac{1}{2} \left\{ (X_T)^2 - (X_0)^2 \right\} + \frac{1}{4} \sum_{n=0}^{n_T-1} (X_{\tau_n} + X_{\tau_{n+1}}) \Delta_n \right\}$$

$$(2.25) \qquad \times \left\{ \frac{1}{2} \sum_{n=0}^{n_T-1} \left[(1 - X_{\tau_{n+1}})^2 X_{\tau_{n+1}} + (1 - X_{\tau_n})^2 X_{\tau_n} \right] \Delta_n \right\}^{-1}.$$

We test this estimator for the true parameter value β which we assume to be known in the following exercise.

Exercise 6.2.3 (PC-Exercise 13.2.2)   *Use the explicit order 1.0 strong scheme (4.2.3) with equidistant step size $\Delta = \Delta_n = 2^{-3}$ to simulate an approximate trajectory of the Ito process (2.24) with $\beta = 2.0$ and initial value $X_0 = 0.5$. Evaluate the approximate estimator $\hat{\beta}_1(T)$ in (2.25) and plot linearly interpolated values of $\hat{\beta}_1(T)$ against T for $T \in [0, 6400]$.*

F. A Two-Dimensional Linear Equation

To illustrate the discretization of maximum likelihood estimators for multi-dimensional processes we study the two-dimensional diffusion process $X = (X^{(1)}, X^{(2)})^T$, generalizing (2.14) by setting in (2.1) $A_t = 0$, $D_t(X_t) = \sigma I$ and

$$B_t(X_t) = \begin{bmatrix} X_t^{(1)} & X_t^{(2)} \\ X_t^{(2)} & X_t^{(1)} \end{bmatrix},$$

that is, we have the system of Ito differential equations

$$(2.26) \qquad dX_t^{(1)} = \left(X_t^{(1)} \theta_1 + X_t^{(2)} \theta_2 \right) dt + \sigma dW_t^1$$

$$dX_t^{(2)} = \left(X_t^{(2)} \theta_1 + X_t^{(1)} \theta_2 \right) dt + \sigma dW_t^2.$$

Thus $\theta = (\theta_1, \theta_2)^T$ is two-dimensional and the process is driven by a two-dimensional Wiener process.

For the two-dimensional linear equation (2.26) we have the estimators

$$(2.27) \qquad \hat{\theta}_{1,T} = \left\{ H_{1,T} \int_0^t |X_t|^2 dt - 2H_{2,T} \int_0^T X_t^{(1)} X_t^{(2)} dt \right\} / N_T$$

and

$$(2.28) \qquad \hat{\theta}_{2,T} = \left\{ -2H_{1,T} \int_0^t X_t^{(1)} X_t^{(2)} dt + H_{2,T} \int_0^T |X_t|^2 dt \right\} / N_T$$

where

$$H_{1,T} = \int_0^T X_t^{(1)} dX_t^{(1)} + \int_0^T X_t^{(2)} dX_t^{(2)},$$

$$H_{2,T} = \int_0^T X_t^{(2)} dX_t^{(1)} + \int_0^T X_t^{(1)} dX_t^{(2)},$$

$$N_T = \left\{ \int_0^T |X_t|^2 dt \right\}^2 - 4 \left\{ \int_0^T X_t^{(1)} X_t^{(2)} dt \right\}^2$$

and

$$|X_t|^2 = \left(X_t^{(1)} \right)^2 + \left(X_t^{(2)} \right)^2 .$$

We can rewrite $H_{1,T}$ and $H_{2,T}$ here as

(2.29) $$H_{1,T} = \frac{1}{2} \left(|X_T|^2 - |X_0|^2 \right) - \sigma^2 T$$

and
(2.30) $$H_{2,T} = X_T^{(1)} X_T^{(2)} - X_0^{(1)} X_0^{(2)} .$$

To estimate the parameters θ_1 and θ_2 from discrete data we first approximate the Riemann integrals in (2.27) and (2.28) and apply the approximate estimators

$$\hat{\theta}_{1,T}^\Delta = \frac{1}{N_T^\Delta} \left[\frac{1}{2} H_{1,T} \sum_{n=0}^{n_T-1} \left\{ \left(X_{t_{n+1}}^{(1)} \right)^2 + \left(X_{t_{n+1}}^{(2)} \right)^2 + \left(X_{t_n}^{(1)} \right)^2 + \left(X_{t_n}^{(2)} \right)^2 \right\} \Delta \right.$$

(2.31) $$\left. - H_{2,T} \sum_{n=0}^{n_T-1} \left\{ X_{t_{n+1}}^{(1)} X_{t_{n+1}}^{(2)} + X_{t_n}^{(1)} X_{t_n}^{(2)} \right\} \Delta \right]$$

and

$$\hat{\theta}_{2,T}^\Delta = \frac{1}{N_T^\Delta} \left[-H_{1,T} \sum_{n=0}^{n_T-1} \left\{ X_{t_{n+1}}^{(1)} X_{t_{n+1}}^{(2)} + X_{t_n}^{(1)} X_{t_n}^{(2)} \right\} \Delta \right.$$

(2.32) $$\left. + \frac{1}{2} H_{2,T} \sum_{n=0}^{n_T-1} \left\{ \left(X_{t_{n+1}}^{(1)} \right)^2 + \left(X_{t_{n+1}}^{(2)} \right)^2 + \left(X_{t_n}^{(1)} \right)^2 + \left(X_{t_n}^{(2)} \right)^2 \right\} \Delta \right]$$

where

$$N_T^\Delta = \frac{1}{4} \left[\sum_{n=0}^{n_T-1} \left\{ \left(X_{t_{n+1}}^{(1)} \right)^2 + \left(X_{t_{n+1}}^{(2)} \right)^2 + \left(X_{t_n}^{(1)} \right)^2 + \left(X_{t_n}^{(2)} \right)^2 \right\} \Delta \right]^2$$

$$- \left[\sum_{n=0}^{n_T-1} \left\{ X_{t_{n+1}}^{(1)} X_{t_{n+1}}^{(2)} + X_{t_n}^{(1)} X_{t_n}^{(2)} \right\} \Delta \right]^2$$

Exercise 6.2.4   *For the two-dimensional linear equation (2.26) simulate in $[0,T]$ with $T = 10$ the corresponding approximation with the parameters $\theta_1 = -1.0$, $\theta_2 = -2.0$ and $\sigma = 0.1$ starting at $\left(X_0^{(1)}, X_0^{(2)} \right) = (1.0, 0.5)$ by the use of the Euler scheme with time step size $\Delta = \Delta_n = 10^{-3}$. Plot the simulated trajectory in the phase plane and both linearly interpolated parameter estimates against time.*

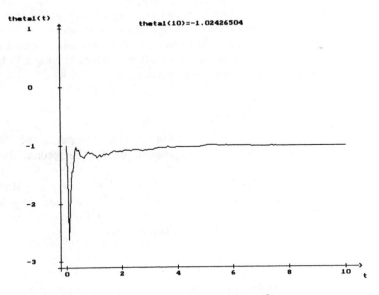

Figure 6.2.1 The estimator $\hat{\theta}_{1,T}^{\Delta}$.

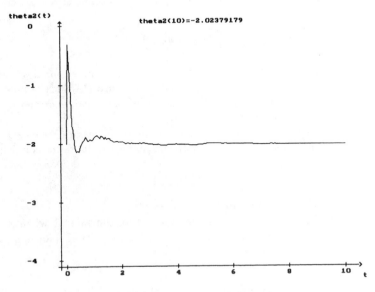

Figure 6.2.2 The estimator $\hat{\theta}_{2,T}^{\Delta}$.

6.3   Filtering

The aim of this section is to construct higher order approximate discrete time filters for continuous time finite-state Markov chains with observations that are perturbed by the noise of a Wiener process. Such higher order explicit and implicit strong order schemes, which involve multiple stochastic integrals of the observation process, provide efficient approximate filters.

A. Markov Chain Filtering

The systematic investigation of filters for Markov chains began in the 1960s, while the construction of discrete time approximations for the optimal filter is more recent.

To begin we first consider filters for continuous time finite state Markov chains. Suppose that the state process $\xi = \{\xi_t, \, t \in [0,T]\}$ is a continuous time, homogeneous Markov chain on the finite state space $\mathcal{S} = \{a_1, a_2, ..., a_d\}$. Its d-dimensional probability vector $p(t)$, with components

$$(3.1) \qquad\qquad p_i(t) = P\left(\xi_t = a_i\right)$$

for each $a_i \in \mathcal{S}$, then satisfies the vector ordinary differential equation

$$(3.2) \qquad\qquad \frac{dp}{dt} = A\,p$$

where A is the intensity matrix. In addition, suppose that the m-dimensional observation process $W = \{W_t, \, t \in [0,T]\}$ is the solution of the stochastic equation

$$(3.3) \qquad\qquad W_t = \int_0^t h\left(\xi_s\right) ds + W_t^*.$$

where $W^* = \{W_t^*, \, t \in [0,T]\}$ with $W_0^* = 0$ is an m-dimensional standard Wiener process with respect to the probability measure P, which is independent of the process ξ. Finally, let $\mathcal{Y}_t$ denote the σ-algebra generated by the observations W_s for $0 \leq s \leq t$.

Our task is to filter as much information about the state process ξ as we can from the observation process W. With this aim we evaluate the conditional expectation

$$(3.4) \qquad\qquad E\left(g\left(\xi_T\right) \mid \mathcal{Y}_T\right)$$

with respect to P for a given function $g : \mathcal{S} \to \Re$.

Let $I_{\{a\}}(x)$ denote the indicator function taking the value 1 when $x = a$ and the value 0 otherwise. It follows from the Kallianpur-Striebel formula, that the conditional probabilities of ξ_t given $\mathcal{Y}_t$ are

$$(3.5) \qquad P\left(\xi_t = a_i \mid \mathcal{Y}_t\right) = E\left(I_{\{a_i\}}\left(\xi_t\right) \mid \mathcal{Y}_t\right) = X_t^i \left/ \sum_{k=1}^d X_t^k \right.$$

for $a_i \in S$ and $t \in [0,T]$, where the d-dimensional process $X_t = \left(X_t^1, \ldots, X_t^d\right)$ of the un-normalized conditional probabilities satifies the Zakai equation,

$$(3.6) \qquad X_t = p(0) + \int_0^t A\, X_s\, ds + \sum_{j=1}^m \int_0^t H_j\, X_s\, dW_s^j$$

for $t \in [0,T]$, which is a homogeneous linear Ito equation. H_j is the $d \times d$ diagonal matrix with iith component $h_j(a_i)$ for $i = 1, \ldots, d$ and $j = 1, \ldots, m$.

The optimal least squares estimate for $g(\xi_t)$ with respect to the observations W_s for $0 \le s \le t$, that is with respect to the σ-algebra $\mathcal{Y}_t$, is given by the conditional expectation

$$(3.7) \qquad \Pi_t(g) = E\left(g\left(\xi_t\right) \mid \mathcal{Y}_t\right) = \sum_{i=1}^d g\left(a_i\right) X_t^i \Big/ \sum_{k=1}^d X_t^k,$$

which we call the optimal filter or the Markov chain filter.

B. Approximate Filters

To compute the optimal filter (3.7) we have to solve the Ito equation (3.6). In practice, however, it is impossible to detect W completely on $[0,T]$. Electronic devices are often used to obtain increments of integral observations over small time intervals, which in the simplest case are the increments of W in integral form

$$\int_{t_0}^{\tau_1} dW_s^j, \ldots, \int_{t_n}^{\tau_{n+1}} dW_s^j, \ldots,$$

for each $j = 1, \ldots, m$, $\tau_n = n\delta$ for $n = 0, 1, 2, \ldots$. We shall see in the next subsection that with such integral observations it is possible to construct strong discrete time approximations Y^δ with time step δ of the solution X of the Zakai equation (3.6). For the given function g we can then evaluate the expression

$$(3.8) \qquad \Pi_t^\delta(g) = \sum_{k=1}^d g\left(a_k\right) Y_t^{\delta,k} \Big/ \sum_{k=1}^d Y_t^{\delta,k}$$

for $t \in [0,T]$, which we define to be the corresponding approximate Markov chain filter.

We shall say that an approximate Markov chain filter $\Pi^\delta(g)$ with step size δ converges on the time interval $[0,T]$ with order $\gamma > 0$ to the optimal filter $\Pi(g)$ for a given function g if there exists a finite constant K, not depending on δ, and a $\delta_0 \in (0,1)$ such that

$$(3.9) \qquad E\left(\left|\Pi_{\tau_n}(g) - \Pi_{\tau_n}^\delta(g)\right|\right) \le K\,\delta^\gamma$$

for all $\delta \in (0,\delta_0)$ and $\tau_n \in [0,T]$.

It has been shown that an approximate Markov chain filter $\Pi^\delta(g)$ with step size δ converges on the time interval $[0,T]$ with order $\gamma > 0$ to the optimal filter $\Pi(g)$ for a given bounded function g if the discrete time approximation Y^δ used in it converges on $[0,T]$ to the solution X of the Zakai equation (3.6) with the same strong order γ.

C. Explicit Filters

Here we describe discrete time approximations converging with a given strong order $\gamma > 0$ to the solution of the Zakai equation (3.6) which can be used in a corresponding approximate filter. Given an equidistant time discretization of the interval $[0, T]$ with step size $\delta = \Delta_n \equiv \Delta = T/N$ for some $N = 1, 2, \ldots$, we define the partition σ-algebra $\mathcal{P}_N^1$ as the σ-algebra of events generated by the increments

$$(3.10) \qquad \Delta W_0^j = \int_0^\Delta dW_s^j, \quad \ldots, \quad \Delta W_{N-1}^j = \int_{(N-1)\Delta}^{N\Delta} dW_s^j$$

for all $j = 1, \ldots, m$. Thus $\mathcal{P}_N^1$ contains the information about the increments of W for this time discretization.

The simplest discrete time approximation for the Zakai equation (3.6) is based on the Euler scheme (4.1.13) and has the form

$$(3.11) \qquad Y_{\tau_{n+1}}^\delta = \left[I + A\,\Delta + G_n \right] Y_{\tau_n}^\delta$$

with

$$(3.12) \qquad G_n = \sum_{j=1}^m H_j\,\Delta W_n^j$$

and initial value $Y_0 = X_0$, where I is the $d \times d$ unit matrix. The scheme (3.11)–(3.12) converges under the given assumptions with strong order $\gamma = 0.5$. However, the special multiplicative noise structure of the Zakai equation allows the strong order $\gamma = 1.0$ to be attained with the information contained in $\mathcal{P}_N^1$. The Milstein scheme (4.1.22) is of strong order $\gamma = 1.0$, which for equation (3.6) has the form

$$(3.13) \qquad Y_{\tau_{n+1}}^\delta = \left[I + \underline{A}\,\Delta + G_n \left(I + \frac{1}{2} G_n \right) \right] Y_{\tau_n}^\delta$$

where

$$(3.14) \qquad \underline{A} = A - \frac{1}{2} \sum_{j=1}^m H_j^2.$$

Newton found that the scheme

$$Y_{\tau_{n+1}}^\delta = \left[I + \underline{A}\,\Delta + G_n + \frac{\Delta^2}{2} A^2 + \frac{\Delta}{2}\,A\,G_n - \frac{\Delta}{2}\,G_n\,A + G_n\underline{A}\,\Delta + \frac{1}{2}\,G_n^2 + \frac{1}{6} G_n^3 \right] Y_{\tau_n}^\delta,$$

$$(3.15)$$

which is said to be asymptotically efficient under $\mathcal{P}_N^1$, is the "best" in the class of strong order 1.0 schemes in the sense that it has asymptotically the smallest leading error coefficient.

The two following Runge-Kutta type schemes due to Newton are also asymptotically efficient under $\mathcal{P}_N^1$. The first scheme for an Ito equation with drift a

and diffusion coefficient b has the form

$$(3.16) \qquad Y_{n+1} \;=\; Y_n + \frac{1}{2}\left(a_0 + a_1\right)\Delta_n$$

$$+ \frac{1}{40}\left(37b_0 + 30b_2 - 27b_3\right)\Delta W_n$$

$$+ \frac{1}{16}\left(8b_0 + b_1 - 9b_2\right)\left(3\Delta_n\right)^{1/2}$$

where

$$a_0 = a\left(Y_n\right), \qquad b_0 = b\left(Y_n\right),$$

$$a_1 = a\left(Y_n + a_0\,\Delta_n + b_0\,\Delta W_n\right), \qquad b_1 = b\left(Y_n - \frac{2}{3}b_0\left\{\Delta W_n + \left(3\Delta_n\right)^{1/2}\right\}\right),$$

$$b_2 = b\left(Y_n + \frac{2}{9}b_0\left\{3\Delta W_n + \left(3\Delta_n\right)^{1/2}\right\}\right),$$

and

$$b_3 = b\left(Y_n - \frac{20}{27}a_0\,\Delta_n + \frac{10}{27}\left(b_1 - b_0\right)\Delta W_n - \frac{10}{27}b_1\left(3\Delta_n\right)^{1/2}\right).$$

The second scheme for a Stratonovich equation with corrected drift $\underline{a}$ is written

$$(3.17) \qquad Y_{n+1} = Y_n + \frac{1}{2}\left(\underline{a}_0 + \underline{a}_1\right)\Delta_n + \frac{1}{6}\left(b_0 + 2b_1 + 2b_2 + b_3\right)\Delta W_n$$

where

$$\underline{a}_0 = \underline{a}\left(Y_n\right), \qquad b_0 = b\left(Y_n\right), \qquad b_1 = b\left(Y_n + \frac{1}{2}b_0\,\Delta W_n\right),$$

$$\underline{a}_1 = \underline{a}\left(Y_n + \frac{1}{2}\underline{a}\left(3\Delta_n - \left(\Delta W_n\right)^2\right) + b_2\,\Delta W_n\right),$$

$$b_2 = b\left(Y_n + \frac{1}{4}\underline{a}\left(3\Delta_n + \left(\Delta W_n\right)^2\right) + \frac{1}{2}b_1\,\Delta W_n\right)$$

and

$$b_3 = b\left(Y_n + \frac{1}{2}\underline{a}\left(3\Delta_n - \left(\Delta W_n\right)^2\right) + b_2\,\Delta W_n\right).$$

We can obtain higher order convergence by exploiting additional information about the observation process such as contained in the integral observations

$$(3.18) \qquad \Delta Z_0^j = \int_0^\Delta \int_0^s dW_r^j\,ds, \ldots, \qquad \Delta Z_{N-1}^j = \int_{(N-1)\Delta}^{N\Delta}\int_{(N-1)\Delta}^s dW_r^j\,ds$$

for $j = 1, \ldots, m$. We define the partition σ-algebra $\mathcal{P}_N^{1.5}$ as the σ-algebra generated by $\mathcal{P}_N^1$ together with the multiple integrals $\Delta Z_0^j, \ldots, \Delta Z_{N-1}^j$ for $j =$

1, ..., m. The order 1.5 strong Taylor scheme (4.1.31) for the Zakai equation (3.6) uses only the information contained in $\mathcal{P}_N^{1.5}$, for which it takes the form

$$Y_{\tau_{n+1}}^\delta = \left[I + \underline{A}\,\Delta + G_n + \frac{\Delta^2}{2}A^2 + A\,M_n - M_n\,A + G_n\,\underline{A}\,\Delta + \frac{1}{2}\,G_n^2 + \frac{1}{6}G_n^3\right]Y_{\tau_n}^\delta$$

(3.19)

where

(3.20)
$$M_n = \sum_{j=1}^m H_j\,\Delta Z_n^j.$$

In order to form a scheme of order $\gamma = 2.0$ we need the information from the observation process expressed in the partition σ-algebra $\mathcal{P}_N^2$ which is generated by $\mathcal{P}_N^{1.5}$ together with the multiple Stratonovich integrals

$$J_{(j_1,j_2,0),n} = \int_{\tau_n}^{\tau_{n+1}}\int_{\tau_n}^{s_3}\int_{\tau_n}^{s_2} odW_{s_1}^{j_1}\circ dW_{s_2}^{j_2}\,ds_3,$$

(3.21)
$$J_{(j_1,0,j_2),n} = \int_{\tau_n}^{\tau_{n+1}}\int_{\tau_n}^{s_3}\int_{\tau_n}^{s_2} odW_{s_1}^{j_1}\,ds_2\circ dW_{s_3}^{j_2},$$

for $n = 0, 1, ..., N-1$ and $j_1, j_2 = 1, ..., m$. Using this information we can apply a slight generalization of the order 2.0 strong Taylor scheme (4.1.33) to the Zakai equation to obtain the approximation

$$(3.22)\quad Y_{\tau_{n+1}}^\delta = \left[I + \underline{A}\,\Delta\left(I + \frac{1}{2}\underline{A}\,\Delta\right) - M_n\underline{A} + \underline{A}M_n\right.$$

$$+ G_n\left(I + \underline{A}\,\Delta + \frac{1}{2}G_n\left(I + \frac{1}{3}G_n\left(I + \frac{1}{4}G_n\right)\right)\right)$$

$$+ \sum_{j_1,j_2=1}^m \left(\underline{A}H_{j_2}H_{j_1}\,J_{(j_1,j_2,0),n} + H_{j_2}\underline{A}H_{j_1}\,J_{(j_1,0,j_2),n}\right.$$

$$\left.+ H_{j_2}H_{j_1}\underline{A}\left(\Delta\,J_{(j_1,j_2)} - J_{(j_1,j_2,0)} - J_{(j_1,0,j_2)}\right)\right)\bigg]Y_{\tau_n}^\delta.$$

D. Implicit Filters

As we have seen in Section 4.3, explicit discrete time approximations are often unstable numerically for stiff stochastic differential equations. Control is lost over the propagation of errors and the approximation is rendered useless. An implicit scheme can nevertheless, still provide a numerically stable approximation. Here we state some of the implicit schemes from Section 4.3 for the Zakai equation. Since the Zakai equation is linear they can all be rearranged algebraically into explicit expressions.

After rearranging the family of implicit Euler schemes (4.3.1) yields

$$(3.23)\quad Y_{\tau_{n+1}}^\delta = (I - \alpha A\,\Delta)^{-1}\left[I + (1-\alpha)\,A\,\Delta + G_n\right]Y_{\tau_n}^\delta$$

where the parameter $\alpha \in [0,1]$ denotes the degree of implicitness and G_n was defined in (3.12). The scheme (3.23) converges with order $\gamma = 0.5$. The family of implicit Milstein schemes (4.3.6), which for all $\alpha \in [0,1]$ converge with order $\gamma = 1.0$, gives us

$$(3.24) \quad Y^\delta_{\tau_{n+1}} = (I - \alpha \underline{A} \Delta)^{-1} \left[I + (1-\alpha)\underline{A}\Delta + G_n \left(I + \frac{1}{2}G_n \right) \right] Y^\delta_{\tau_n}.$$

In principle, to each explicit scheme there corresponds a family of implicit schemes by making implicit the terms involving the nonrandom multiple stochastic integrals such as Δ or $\frac{1}{2}\Delta^2$. As a final example we mention the order 1.5 implicit Taylor scheme (4.3.10), which for the Zakai equation (3.6) can be rewritten as

$$(3.25) \quad Y^\delta_{\tau_{n+1}} = \left(I - \frac{1}{2}\underline{A}\Delta \right)^{-1} \left[I + \frac{1}{2}\underline{A}\Delta + G_n \underline{A}\Delta - M_n \underline{A} + \underline{A}M_n \right.$$

$$\left. + G_n \left(I + \frac{1}{2}G_n \left(I + \frac{1}{3}G_n \right) \right) \right] Y^\delta_{\tau_n}.$$

E. A Numerical Example

The random telegraphic noise process is a two state continuous time Markov chain ξ on the state space $\mathcal{S} = \{-1, +1\}$ with intensity matrix

$$A = \begin{bmatrix} -0.5 & 0.5 \\ 0.5 & -0.5 \end{bmatrix}.$$

We consider the initial probability vector $p(0) = (0.9, 0.1)$ and suppose that

$$h(a_k) = \begin{cases} 5 & : \quad a_k = +1 \\ 0 & : \quad a_k = -1 \end{cases}$$

in the observation equation (3.3)

On the basis of the observations of W, we want to determine the actual state of the chain, that is to compute the conditional probability vector

$$\hat{p}(t) = \left(P\left(\xi_t = -1 \,|\, \mathcal{Y}_t \right), \; P\left(\xi_t = +1 \,|\, \mathcal{Y}_t \right) \right).$$

We shall say that ξ_t has most likely the value $+1$ if $P\left(\xi_t = +1 \,|\, \mathcal{Y}_t \right) \geq 0.5$, and the value -1 otherwise. For this we need to evaluate the conditional probability

$$P\left(\xi_t = +1 \,|\, \mathcal{Y}_t \right) = E\left(I_{\{+1\}}\left(\xi_t \right) \,|\, \mathcal{Y}_t \right) = \Pi_t \left(I_{\{+1\}} \right),$$

which is the optimal filter here, using a filter $\Pi^\delta_t \left(I_{\{+1\}} \right)$ based on a discrete time strong approximation to obtain a good approximation of $\Pi_t \left(I_{\{+1\}} \right)$.

Exercise 6.3.1 (PC-Exercise 13.3.2)   *Suppose that we have a realization of the above Markov chain on the interval* [0, 4] *with*

$$\xi_t = \begin{cases} +1 & : \quad 0 \le t < \frac{1}{2} \\ -1 & : \quad \frac{1}{2} \le t \le 4. \end{cases}$$

Compute the approximate filters $\Pi_t^\delta \left(I_{\{+1\}} \right)$ *for the same realization of the Wiener process* W^* *using the Euler, Milstein, order 1.5 strong Taylor and order 1.5 asymptotically efficient schemes with equidistant step size* $\delta = \Delta_n = 2^{-7}$. *Plot and compare the calculated sample paths.*

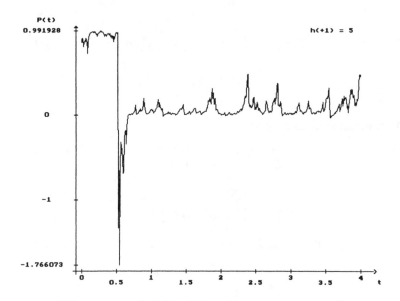

Figure 6.3.1   Results of the Euler scheme in Exercise 6.3.1.

We see from Figure 6.3.1 that the Euler scheme succeeds in detecting the jump in the hidden Markov chain at time $t = 0.5$, but also computes meaningless negative "probabilities". The printout for the order 1.5 strong Taylor scheme would show better numerical stability.

Exercise 6.3.2 (PC-Exercise 13.3.3)   *Simulate the observation process of the above example using a realization of the telegraphic noise process as in Exercise 6.3.1. Apply each of the four approximate Markov chain filters in Exercise 6.3.1 to detect the state of the Markov chain at time* $T = 1$ *and estimate the frequency of failure of the filter. Take equidistant step sizes* $\delta = \Delta_n = 2^{-3}$, 2^{-4} *and* 2^{-5} *and generate* $M = 20$ *batches with* $N = 100$ *simulations. Plot the 90%-confidence intervals against* δ.

Exercise 6.3.3 (PC-Exercise 13.4.2)   *Repeat Exercise 6.3.2 using the order 1.0 asymptotically efficient schemes (3.15), (3.16) and (3.17).*

F. A Markov Chain with High Jump Intensities

For Markov chains with high jump intensities the resulting equations can be interpreted as stiff stochastic differential equations, so usually an implicit scheme must be used to ensure that the numerical results are reliable.

To illustrate this we change the intensity matrix in our example to

$$
A = \left[\begin{array}{cc} -50.0 & 50.0 \\ 50.0 & -50.0 \end{array} \right],
$$

so the state process now typically jumps very frequently between $+1$ and -1.

Exercise 6.3.4 (PC-Exercise 13.3.5)   *Repeat Exercise 6.3.1 for the new intensity matrix using the explicit and implicit Euler and Milstein schemes with implicitness parameter $\alpha = 1.0$ in the implicit versions.*

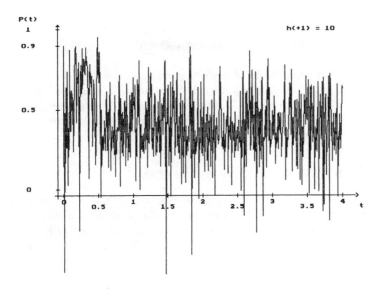

Figure 6.3.2   Results of the explicit Milstein scheme in Exercise 6.3.4.

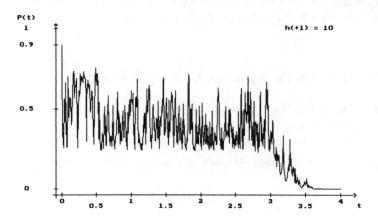

Figure 6.3.3   Results of the implicit Milstein scheme in Exercise 6.3.4.

It is apparent from Figures 6.3.2 and 6.3.3 that the implicit filter detects the state of the system better than its explicit counterpart in this example.

Exercise 6.3.5 (PC-Exercise 13.3.6)   *Repeat Exercise 6.3.2 for the new intensity matrix to compare the explicit approximate filters with the implicit Markov chain filters based on the explicit and implicit Euler and Milstein schemes, respectively.*

6.4   Functional Integrals and Invariant Measures

Two important applications of weak approximations will be considered in this section, the approximate computation of functional integrals and the determination of invariant measures of diffusion processes. The former often arise in physics, while the latter are used, for instance, to compute Lyapunov exponents in stochastic dynamical systems.

A. Wiener Function Space Integrals

Integrals on a Wiener function space play an important role in mathematical physics, where they are often called functional integrals. They occur, for example, in representations of solutions of the Schrödinger equation and in the analysis of wave scattering in random media. Typically one is interested in computing functionals of the form

$$(4.1) \qquad u(t,x) = E\left(f(T, W_T) \exp\left(\int_t^T V(W_s)\, ds \right) \; \middle| \; W_t = x \right),$$

where $W = \{W_s. \; s \geq t\}$ is an m-dimensional Wiener process starting at a point $x = (x^1, \ldots, x^m) \in \Re^m$ at time t and the expectation here is with respect to the probability measure of the Wiener process W. The functions V and f are given.

In view of the Feynman-Kac formula, evaluating $u(t, x)$ in (4.1) is the same as solving the parabolic equation

$$(4.2) \qquad \frac{\partial}{\partial t} u(t, x) + \frac{1}{2} \sum_{k=1}^{m} \frac{\partial^2}{\partial x^k \partial x^k} u(t, x) + V(t, x) u(t, x) = 0$$

for $0 \leq t \leq T$ and $x \in \Re^m$ with the final time condition

$$u(T, x) = f(T, x).$$

Weak approximations for stochastic differential equations provide a powerful tool to evaluate functionals of the type (4.1). In higher dimensions this can be an effective way of solving second order partial differential equations such as (4.2), particularly on parallel computers.

For given, sufficiently smooth functions f, ϕ and a we introduce the functional

$$(4.3) \qquad F = E\left(f(T, W_T) \phi \left(\int_0^T a(s, W_s) \, ds \right) \right)$$

with W as above. Using the $(m+1)$-dimensional Ito process $X = \{X_t = (X_t^1, \ldots, X_t^{m+1}), \; t \geq 0\}$ satisfying the stochastic equation

$$(4.4) \qquad X_t^k = x^k + \int_0^t dW_s^k, \qquad \text{for} \quad k = 1, \ldots, m,$$

$$X_t^{m+1} = \int_0^t a\left(s, \tilde{X}_s\right) ds,$$

where $\tilde{X}_s = (X_s^1, \ldots, X_s^m)$, we can rewrite the functional (4.3) in the form

$$(4.5) \qquad F = E\left(g(X_T) \right),$$

where

$$g(y) = f\left(T, (y^1, \ldots, y^m)\right) \phi\left(y^{m+1}\right).$$

The expression (4.5) is now a functional of an $(m+1)$-dimensional Ito process and can be approximated by the weak schemes discussed in Chapter 5. The structure of the stochastic equation (4.4) is relatively simple and will lead to simple weak schemes.

The Euler scheme (5.1.1) for (4.4) has the form

$$(4.6) \qquad Y_{n+1}^k = Y_n^k + \Delta \hat{W}_n^k, \qquad \text{for} \quad k = 1, \ldots, m,$$

$$Y_{n+1}^{m+1} = Y_n^{m+1} + a\left(\tau_n, \tilde{Y}_n\right) \Delta$$

with $\tilde{Y}_0 = x$ and $Y_0^{m+1} = 0$, where $\tilde{Y}_n = \left(Y_n^1, \ldots, Y_n^m\right)$. Here the $\Delta\hat{W}_n^k$ are independent $N(0; \Delta)$ Gaussian random variables or two-point distributed random variables with

$$P\left(\Delta\hat{W}_n^k = \pm\sqrt{\Delta}\right) = \frac{1}{2}$$

for $k = 1, \ldots, m$ and $n = 0, 1, \ldots$. Using the solution of (4.6) with a step size $\delta = \Delta$, we can define the random variable

$$(4.7) \qquad \tilde{F}_1^\delta = f\left(T, \tilde{Y}_{n_T}\right) \phi\left(\Delta \sum_{n=0}^{n_T-1} a\left(\tau_n, \tilde{Y}_n\right)\right)$$

where $n_T = T/\Delta$. Its expectation approximates the functional

$$(4.8) \qquad F = E\left(f\left(T, \tilde{X}_T\right) \phi\left(\int_0^T a\left(s, \tilde{X}_s\right) ds\right)\right)$$

with weak order $\beta = 1.0$, that is

$$(4.9) \qquad \left|F - E\left(\tilde{F}_1^\delta\right)\right| \le K\,\delta.$$

To illustrate the approximation of specific functional integrals we consider the following example of (4.4)–(4.5) with $m = 1$, $T = 1$,

$$(4.10) \quad f(t, y^1) \equiv 1, \qquad \phi\left(y^2\right) = \exp\left(y^2\right), \qquad a\left(t, y^1\right) = -\frac{1}{2}\left(y^1\right)^2$$

and $\tilde{X}_0 = 0$ for which the functional (4.3) can be evaluated exactly, namely

$$(4.11) \qquad F = E\left(\exp\left(-\frac{1}{2}\int_0^T (W_s)^2\, ds\right)\right) = \left(\frac{2e}{1+e^2}\right)^{\frac{1}{2}}.$$

Exercise 6.4.1 (PC-Exercise 17.1.3)   *Apply the Euler scheme with step size δ $= \Delta = 2^{-3}$ for the example (4.10)–(4.11) to simulate $M = 20$ batches each of N $= 100$ realizations of the random variable $\tilde{F}_1^\delta$ and evaluate the 90% confidence interval for the mean error $\mu = F - E(\tilde{F}_1^\delta)$. Repeat the calculations for step sizes $\delta = \Delta = 2^{-4}$, 2^{-5} and 2^{-6}. Plot the confidence intervals on μ versus δ axes and also $\log_2 |\mu|$ versus $\log_2 \delta$ axes.*

B. Higher Order Approximation of Functional Integrals

The scheme used in the above subsection is the simplest and may thus be inefficient and even numerically unstable in some cases. Here we shall approximate the functional integral (4.3) with a second order weak scheme, for which we take an implicit one in anticipation of those cases where the underlying stochastic differential equations are stiff.

The implicit order 2.0 weak scheme (5.3.7) for the system (4.4) simplifies to

$$(4.12) \qquad Y_{n+1}^k = Y_n^k + \Delta \hat{W}_n^k, \qquad \text{for} \quad k = 1, \ldots, m,$$

$$Y_{n+1}^{m+1} = Y_n^{m+1} + \frac{1}{2} \left\{ a\left(\tau_{n+1}, \tilde{Y}_{n+1}\right) + a\left(\tau_n, \tilde{Y}_n\right) \right\} \Delta$$

with $\tilde{Y}_0 = x$ and $Y_0^{m+1} = 0$, where $\tilde{Y}_n = (Y_n^1, \ldots, Y_n^m)$. It is, in fact, explicit because $\tilde{Y}_{n+1}$ does not include the Y_{n+1}^{m+1} component. The $\Delta \hat{W}_n^k$ here can be chosen as independent $N(0; \Delta)$ Gaussian random variables or as three-point distributed random variables with

$$P\left(\Delta \hat{W}_n^k = \pm\sqrt{3\Delta}\right) = \frac{1}{6}, \qquad P\left(\Delta \hat{W}_n^k = 0\right) = \frac{2}{3}$$

for $k = 1, \ldots, m$ and $n = 0, 1, \ldots$. Using (4.12) with step size $\delta = \Delta$ we define the random variable

$$(4.13) \quad \tilde{F}_2^\delta = f\left(T, \tilde{Y}_{n_T}\right) \phi\left(\Delta \sum_{n=0}^{n_T} a\left(\tau_n, \tilde{Y}_n\right)\right.$$

$$\left. - \frac{1}{2}\Delta \left\{ a\left(\tau_0, x\right) + a\left(\tau_{n_T}, \tilde{Y}_{n_T}\right) \right\} \right)$$

where $n_T = T/\Delta$. We can interpret (4.13) as a stochastic generalization of the trapezoidal formula for Riemann integrals. Its expectation $E(\tilde{F}_2^\delta)$ approximates F with weak order $\beta = 2.0$.

Exercise 6.4.2 (PC-Exercise 17.1.4)   *Repeat Exercise 6.4.1 with $\tilde{F}_2^\delta$ defined in (4.13), for $\Delta = \delta = 2^{-1}, \ldots, 2^{-4}$ and the implicit order 2.0 weak scheme (4.12). Plot the logarithm of the necessary computer time versus $-\log_2 |\mu|$.*

Much can be gained in efficiency by using an extrapolation method. The order 4.0 weak extrapolation method given in (5.2.13) with step sizes $\Delta = \delta$, 2δ and 4δ, where $T/4\delta$ is an integer, generates the random variable

$$(4.14) \qquad\qquad \tilde{F}_3^\delta = \frac{1}{21}\left[32\tilde{F}_2^\delta - 12\tilde{F}_2^{2\delta} + \tilde{F}_2^{4\delta}\right].$$

The expectation $E(\tilde{F}_3^\delta)$ approximates F with weak order 4.0.

Exercise 6.4.3 (PC-Exercise 17.1.5)   *Repeat Exercise 6.4.2 with $\tilde{F}_3^\delta$ defined in (4.14) with the $\tilde{F}_2^\Delta$ described in (4.13) based on the implicit order 2.0 weak scheme (4.12).*

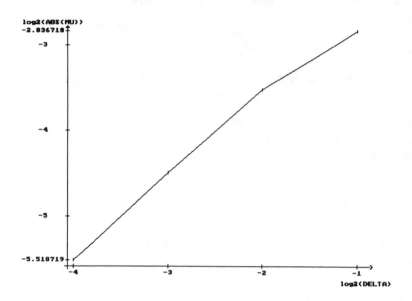

Figure 6.4.1 $\log_2$ (computer time) versus $-\log_2 |\mu|$ for Exercise 6.4.2.

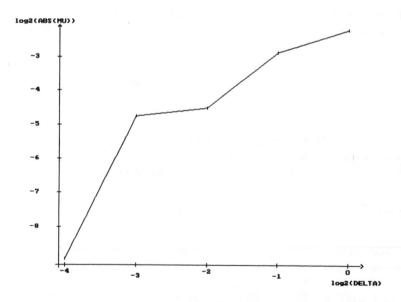

Figure 6.4.2 $\log_2$ (computer time) versus $-\log_2 |\mu|$ for Exercise 6.4.3.

C. Variance Reduction for an Exponential Functional

The special case with $f \equiv 1$ and $\phi(y^{m+1}) = \exp(y^{m+1})$ in (4.3) gives an exponential functional of the form

$$(4.15) \qquad \hat{F} = E\left(\exp\left(\int_0^T a\left(s, W_s\right) ds\right)\right),$$

where $W = \{W_t, t \geq 0\}$ is an m-dimensional Wiener process starting at a point $x = (x^1, \ldots, x^m) \in \Re^m$. It can be shown that

$$(4.16) \qquad \hat{F} = E\left(X_T^{m+1}\right)$$

for the Ito process $X = \{X_t = (X_t^1, \ldots, X_t^{m+1}), t \geq 0\}$ satisfying the stochastic equation

$$(4.17) \qquad X_t^k = x^k + \int_0^t dW_s^k, \qquad \text{for} \quad k = 1, \ldots, m,$$

$$X_t^{m+1} = 1 + \int_0^t a\left(s, \tilde{X}_s\right) X_s^{m+1} ds,$$

where once again $\tilde{X}_s = (X_s^1, \ldots, X_s^m)$.

It is known that estimators based on weak approximations may have rather large variances. This problem can be handled with the variance reduction techniques discussed in Chapter 5, although the simple structure of the stochastic equation (4.4) may then be lost. The corresponding stochastic equations that arise when such variance reduction techniques are used can, nevertheless, still be simulated with the general weak approximations from Chapter 5.

Let us apply the measure transformation method described in Section 5 of Chapter 5 with the parameter function

$$(4.18) \qquad d^1(t, y^1) = d(t, y^1) = \frac{T - t}{1 + T - t} y^1$$

to the example given by (4.10)–(4.11). Instead of (4.17), from (5.5.7) we then obtain the system of stochastic equations

$$(4.19) \qquad Z_t^1 = x + \int_0^t dW_s - \int_0^t d\left(s, Z_s^1\right) ds$$

$$Z_t^2 = 1 - \frac{1}{2} \int_0^t \left(Z_s^1\right)^2 Z_s^2 ds$$

$$Z_t^3 = 1 + \int_0^t d\left(s, Z_s^1\right) Z_s^3 dW_s.$$

According to (5.5.8) we have

$$(4.20) \qquad \hat{F} = u(t, y) = E\left(Z_T^2 Z_T^3\right).$$

We can use a weak approximation $Y = (Y^1, Y^2, Y^3)$ to approximate the solution of the stochastic system (4.19) in (4.20) and expect the functional

$$(4.21) \qquad\qquad \tilde{F}_4^\delta = Y_{n_T}^2 \, Y_{n_T}^3$$

to have a relatively small variance.

Exercise 6.4.4 (PC-Exercise 17.1.8)   *Repeat Exercise 6.4.2 with variance reduction using (4.19)–(4.21) and applying the implicit order 2.0 weak scheme (5.3.7).*

A comparison of Figure 6.4.2 and Figure 6.4.3 shows that the variance reduction technique yields a significant reduction in required computer time to approximate the functional integral to a given accuracy.

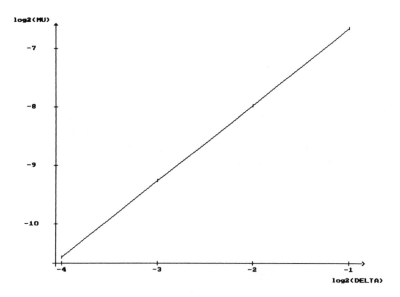

Figure 6.4.3 $\log_2$ (computer time) versus $-\log_2 |\mu|$ for Exercise 6.4.4.

In principle, we have solved specific second order partial differential equations in this section. It is also possible to use this approach in a similar manner involving higher order weak schemes and variance reduction techniques to approximate higher dimensional linear second order parabolic equations. We note that the underlying algorithms are well suited to parallel computers. Furthermore we have the crucial advantage that the proposed weak approximation methods work also in degenerate situations, that is when the diffusion coefficient becomes sometimes zero.

D. Approximation of Invariant Measures

Until now we have considered problems which only require a weak or strong approximation on some bounded time interval $[0, T]$. Here we consider a situation which requires weak approximations to work equally well on $[0, \infty)$.

We suppose that we have a d-dimensional ergodic Ito process

$$(4.22) \qquad X_t = X_0 + \int_0^t a(X_s)\, ds + \sum_{j=1}^{m} \int_0^t b^j(X_s)\, dW_s^j$$

for $t \geq 0$ driven by an m-dimensional Wiener process $W = (W^1, \ldots, W^m)$, that is, with a unique invariant probability law μ such that

$$(4.23) \qquad \lim_{t \to \infty} \frac{1}{t} \int_0^t f(X_s)\, ds = \int_{\Re^d} f(y)\, d\mu(y),$$

w.p.1, for all μ–integrable functions $f : \Re^d \to \Re$ and any admissible initial value X_0.

In many applications we know on theoretical grounds that an Ito process X has a unique invariant probability law μ and are interested in evaluating a functional of the form

$$(4.24) \qquad F = \int_{\Re^d} f(y)\, d\mu(y)$$

for a given function $f : \Re^d \to \Re$, but we do not know μ explicitly. For instance, when $f(y) \equiv y^2$ the functional (4.24) is simply the asymptotic second moment, $\lim_{t\to\infty} E\left((X_t)^2\right)$.

A simple example of an ergodic Ito process is the Ornstein-Uhlenbeck process

$$(4.25) \qquad X_t = X_0 - \int_0^t X_s\, ds + \int_0^t \sqrt{2}\, dW_s,$$

which has the standard Gaussian law $N(0; 1)$ with density

$$(4.26) \qquad p(y) = \frac{1}{\sqrt{2\pi}} \exp\left(-\frac{1}{2} y^2\right)$$

giving its invariant measure μ. Using (4.23) with $f(y) = y^2$, we see that the asymptotic second moment of this Ornstein–Uhlenbeck process satisfies

$$(4.27) \quad \lim_{t \to \infty} \frac{1}{t} \int_0^t (X_s)^2\, ds = \int_{-\infty}^{\infty} y^2 p(y)\, dy = \lim_{t \to \infty} E\left((X_t)^2\right) = F = 1.$$

We could thus use the expression

$$(4.28) \qquad F_T = \frac{1}{T} \int_0^T (X_s)^2\, ds$$

for large T as an estimate for the asymptotic second moment of X, if we did not already know it. Note that F_T here involves just one sample path of the

Ito process X, though taken over a rather long time interval $[0, T]$. We can determine it approximately by evaluating the sum

$$(4.29) \qquad F_T^\delta = \frac{1}{n_T} \sum_{n=0}^{n_T - 1} \left(Y_n^\delta \right)^2$$

for a single trajectory of a discrete time approximation Y^δ with equidistant step size $\delta = \Delta = T/n_T$.

This method of using a single simulated trajectory to approximate a functional of the form (4.24) numerically, proposed by Talay, can also be used for other functions f and more general ergodic Ito processes X. If the invariant measure μ is absolutely continuous with respect to the Lebesgue measure, then its density p satisfies the stationary Kolmogorov equation

$$(4.30) \qquad \mathcal{L}p = \sum_{k=1}^{d} a^k \frac{\partial p}{\partial y^k} + \frac{1}{2} \sum_{i,j=1}^{d} b^i b^j \frac{\partial^2 p}{\partial y^i \partial y^j} = 0.$$

The simulation method discussed above thus offers an alternative means for calculating the stationary solution of a partial differential equation, and in higher dimensional cases may sometimes be the only efficient method available.

Talay proposed a criterion for assessing discrete time approximations Y^δ which are used to calculate limits of the form

$$(4.31) \qquad F^\delta = \lim_{T \to \infty} \frac{1}{n_T} \sum_{n=0}^{n_T - 1} f \left(Y_n^\delta \right),$$

where $\delta = T/n_T$ is kept fixed. A discrete time approximation Y^δ is said to *converge with respect to the ergodic criterion with order* $\beta > 0$ to an ergodic Ito process X as $\delta \to 0$ if for each polynomial f there exist a positive constant C_f, which does not depend on δ, and a $\delta_0 > 0$ such that

$$(4.32) \qquad \left| F^\delta - F \right| \leq C_f \, \delta^\beta$$

for $\delta \in (0, \delta_0)$.

In view of the ergodicity relationship (4.23) we can interpret F as $E(f(X_\infty))$ and F^δ as $E(f(Y_\infty^\delta))$. The ergodic convergence criterion (4.32) can thus be considered to be an extension of the weak convergence criterion (3.4.16) to the infinite time horizon $T = \infty$.

Under appropriate assumptions, it follows that nearly all the schemes of a specific weak order mentioned in Chapter 5 converge with the same order with respect to the ergodic criterion (4.32).

Let us now compute some functionals of the invariant measure μ of the Ornstein-Uhlenbeck process (4.25) with density p given in (4.26).

Exercise 6.4.5 (PC-Exercise 17.2.2)   *Use the Euler scheme with step size $\delta = \Delta = 2^{-3}$ to simulate the sum F^δ in (4.29) with $T \in [0, 25600]$ for the*

Ornstein-Uhlenbeck process (4.25) with $X_0 = 0$ to estimate the second moment of the invariant law. Plot the linearly interpolated values of F^δ against T.

Exercise 6.4.6 (PC-Exercise 17.2.3)   *Repeat Exercise 6.4.5 using the explicit order 2.0 weak scheme (5.2.1).*

Exercise 6.4.7 (PC-Exercise 17.2.4)   *Consider Exercise 6.4.5, but now estimating the probabilities*

$$\mu\left([r\epsilon, (r+1)\epsilon]\right) = \int_{r\epsilon}^{(r+1)\epsilon} d\mu(y)$$

for $r = -10, -9, \ldots, 8, 9$ and $\epsilon = 0.3$. Plot the results in a histogram.

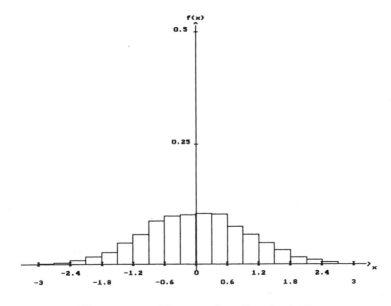

Figure 6.4.4   Histogram from Exercise 6.4.7.

6.5   Stochastic Stability and Bifurcation

The inclusion of noise in a dynamical system can modify its behaviour, particularly the stability of its stationary solutions and their bifurcation to new stationary or time-periodic solutions as a crucial systems parameter is varied. While the effect of low intensity, additive noise is often slight, new forms of behaviour not present in the underlying deterministic system may arise when the noise is multiplicative or strong, as we saw in Section 6.1. Of course, there are also many stochastic dynamical systems that are not based on an underlying deterministic system. In both cases, the asymptotic stability of a stationary solution is of interest.

Asymptotic stability can be characterized by the Lyapunov exponents of the system, with the change in sign of the largest Lyapunov exponent from negative to positive as a parameter is varied indicating the loss of stability and the possible occurrence of a stochastic bifurcation. Since they can rarely be determined explicitly, Lyapunov exponents usually have to be evaluated numerically. In addition, while the theory of stochastic stability is well established, that of stochastic bifurcation is far from complete, so numerical simulations are a practical way of discovering what may happen in specific examples. Here we first consider a simple example for which the Lyapunov exponents are known exactly. Then we investigate the asymptotic stability of the zero solution of the noisy Brusselator equations and its apparent bifurcation to a noisy limit cycle.

A. Lyapunov Exponents

Lyapunov exponents measure the asymptotic rates of expansion or contraction of the solutions of a linear system, thus generalizing the real parts of the eigenvalues of an autonomous deterministic linear system. For a d-dimensional linear Stratonovich stochastic differential equation

$$(5.1) \qquad dZ_t = AZ_t\, dt + \sum_{k=1}^{m} B^k Z_t \circ dW_t^k,$$

where $d \geq 2$ and W is an m-dimensional Wiener process the *Lyapunov exponents* are defined as

$$(5.2) \qquad \lambda(z,\omega) = \limsup_{t\to\infty} \frac{1}{t} \ln |Z_t^z(\omega)|$$

for each solution with initial value $Z_0^z = z$. Under appropriate assumptions on the matrices $A, B^1, \ldots, B^m$ in (5.1), the Multiplicative Ergodic Theorem of Oseledecs assures the existence of just d *nonrandom* Lyapunov exponents

$$\lambda_d \leq \ldots \leq \lambda_2 \leq \lambda_1,$$

some of which may coincide, as well as the existence of the limits rather than only the upper limits in (5.2). The counterparts of the eigenspaces for these Lyapunov exponents are, however, generally random.

There is an explicit formula for the top Lyapunov exponent λ_1. To determine it we change to spherical coordinates $r = |z|$ and $s = z/|z|$ for $z \in \Re^d \setminus \{0\}$, in which case the linear Stratonovich system (5.1) transforms into the system

$$(5.3) \qquad dR_t = R_t\, q(S_t)\, dt + \sum_{k=1}^{m} R_t\, q^k(S_t) \circ dW_t^k$$

$$(5.4) \qquad dS_t = h(S_t, A)\, dt + \sum_{k=1}^{m} h(S_t, B^k) \circ dW_t^k$$

on S^{d-1}, where S^{d-1} is the unit sphere in $\Re^d$ and

$$q(s) = s^\top A s + \sum_{k=1}^m \left(\frac{1}{2} s^\top \left(B^k + (B^k)^\top \right) s - \left(s^\top B^k s \right)^2 \right),$$

$$q^k(s) = s^\top B^k s, \quad h(s, A) = \left(A - (s^\top A s) I \right) s.$$

Equation (5.4) does not involve the radial variable R_t and is an example of a stochastic differential equation on a compact manifold. The advantage of using the Stratonovich interpretation here is that the transformation laws of classical calculus apply, and keep the formulae simple. They are in fact used to derive (5.3) and (5.4) from (5.1). From (5.3) we have

$$\ln R_t = \ln |Z_t| = \int_0^t q(S_u)\, du + \sum_{k=1}^m \int_0^t q^k(S_u) \circ dW_u^k,$$

and hence

(5.5) $$\lim_{t \to \infty} \frac{1}{t} \ln |Z_t| = \lim_{t \to \infty} \frac{1}{t} \int_0^t q(S_u)\, du.$$

Consequently, the top Lyapunov exponent λ_1 of system (5.1) is given by

(5.6) $$\lambda_1 = \int_{S^{d-1}} q(s)\, d\bar{\mu}(s)$$

where $\bar{\mu}$ is the invariant probability measure of an ergodic solution process $S = \{S_t, t \geq 0\}$ of (5.1) on S^{d-1}. In most cases, however, it is not easy to determine $\bar{\mu}$, so we shall use the limit (5.5) to approximate the λ_1 instead.

For the stochastic differential equation

(5.7) $$dZ_t = A Z_t\, dt + B Z_t \circ dW_t$$

with coefficient matrices

$$A = \begin{bmatrix} a & 0 \\ 0 & b \end{bmatrix}, \qquad B = \begin{bmatrix} 0 & -\sigma \\ \sigma & 0 \end{bmatrix}$$

and real-valued parameters a, b, σ. Baxendale has shown that the top Lyapunov exponent λ_1 is

(5.8) $$\lambda_1 = \frac{1}{2}(a + b) + \frac{1}{2}(a - b) \frac{\displaystyle\int_0^{2\pi} \cos(2\theta) \exp\left(\frac{a - b}{2\sigma^2} \cos(2\theta) \right) d\theta}{\displaystyle\int_0^{2\pi} \exp\left(\frac{a - b}{2\sigma^2} \cos(2\theta) \right) d\theta}.$$

The projected process S_t lives on the unit circle S^1 since $d = 2$ here, so we can represent it in terms of polar coordinates, specifically, the polar angle ϕ. The resulting process

$$\phi_t = \arctan\left(\frac{S_t^2}{S_t^1} \right)$$

satisfies the stochastic equation

(5.9)
$$d\phi_t = \frac{1}{2}(b - a)\sin(2\phi_t)\,dt + \sigma dW_t,$$

which we interpret modulo 2π, and

(5.10)
$$q(S_t) = \tilde{q}(\phi_t) = a(\cos\phi_t)^2 + b(\sin\phi_t)^2.$$

In view of the remark following formula (5.6) we shall use the functional

(5.11)
$$F_T^\delta = \frac{1}{n_T}\sum_{n=0}^{n_T-1}\tilde{q}(Y_n^\delta)$$

for a discrete time approximation Y^δ of the solution of equation (5.9) to approximate the top Lyapunov exponent λ_1, noting from (5.8) that $\lambda_1 \simeq -0.489\ldots$ for the choice of parameters $a = 1.0$, $b = -2.0$ and $\sigma = 10$. Recall that n_T is the largest integer n such that $n\delta \leq T$.

Exercise 6.5.1 (PC-Exercise 17.3.2)   *Simulate the Lyapunov exponent λ_1 of (5.7) for the parameters $a = 1.0$, $b = -2.0$ and $\sigma = 1.0$ by (5.11) using the explicit order 2.0 weak scheme (5.2.1) with $\delta = \Delta = 2^{-9}$, $T = 512$ and $Y_0 = \phi_0 = 0.0$. Plot the linearly interpolated values of F_t^δ against t for $0 \leq t \leq T$.*

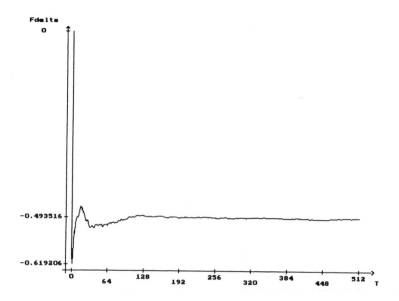

Figure 6.5.1   The top Lyapunov exponent approximated with the explicit order 2.0 weak scheme.

We see from Figure 6.5.1 that F_t^δ tends to the true value of λ_1 as $t \to \infty$. The performance of the numerical scheme here is enhanced by the asymptotic stability of the null solution of the system of equations (5.7) as indicated by the negativity of its Lyapunov exponents. For stiff and unstable systems an implicit weak scheme may be required for satisfactory results.

In higher dimensions the stochastic differential equation (5.4) for S_t on S^{d-1} will not simplify so nicely and difficulties may be encountered in trying to solve it numerically, particularly in ensuring that the successive iterates remain on S^{d-1}. To circumvent these difficulties the first limit in (5.5) could be approximated directly, that is by the functional

$$(5.12) \qquad \tilde{L}_T^\delta = \frac{1}{n_T \delta} \ln |Y_{n_T}^\delta|$$

where Y^δ is now a discrete time approximate solution of the original stochastic differential equation (5.1). Without loss of generality the initial value Y_0^δ can be chosen here to satisfy $|Y_0^\delta| = 1$. Actually, it is computationally preferable to evaluate the functional

$$(5.13) \qquad L_T^\delta = \frac{1}{n_T \delta} \sum_{n=1}^{n_T} \ln \left(\frac{|Y_n^\delta|}{|Y_{n-1}^\delta|} \right)$$

here since the logarithms in (5.12) will become very large in magnitude as $|Y_n^\delta|$ tends to zero or becomes very large.

Exercise 6.5.2 *Repeat Exercise 6.5.1 using the functional (5.12) and the explicit order 2.0 weak scheme (5.2.1) applied to the stochastic differential equation (5.7) with $Z_0 = (1.0, 0.0)$ and parameters as in Exercise 6.5.1.*

B. Stochastic Stability

Given the variety of convergences for stochastic processes there are many different ways of defining stability in stochastic systems. Nevertheless the idea is the same in all cases: solutions will always remain close to a reference solution if they start sufficiently close to it. This is a strengthening of the property of continuity in initial values to hold over all future time. The reference solution is typically a statistically stationary solution, which, to simplify matters, we shall suppose to be the zero solution $X_t \equiv 0$. Consequently, the coefficients of the stochastic differential equation under consideration, say,

$$(5.14) \qquad dX_t = \underline{a}(X_t)\, dt + \sum_{j=1}^{m} b^k(X_t) \circ dW_t^k$$

need to satisfy

$$\underline{a}(0) = b^1(0) = \ldots = b^m(0) = 0.$$

Sufficient regularity to ensure the existence of a unique solution $X_t = X_t^{x_0}$ of (5.14) for all $t \geq 0$ and all appropriate deterministic initial values $X_0 = x_0$

will also be assumed. A widely accepted definition of stochastic stability is a probabilistic one due to Hasminski. The steady solution $X_t \equiv 0$ is called *stochastically stable* if for any $\epsilon > 0$ and $t_0 \geq 0$

$$\lim_{x_0 \to 0} P\left(\sup_{t \geq t_0} \left|X_t^{t_0, x_0}\right| \geq \epsilon\right) = 0,$$

and *stochastically asymptotically stable* if, in addition,

$$\lim_{x_0 \to 0} P\left(\lim_{t \to \infty} \left|X_t^{t_0, x_0}\right| \to 0\right) = 1.$$

An extensive array of tests for stochastic stability and stochastic asymptotic stability is now available. Often the stability follows from that of the zero solution $Z_t \equiv 0$ of the corresponding linearized stochastic differential equation

$$(5.15) \qquad dZ_t = A\, Z_t\, dt + \sum_{k=1}^{m} B^k Z_t \circ dW_t^k$$

where $A = \nabla \underline{a}(0)$, $B^1 = \nabla b^1(0)$, ..., $B^m = \nabla b^m(0)$ with ∇f denoting the gradient vector of f. From the previous subsection it is apparent that the asymptotic stability of the zero solution of (5.15) is characterized by the negativity of its top Lyapunov exponent λ_1. When the coefficients of (5.15) depend on parameters, so too will λ_1 and its sign may change as the parameters change, thus changing the stability of the zero solution.

The system of ordinary differential equations

$$(5.16) \qquad \frac{dx_1}{dt} = (a-1)\,x_1 + a\,x_1^2 + (x_1+1)^2 x_2$$

$$\frac{dx_2}{dt} = -a\,x_1 - a\,x_1^2 - (x_1+1)^2 x_2$$

with parameter a is based on the Brusselator equations which model unforced periodic oscillations in certain chemical reactions. When $a < 2$ the zero solution $(x_1, x_2) \equiv (0,0)$ is globally asymptotically stable, but loses stability in a Hopf bifurcation at $a = 2$ to give rise to a limit cycle for $a > 2$. Supposing that the parameter a is noisy, that is of the form $\alpha + \sigma \xi_t$ where ξ_t is Gaussian white noise, leads to the system of Ito stochastic differential equations with scalar multiplicative noise

$$(5.17)\, dX_t^1 = \left\{(\alpha-1)X_t^1 + \alpha(X_t^1)^2 + (X_t^1+1)^2 X_t^2\right\} dt + \sigma\, X_t^1(1+X_t^1)dW_t$$

$$dX_t^2 = \left\{-\alpha\, X_t^1 - \alpha(X_t^1)^2 - (X_t^1+1)^2 X_t^2\right\} dt - \sigma\, X_t^1(1+X_t^1)dW_t$$

for which $(X_t^1, X_t^2) \equiv (0,0)$ is a solution. The corresponding linearized system is then

$$(5.18) \quad d\begin{pmatrix} Z_t^1 \\ Z_t^2 \end{pmatrix} = \begin{bmatrix} \alpha-1 & 1 \\ -\alpha & -1 \end{bmatrix} \begin{pmatrix} Z_t^1 \\ Z_t^2 \end{pmatrix} dt + \sigma \begin{bmatrix} 1 & 0 \\ -1 & 0 \end{bmatrix} \begin{pmatrix} Z_t^1 \\ Z_t^2 \end{pmatrix} dW_t$$

in its Ito version, with the Stratonovich version having the same form with α replaced by $\alpha - \sigma^2/2$. Its top Lyapunov exponent $\lambda_1 = \lambda_1(\alpha, \sigma)$ will depend on the two parameters α and σ. In the deterministic case $\sigma = 0$ we have $\lambda_1(\alpha, 0) = \frac{1}{2}(\alpha - 2)$, being the real part of the complex conjugate eigenvalues of the drift coefficient matrix. When noise is present we need to evaluate $\lambda_1(\alpha, \sigma)$ numerically.

Exercise 6.5.3 *Evaluate the functional (5.13) using the explicit order 2.0 weak scheme (5.2.1) with step size $\delta = \Delta_n = 2^{-9}$, initial value $(Z_0^1, Z_0^2) = (1.0, 0.0)$ and $T = 512$ for the linear system (5.18). Consider the parameter values $0 < \alpha \leq 3$ and $\sigma = 0.0, 0.4, 0.8, 1.2, 1.6$, plotting the functional value against α for each choice of noise intensity. Use 90% confidence intervals and 20 batches of 20 samples in each case.*

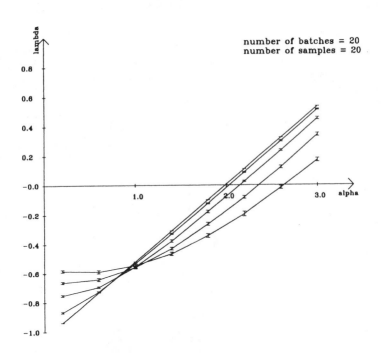

Figure 6.5.2 The top Lyapunov exponent $\lambda_1(\alpha, \sigma)$.

In Figure 6.5.2 the values of the Lyapunov exponents decrease for $\alpha = 3.0$ for increasing $\sigma = 0.0, 0.4, 0.8, 1.2$ and 1.6. We see that the presence of noise stabilizes the system for $\alpha > 1$. In general it loses stability for larger α values. The same effect will also occur in the original nonlinear system (5.17).

C. An Example of Stochastic Bifurcation

The zero solution of the deterministic Brusselator equations (5.16) loses stability at $\alpha = 2$ in a Hopf bifurcation to yield a limit cycle for $\alpha > 2$. This limit cycle encloses the now unstable zero solution and, from bifurcation theory, has magnitude and period proportional to $\sqrt{\alpha - 2}$. In the preceding subsection we saw that the zero solution of the linearized noisy Brusselator equations (5.18) loses stability as α increases in the vicinity of 2 with the critical value depending on the noise intensity σ. Stochastic stability theory implies a similar loss of stability in the nonlinear noisy Brusselator equations (5.17). At present there is no complete stochastic bifurcation theory to tell us what may happen thereafter. From our simulations of the noisy Bonhoeffer-Van der Pol equations (1.2) in Section 6.1 we might suspect that there will be a noisy limit cycle about the deterministic limit cycle, for low noise intensity at least. The noise is, however, now multiplicative rather than additive, which may have a profound effect on what may happen. Here we shall use numerical simulations to investigate the situation.

Exercise 6.5.4   *Repeat Exercise 6.1.1 to compute the phase diagram for the noisy Brusselator equations (5.17) with initial value $X_0^1 = -0.1$, $X_0^2 = 0.0$ over the time interval $0 \leq t \leq T = 1$ for the parameters $\alpha = 2.1$ and $\sigma = 0.1$. Repeat for $\alpha = 2.2, 2.3$ and 2.5.*

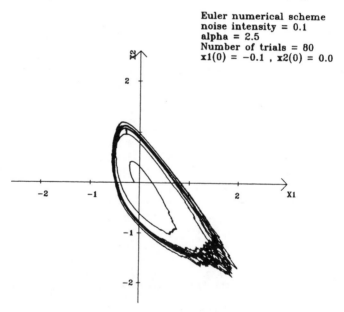

Figure 6.5.3   Noisy Brusselator phase diagram.

From Figure 6.5.3 there appears to be a noisy limit cycle which increases in size as the parameter α increases. This is confirmed by the frequency histograms calculated in the next exercise.

Exercise 6.5.5 *Repeat Exercise 6.1.2 to compute the frequency histogram for the noisy Brusselator equations (5.17) with initial value $X_0^1 = -0.1$, $X_0^2 = 0.0$ over the time interval $0 \leq t \leq T = 10$ for parameters $\alpha = 2.1$ and $\sigma = 0.1$. Repeat for $\alpha = 2.2, 2.3$ and 2.5.*

A noisy limit cycle is obviously not a closed, periodic curve as in the deterministic case. Nevertheless it appears to have a characteristic mean radius, or magnitude, and period. We can estimate these quantities to investigate their dependence on the bifurcation parameter α and noise intensity σ. To be specific we can evaluate the mean radius

$$(5.19) \qquad R_T^\delta = \frac{1}{n_T} \sum_{n=0}^{n_T-1} |Y_n^\delta|$$

and the mean period

$$(5.20) \qquad P_T^\delta = \frac{\delta}{n_T} \# \left\{ n \in \{0, 1, \ldots, n_T - 1\}; \; Y_n^{1,\delta} < 0 < Y_{n+1}^{1,\delta} \right\}$$

for a time-discretization $Y^\delta = (Y^{1,\delta}, Y^{2,\delta})$ with equal time steps δ for the noisy Brusselator equations (5.17). Here $\#A$ denotes the cardinality of the set A. Note that a strong scheme should be used for P_T^δ, while a weak scheme could be used for R_T^δ here.

Exercise 6.5.6 *Apply the Milstein scheme (4.1.21) with equal step size $\delta = \Delta_n = 2^{-7}$ to the noisy Brusselator equations (5.17) with initial value $X_0^1 = 1.0$, $X_0^2 = 0.0$ over the time interval $0 \leq t \leq T = 512$ for the parameters $\alpha = 2.1$ and $\sigma = 0.1$. Plot the mean radius R_t^δ and mean period P_t^δ against t for $0 \leq t \leq 512$. Repeat for $\alpha = 2.2, 2.3, 2.4$ and 2.5. Then plot R_T^δ and P_T^δ against α. Repeat this for $\sigma = 0.4$ and $\sigma = 0.8$.*

In Figure 6.5.4 the upper curve relates to $\sigma = 0.1$, the middle to $\sigma = 0.4$ and the lower to $\sigma = 0.8$. There seems to be a square root relationship

$$R_T^\delta \sim \sqrt{\alpha - \bar{\alpha}(\sigma)}$$

for $\sigma = 0.1$ with some $\bar{\alpha}(\sigma)$ similar to that predicted theoretically in the deterministic case $\sigma = 0.0$.

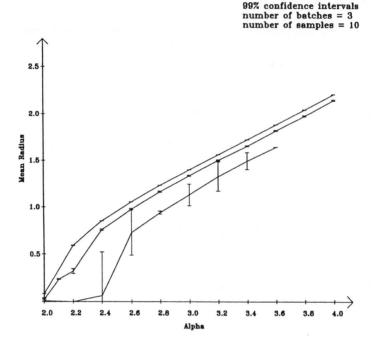

Figure 6.5.4   Mean radius versus α.

6.6   Simulation in Finance

In this final section we would like to indicate that the preceding numerical methods are also readily applicable to stochastic differential equations in finance, for example in option pricing and hedging. Consequently, it is no longer necessary to restrict attention to models that have explicitly known solutions.

A. The Black and Scholes Model

In 1973 Black and Scholes introduced a model describing the evolution of a risky asset $S = \{S_t, 0 \le t \le T\}$ such as the price of a stock, an exchange rate or an index. The model also includes a bond $B = \{B_t, 0 \le t \le T\}$ which evolves according to an interest rate r. The dynamics of the bond and the risky asset are described by the following linear system of stochastic differential equations

$$(6.1) \qquad\qquad dB_t \;=\; r\, B_t\, dt$$

$$dS_t \;=\; r\, S_t\, dt + \sigma S_t\, dW_t,$$

$t \in [0, T]$, with given initial values B_0 and S_0. Here $W = \{W_t, 0 \le t \le T\}$ is a Wiener process under the given probability measure P, called the probability measure of the risk-neutral world, and the parameter σ denotes the volatility of the risky asset.

Let us now consider an option on the risky asset S. A European call option with strike price K gives the right to buy the stock at time T at the fixed price K. The resulting payoff is thus

$$(6.2) \qquad f(S_T) = (S_T - K)^+ = \begin{cases} S_T - K & : \quad S_T > K \\ 0 & : \quad S_T \leq K \end{cases},$$

where a^+ denotes the positive part of a.

Suppose that we apply a dynamical portfolio strategy $\phi = (\zeta_t, \eta_t)_{t \in [0,T]}$, where at time $t \in [0, T]$ we hold the amount η_t in the riskless asset B_t and the amount ζ_t in the risky asset S_t. Then the value V_t of the portfolio at time t is

$$(6.3) \qquad V_t = \zeta_t S_t + \eta_t B_t.$$

Black and Scholes suggested that the fair option price at time $t \in [0, T]$ is the conditional expectation

$$(6.4) \qquad V_t = u(t, S_t) = e^{-r(T-t)} E(f(S_T)| \mathcal{F}_t),$$

where the σ-algebra $\mathcal{F}_t$ expresses the information about the risky asset up to time t. The corresponding hedging strategy $\phi = (\zeta_t, \eta_t)_{t \in [0,T]}$ has the form

$$(6.5) \qquad \zeta_t = \frac{\partial}{\partial S} u(t, S_t)$$

with

$$(6.6) \qquad \eta_t = \frac{V_t - \zeta_t S_t}{B_t},$$

the application of which leads to a perfect replication of the claim at maturity T, that is

$$(6.7) \qquad V_T = f(S_T).$$

This theoretical model allows the replication of the payoff, the claim at time T, without any extra cost just by continuously hedging the portfolio. Therefore the dynamical portfolio strategy is also called self financing. Black and Scholes were able to derive explicit formulae for the computation of V_t and ζ_t because of the linear structure of their model.

B. A Model with Stochastic and Past Dependent Volatility

The assumptions imposed on the behaviour of the underlying stock price, in particular that of a constant volatility, in the Black and Scholes model have been the cause of much critical comment.

The availability of stochastic numerical methods for solving multi-dimensional Markovian models described by stochastic differential equations now provides a possibility to specify very general patterns of stochastic volatility as well as stochastic interest rates. Our aim here is to present an example of how this works in a specific case with stochastic and past dependent volatility.

We consider a process $X = (X^0, X^1, X^2, X^3)^\top = (B, S, \sigma, \xi)^\top$ satisfying the following system of stochastic differential equations:

$$(6.8) \qquad\qquad dB_t = r(t, X_t) B_t \, dt$$

$$dS_t = r(t, X_t) S_t \, dt + \sigma_t S_t \, dW_t^1$$

$$d\sigma_t = -q(\sigma_t - \xi_t) \, dt + p\sigma_t \, dW_t^2$$

$$d\xi_t = \frac{1}{\alpha}(\sigma_t - \xi_t) \, dt$$

with $p > 0$, $q > 0$, $\alpha > 0$, where W^1, W^2 are independent Wiener processes under P.

Here the bond price B has the usual structure with a Markovian instantaneous interest rate r and the stock price S follows a generalized geometric Brownian motion since drift and volatility are not constant. Taking the drift to be $r(t, X_t)$ means that the discounted stock price process S/B is a martingale under the measure P which we use for pricing. The processes σ and ξ are interpreted as the instantaneous and the weighted average volatility of the stock, respectively. The equation for σ shows that the instantaneous volatility σ_t is disturbed by external noise with an intensity parameter p and at the same time is continuously pulled back towards the average volatility ξ_t, which is known as mean reversion. The parameter q measures the strength of the restoring force or speed of adjustment.

The equation for the average volatility ξ can be solved explicitly to give

$$\xi_t = \xi_0 \exp\left(-\frac{t}{\alpha}\right) + \frac{1}{\alpha} \int_0^t \exp\left(-\frac{t-s}{\alpha}\right) \sigma_s ds, \qquad 0 \le t \le T,$$

which shows that ξ_t is an average of the values σ_s $(0 \le s \le T)$, weighted with an exponential factor. For very large α we obtain $\xi_t \approx \xi_0$, while a very small value of α yields $\xi_t \approx \sigma_t$. Thus, the parameter α measures the strength of the past-dependence of the average volatility.

The stochastic volatility in our model yields an intrinsic risk which does not allow in general replication of the payoff without extra cost. Following the approach of Föllmer and Schweizer for a European call option we obtain as option price

$$(6.9) \qquad V_t = v(t, X_t) = E\left(\exp\left\{-\int_t^T r(s, X_s) \, ds\right\} f(S_T) \Big| \mathcal{F}_t\right),$$

where $\mathcal{F}_t$ denotes the σ-algebra generated by X up to time t. Here the hedging strategy is similar to that in the Black and Scholes case, having

$$(6.10) \qquad\qquad \zeta_t = \frac{\partial}{\partial S} v(t, X_t)$$

and

$$(6.11) \qquad \eta_t = \frac{V_t - \zeta_t S_t}{B_t}.$$

Except in special cases, however, the above strategy $\phi = (\zeta_t, \eta_t)_{t \in [0,T]}$ will not be self-financing, but it is still self-financing on average. In general, there is no explicit solution for the option price or hedging strategy for this model, so a numerical approximation must be used.

C. Numerical Example

We consider the model (6.8) with its stochastic and past-dependent volatility. For simplicity, we take the interest rate $r(t, x)$ to be identically 0. The other parameters are $p = 0.3$ and $q = 1.0$ with expiration time $T = 1$. We thus have the three-dimensional stochastic differential equation

$$
\begin{aligned}
(6.12) \qquad dS_t &= \sigma_t S_t \, dW_t^1 \\
d\sigma_t &= -(\sigma_t - \xi_t) \, dt + 0.3 \sigma_t \, dW_t^2 \\
d\xi_t &= \frac{1}{\alpha} (\sigma_t - \xi_t) \, dt,
\end{aligned}
$$

for which we take initial values $S_0 = 1.0$, $\sigma_0 = 0.1$ and $\xi_0 = 0.1$. The claim to be considered will be a European call option with strike price $K = 1.0$.

We shall first generate an approximate solution path of the three-dimensional stochastic differential equation (6.12) using a strong scheme because we are interested in simulating a sample path. Along this trajectory we then compute the price of the option and the hedging strategy which we represent as functionals and we can use weak approximations for their evaluation.

Exercise 6.6.1 *Simulate a trajectory of (6.12) with $\alpha = 0.1$ by applying the strong scheme (4.2.8) with step size $\delta = \Delta_n = 2^{-9}$ and double Wiener integrals approximated by (2.3.32) with $p = 10$. At each of these discretization points compute the option price (6.9), the hedging ratio (6.10) and the amount to hold in the bond (6.11). Use the weak method (5.2.1) with step size $\delta = \Delta_n = 2^{-2}$ simulating $M = 20$ batches with $N = 50$ trajectories. Plot σ_t and ξ_t, S_t and ζ_t, V_t and $f(S_t)$ against $t \in [0, T]$.*

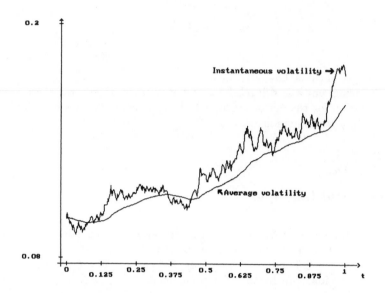

Figure 6.6.1   Instantaneous and average volatility σ_t and ξ_t, respectively.

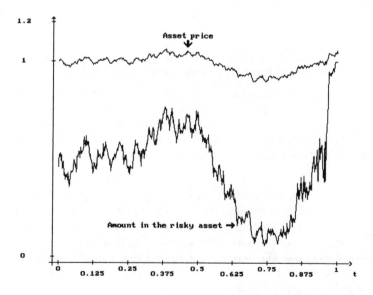

Figure 6.6.2   Asset price and amount in the risky asset ζ_t.

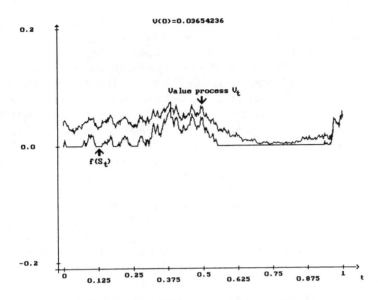

Figure 6.6.3 Value process V_t and $f(S_t)$.

We observe from Figure 6.6.1 that the average volatility ξ smoothens the instantaneous volatility σ, but with some delay. We also note from Figure 6.6.2 that our simulated trajectory of the risky asset ends up with a value $S_T > K = 1.0$ which means it has made a profit. Further, the stock component ζ of our hedging strategy, denoting the amount we have to hold in the risky asset, shows how we have finally to invest all our money, $\zeta_T = 1$, in the risky asset to obtain the payoff $(S_T - K)^+$. This is just what we would expect to happen here since the exercise of the option yields a profit. Figure 6.6.3 shows on the top the option price V_0. Further, it contains the value process V_t and for comparison the inner value $(S_t - K)^+$ of the option along the trajectory of S. As expected, the value of the hedging portfolio approaches the inner value of the option as the time to expiration goes to 0. We emphasize that all three figures are based on the same sample path solution of the stochastic differential equation (6.12). Repeating the simulation may result in an asset price $S_T < K$, which should result ζ_T being close to 0.

In the above calculations we used a weak past dependence factor α. The following exercise considers stronger past dependence.

Exercise 6.6.2   *Repeat Exercise 6.6.1 for past dependence parameter $\alpha = 1.0$.*

The above computations show that numerical methods for stochastic differential equations provide an invaluable tool for pricing options in situations where explicit solutions are not available.

Literature for Chapter 6

Treutlein & Schulten (1985) used Monte-Carlo methods to simulate the noisy
Bonhoeffer-Van der Pol equations, see also Kloeden, Platen & Schurz (1991).
The theory of stochastic flows is presented in Kunita (1990), while the examples
of stochastic flows on the circle and the torus were taken from Carverhill, Chappel & Elworthy (1986), from Baxendale (1986) and from Baxendale & Harris
(1986), respectively. For surveys of parametric estimation with continuous sample of a diffusion see Basawa & Prakasa Rao (1980) and Kutoyants (1984). The
discrete sampling case is treated in Le Breton (1976), Dacunha-Castelle & Flo-
rens Zmirou (1986), Florens-Zmirou (1989), Campillo & Le Gland (1989) and
Genon-Catalot (1990). The model considered in Subsection 6.4.A is taken from
Kloeden, Platen, Schurz & Sørensen (1992). Exponential families of processes
are treated in Küchler & Sørensen (1989) and quasi-likelihood estimators in
Hutton & Nelson (1986), Godambe & Heyde (1987) and Sørensen (1990). See
Shiga (1985) for population models. Filtering of Markov chains goes back to
Wonham (1965), Zakai (1969) and Fujisaki, Kallianpur & Kunita (1972). Dis-
crete time approximations of optimal filters have been considered by Clark
& Cameron (1980), Newton (1986) and Kloeden, Platen & Schurz (1993).
Functional integrals like those in Subsection 6.4.A have been investigated by
Blankenship & Baras (1981), amongst others. The ergodic convergence crite-
rion is due to Talay (1987). See also Talay (1990) for his order $\beta = 2.0$ weak
scheme. The explicit formula for the top Lyapunov exponent in Subsection
6.5.A is from Baxendale (1986). See Talay (1989) for the numerical approxima-
tion of Lyapunov exponents. The theory of stochastic stability is developed in
Hasminski (1980) and, from the perspective of Lyapunov exponents, in Arnold
& Wihstutz (1986). See Ehrhardt (1983) for the noisy Brusselator equations.
For finance models see Black & Scholes (1973), Hull & White (1987), Johnson &
Shanno (1987), Scott (1987), Wiggins (1987) and Föllmer & Schweizer (1991).
The model in Subsection 6.6.B is taken from Hofmann, Platen & Schweizer
(1992).

References

Anderson, S.L.
1990   Random number generators on vector supercomputers and other advanced structures. SIAM Review **32**, 221-251.

Arnold, L.
1974   Stochastic differential equations. Wiley, New York.

Arnold, L. and Wihstutz, V. (editors)
1986   Lyapunov exponents. Springer Lecture Notes in Mathematics Vol. 1186.

Azencott, R.
1982   Formule de Taylor stochastique et développement asymptotique d' intégrales de Feynman. Springer Lecture Notes in Mathematics Vol. 921, pp. 237-285.

Basawa, I.V. and Prakasa Rao, B.L.S.
1980   Statistical inference for stochastic processes. Academic Press, London.

Baxendale, P.
1986   Asymptotic behaviour of stochastic flows of diffeomorphisms. Springer Lecture Notes in Mathematics Vol. 1203.

Baxendale, P. and Harris, T.
1986   Isotropic stochastic flows. Annals Probab. **14**, 1155-1179.

Black, F. and Scholes, M.
1973   The pricing of options and corporate liabilities. J. Pol. Econ. **81**, 637-659.

Blankenship, G.L. and Baras, J.S.
1981   Accurate evaluation of stochastic Wiener integrals with applications to scattering in random media and nonlinear filtering. SIAM J. Appl. Math. **41**, 518-552.

Butcher, J.C.
1987   The numerical analysis of ordinary differential equations. Runge-Kutta and General Linear Methods. Wiley, Chichester.

Campillo, F. and Le Gland, F.
1989   MLE for partially observed diffusions. Direct maximization vs. the EM algorithm. Stoch. Proc. Appl. **33**, 245-274.

Carverhill, A., Chappel, M. and Elworthy, K.D.
1986   Characteristic exponents for stochastic flows. Springer Lecture Notes in Mathematics Vol. 1158.

Chang, C.C.
1987   Numerical solution of stochastic differential equations with constant diffusion coefficients. Math. Comput. **49**, 523-542.

Chung, K.L.
   1975   Elementary probability with stochastic processes. Springer.

Clark, J.M.C.
   1978   The design of robust approximations to the stochastic differential equations
          of nonlinear filtering. In Communication Systems and Random Process
          Theory (J.K. Skwirzynski, editor). Sijthoff& Noordhoff, Alphen naan den
          Rijn, pp. 721-734.

Clark, J.M.C. and Cameron, R.J.
   1980   The maximum rate of convergence of discrete approximations for stochastic
          differential equations. Springer Lecture Notes in Control and Inform. Sc.
          Vol. 25, pp. 162-171.

Dacunha-Castelle, D. and Florens-Zmirou, D.
   1986   Estimators of the coefficients of a diffusion from discrete observations.
          Stochastics 19, 263-284.

Dahlquist, G. and Bjorck, A.
   1974   Numerical methods. Prentice-Hall, New York.

Deuflhard, P.
   1985   Recent progress in extrapolation methods for ordinary differential equa-
          tions. SIAM Rev. 27, pp. 505-535.

Drummond, P.D. and Mortimer, I.K.
   1991   Computer simulation of multiplicative stochastic differential equations. J.
          Comput. Phys. 93, No 1, 144-170.

Ehrhardt, M.
   1983   Invariant probabilities for systems in a random environment — with ap-
          plications to the Brusselator. Bull. Math. Biol. 45, 579-590.

Florens-Zmirou, D.
   1989   Approximate discrete time schemes for statistics of diffusion processes.
          Statistics 20, 547-557.

Föllmer, H. and Schweizer, M.
   1991   Hedging of contingent claims under incomplete information. In Applied
          Stochastic Analysis, eds. M.H.A. Davis and R.J. Elliott, Stochastics Mono-
          graphs vol 5. Gordon and Breach, London/New York, 389-414.

Fujisaki, M., Kallianpur, G. and Kunita, H.
   1972   Stochastic differential equations for the nonlinear filtering problem. Osaka
          J. Math. 9, 19-40.

Gard, T.C.
   1988   Introduction to stochastic differential equations.   Marcel Dekker, New
          York.

Gear, C.W.
   1971   Numerical initial value problems in ordinary differential equations.
          Prentice-Hall, Englewood Cliffs, N.J.

Genon-Catalot, V.
   1990   Maximum contrast estimation for diffusion processes from discrete obser-
          vations. Statistics 21, 99-116.

Gikhman, I.I. and Skorokhod, A.V.
1972   Stochastic differential equations, Springer.
1979   The theory of stochastic processes, Vol. I-III. Springer.

Godambe, V.P. and Heyde, C.C.
1987   Quasi-likelihood and optimal estimation. Int. Statist. Rev. **55**, 231-244.

Groeneveld, R.A.
1979   An introduction to probability and statistics using BASIC. Marcel-Dekker, New York.

Hairer, E., Nørsett, S.P. and Wanner, G.
1987   Solving differential equations I: Nonstiff problems. Springer Series in Computational Mathematics, Vol. **8**, Springer.

Hairer, E. and Wanner, G.
1991   Solving differential equations II: Stiff and differential-algebraic problems. Springer Series in Computational Mathematics, Vol. **14**, Springer.

Henrici, P.
1962   Discrete variable methods in ordinary differential equations. Wiley, New York.

Hernandez, D.B. and Spigler, R.
1991   Numerical stability for stochastic implicit Runge-Kutta methods. Proceedings of ICIAM'91, July 8-12, Washington.

Hida, T.
1980   Brownian motion. Springer.

Hofmann, N., Platen, E. and Schweizer, M.
1992   Option pricing under incompleteness and stochastic volatility. Mathematical Finance, Vol. **2**, No. 3, 153-187.

Hull, J. and White, A.
1987   The pricing of options as assets with stochastic volatilities. Journal of Finance, **42**, 281-300.

Hutton, J.E. and Nelson, P.I.
1986   Quasi-likelihood estimation for semimartingales. Stoch. Proc. Appl. **22**, 245-257.

Ikeda, N. and Watanabe, S.
1981   Stochastic differential equations and diffusion processes. North-Holland, Amsterdam.
1989   2nd edition.

Johnson, H. and Shanno, D.
1987   Option pricing when the variance is changing. J. Finan. Quant. Anal. **22**, 143-151.

Kalos, M.H. and Whitlock, P.A.
1986   Monte-Carlo methods. Vol. 1, Basics. Wiley, New York.

Karatzas, I. and Shreve, S.E.
1988   Brownian motion and stochastic calculus. Springer.

Karlin, S. and Taylor, H. M.
   1970   A first course in stochastic processes, 2nd edition. Academic Press, New
           York

   1981   A second course in stochastic processes. Academic Press, New York

Klauder, J.R. and Petersen, W.P.
   1985   Numerical integration of multiplicative-noise stochastic differential equa-
           tions. SIAM J. Numer. Anal. **22**, 1153-1166.

Kloeden, P.E. and Pearson, R.A.
   1977   The numerical solution of stochastic differential equations. J. Austral.
           Math. Soc., Series B, **20**, 8-12.

Kloeden, P.E. and Platen, E.
   1989   A survey of numerical methods for stochastic differential equations. J.
           Stoch. Hydrol. Hydraul. **3**, 155-178.

   1991a   Stratonovich and Ito Taylor expansions. Math. Nachr. **151**, 33-50.

   1991b   Relations between multiple Ito and Stratonovich integrals. Stoch. Anal.
           Appl. Vol. **IX**, No 3, 86-96.

   1992a   Numerical solution of stochastic differential equations. Springer, Applica-
           tions of Mathematics, Vol. **23**. (Corrected reprinting 1994).

   1992b   Higher-order implicit strong numerical schemes for stochastic differential
           equations. J. Statist. Physics, Vol. **66**, No 1/2, 283-314.

Kloeden, P.E., Platen, E. and Hofmann, N.
   1993   Extrapolation methods for the weak approximation of Ito diffusions. SIAM
           J. Numer. Anal. **32** (1995), 1519-1534.

Kloeden, P.E., Platen, E. and Schurz, H.
   1991   The numerical solution of nonlinear stochastic dynamical systems: a brief
           introduction. J. Bifur. Chaos Vol. **1**, No 2, 277-286.

   1993   Higher order approximate Markov chain filters. A Festschrift in Honour of
           G. Kallianpur. ed. S. Cambanis et. al., Springer, 181-190.

Kloeden, P.E., Platen, E., Schurz, H. and Sørenson, M.
   1992   On the effects of discretization on estimators of diffusions. J. Appl.
           Probab. **33** (1996), 1061-1076.

Kloeden, P.E., Platen, E. and Wright, I.
   1992   The approximation of multiple stochastic integrals. J. Stoch. Anal. Appl.,
           10 (4), 431-441.

Küchler, U. and Sørensen, M.
   1989   Exponential families of stochastic processes: A unifying semimartingale
           approach. Intern. Statist. Rev. **57**, 123-144.

Kunita, H.
   1990   Stochastic flows and stochastic differential equations. Cambridge Univer-
           sity Press, Cambridge.

Kushner, H.J. and Dupuis, P.G.
   1992   Numerical methods for stochastic control problems in continuous time.
           Springer, Applications of Mathematics, Vol. **24**.

Kutoyants, Y.A.
   1984   Parameter estimation for stochastic processes. Heldermann Verlag, Berlin.

Le Breton, A.
   1976   On continuous and discrete sampling for parameter estimation in diffusion
          type processes. Math. Prog. Stud. **5**, 124-144.

Liske, H. and Platen, E.
   1987   Simulation studies on time discrete diffusion approximations.   Math.
          Comp. Simul. **29**, 253-260.

Liske, H., Platen, E. and Wagner, W.
   1982   About mixed multiple Wiener integrals. Preprint P-Math-23/82, IMath,
          Akad. der Wiss. der DDR, Berlin.

Marsaglia, G.
   1985   A current view of random number generators.   Computer Science and
          Statistics: The Interface (ed. by L. Billard), North Holland, 3-10.

Maruyama, G.
   1955   Continuous Markov processes and stochastic equations. Rend. Circolo
          Math. Palermo 4, 48-90.

McShane, E.J.
   1974   Stochastic calculus and stochastic models. Academic Press, New York.

Mikhailov, G.A.
   1992   Optimization of weighted Monte-Carlo methods. Springer.

Mikulevicius R. and Platen, E.
   1991   Rate of convergence of the Euler approximation for diffusion processes.
          Math. Nachr. **151**, 233-239.

   1988   Time discrete Taylor approximations for Ito processes with jump compo-
          nent. Math. Nachr. **138**, 93-104.

Milstein, G.N.
   1974   Approximate integration of stochastic differential equations. Theor. Prob.
          Appl. **19**, 557-562.

   1978   A method of second-order accuracy integration of stochastic differential
          equations. Theor. Prob. Appl. **23**, 396-401.

   1985   Weak approximation of solutions of systems of stochastic differential equa-
          tions. Theor. Prob. Appl. **30**, 750-766.

   1988a The numerical integration of stochastic differential equations. Urals Univ.
          Press, Sverdlovsk. (In Russian. English Translation: Kluwer, 1995).

   1988b A theorem on the order of convergence of mean-square approximations
          of solutions of systems of stochastic differential equations. Theor. Prob.
          Appl. **32**, 738-741.

Morgan, B.J.J.
   1984   Elements of simulation. Chapman & Hall, London.

Nakazawa, H.
   1990   Numerical procedures for sample structures on stochastic differential equa-
          tions. J. Math. Phys. **31**, 1978-1990.

Newton, N.J.
  1986   An asymptotically efficient difference formula for solving stochastic differential equations. Stochastics **19**, 175-206.

  1991   Asymptotically efficient Runge-Kutta methods for a class of Ito and Stratonovich equations. SIAM J. Appl. Math. **5**, 542-567. *

  1992   Variance reduction for simulated diffusions. SIAM J. Appl. Math. **54** (1994), 1780-1805.

Niederreiter, H.
  1992   Random number generation and quasi-Monte-Carlo methods. SIAM, Philadelphia, Pennsylvania.

Øksendal, B.
  1985   Stochastic differential equations. Springer, Universitext Series.

Pardoux, E. and Talay, D.
  1985   Discretization and simulation of stochastic differential equations. Acta Appl. Math. **3**, 23-47.

Petersen, W.P.
  1988   Some vectorized random number generators for uniform, normal and Poisson distributions for CRAY X-MP. J. Supercomputing **1**, 318-335.

  1990   Stability and accuracy of simulations for stochastic differential equations. IPS Research Report No. 90-02, ETH Zürich.

Platen, E.
  1980   Approximation of Ito integral equations. Springer Lecture Notes in Control and Inform. Sc. Vol. 25, pp. 172-176.

  1981   An approximation method for a class of Ito processes. Lietuvos Matem. Rink. **21**, 121-133.

  1984   Zur zeitdiskreten Approximation von Itoprozessen. Diss. B., IMath, Akad. der Wiss. der DDR, Berlin.

Platen, E. and Wagner, W.
  1982   On a Taylor formula for a class of Ito processes. Prob. Math. Statist. **3**, 37-51.

Protter, P.
  1990   Stochastic integration and differential equations. Springer, Applications in Mathematics, Vol **21**.

Rao, N.J., Borwankar, J.D. and Ramakrishna, D.
  1974   Numerical solution of Ito integral equations. SIAM J. Control **12**, 124-139.

Ripley, B.D.
  1983   Stochastic simulation. Wiley, New York.

Rubinstein, R.Y.
  1981   Simulation and the Monte-Carlo method. Wiley, New York.

Rümelin, W.
  1982   Numerical treatment of stochastic differential equations. SIAM J. Numer. Anal. **19**, 604-613.

Saito, Y. and Mitsui, T.
  1992   Discrete approximations for stochastic differential equations. Trans. Japan SIAM, **2**, 1-16.

Scott, L.O.
   1987   Option pricing when the variance changes randomly: Theory, estimation and an application. Journal of Financial and Quantitative Analysis **22**, 419-438.

Shiga, T.
   1985   Mathematical results on the stepping stone model of population genetices. In Population Genetics and Molecular Evolution. (T. Ohta and K. Aoki, editors). Springer, pp. 267-279.

Shiryayev, A. N.
   1984   Probability theory. Springer.

Smith, A.M. and Gardiner, C.W.
   1988   Simulation of nonlinear quantum damping using the positive representation. Univ. Waikato Research Report.

Sørensen, M.
   1990   On quasi-likelihood for semi-martingales. Stoch. Processes Appl. **35**, 331-346.

Stoer, J. and Bulirsch, R.
   1980   Introduction to numerical analysis. Springer.

Talay, D.
   1982   Analyse numérique des equations différentielles stochastiques. Thèse 3ème cycle, Univ. Provence.

   1984   Efficient numerical schemes for the approximation of expectations of functionals of the solution of an SDE and applications. Springer Lecture Notes in Control and Inform. Sc. Vol. 61, pp. 294-313.

   1987   Classification of discretization schemes of diffusions according to an ergodic criterium. Springer Lecture Notes in Control and Inform. Sc. Vol. 91, pp. 207-218.

   1989   Approximation of upper Lyapunov exponents of bilinear stochastic differential equations. SIAM J. Numer. Anal. **28** (1991), 1141-1164.

   1990   Second order discretization schemes of stochastic differential systems for the computation of the invariant law. Stochastics Stoch. Rep. **29**, 13-36.

Talay, D. and Tubaro, L.
   1990   Expansions of the global error for numerical schemes solving stochastic differential equations. Stoch. Analysis Appl. Vol. **8**, No 4, 483-509.

Treutlein, H. and Schulten, K.
   1985   Noise induced limit cycles of the Bonhoeffer-Van der Pol model of neural pulses. Ber. Bunsenges. Phys. Chem. **89**, 710-718.

Wagner, W.
   1988a   Unbiased multi-step estimators for the Monte-Carlo evaluation of certain
            functionals. J. Comput. Physics **79**, 336-352.

   1988b   Monte-Carlo evaluation of functionals of solutions of stochastic differential
            equations. Variance reduction and numerical examples. Stoch. Anal.
            Appl. **6**, 447-468.

   1989a   Unbiased Monte-Carlo estimators for functionals of weak solutions of sto-
            chastic differential equations. Stochastics Stoch. Rep. **28**, 1-20.

   1989b   Stochastische numerische Verfahren zur Berechnung von Funktionalinte-
            gralen. Report R-Math 02/89 Inst. Math. Akad. d. Wiss. DDR. 149
            pp.

Wagner, W. and Platen, E.
   1978   Approximation of Ito integral equations. Preprint ZIMM, Akad. der Wiss.
            der DDR, Berlin.

Wiggins, J.B.
   1987   Option values under stochastic volatilities. Journal of Financial Economics
            19, 351-372.

Wonham, W.M.
   1965   Some applications of stochastic differential equations to optimal nonlinear
            filtering. SIAM J. Control **2**, 347-369.

Zakai, M.
   1969   On the optimal filtering of diffusion processes. Z. Wahrsch. verw. Gebiete
            11, 230-343.

Subject Index

List of PC-Exercises

PC-Exercises in
*) Kloeden, P.E. and Platen, E.: Numerical Solution of Stochastic Differential Equations. Springer-Verlag, Applications of Mathematics, Vol. **23** (1992).

Frequently Used Notations

$\underline{a}(\cdot)$	Stratonovich corrected drift	75, 140
$\frac{\partial}{\partial x} a$	partial derivative of a with respect to x	73
$\frac{\partial^2}{\partial x_i\, \partial x_j} a$	second order partial derivative of a	73
$a'(x)$	derivative with respect to x	
$\dot{x}$	time derivative of x	
$l!$	factorial of l	77
L^0, L^1, L^j	differential operators	78, 139
$\underline{L}^0, \underline{L}^1, \underline{L}^j$	differential operators	80, 139
$J_{(j_1,j_2,\ldots,j_l),t}$	multiple Stratonovich integral	82, 140
$J^p_{(j_1,j_2,\ldots,j_l),t}$	approximate multiple Stratonovich integral	82
$\max\{\cdot,\cdot\}$	maximum	65
$\Delta W_n, \Delta W_n^j$	Wiener process increment	110, 140
$\Delta Z_n, \Delta Z_n^j$	multiple stochastic integral	146
$I_{(j_1,j_2,\ldots,j_l),t}$	multiple Ito integral	89, 140
τ_n	discretization time	110
Y_n	approximation Y at discretization time τ_n	110
n_t	largest integer for which $\tau_n \le t$	111
ϵ	absolute error	115
$\hat{\epsilon}$	estimate for absolute error ϵ	115, 118
μ	mean error	124
$\hat{\mu}$	estimate for mean error μ	124
$a \ll b$	b is much larger than a	132
ι	square root of -1	
$\mathrm{Re}(\lambda)$	real part of a complex number λ	132
$\mathrm{Im}(\lambda)$	imaginary part of a complex number λ	132
d	dimension of Ito process X	139
m	dimension of Wiener process W	139
$a, b, \ldots$	abbreviation in schemes for $a(Y_n), b(Y_n), \ldots$	140
I	unit matrix	158
V_{j_1,j_2}	random variable for double Wiener integrals	182

$\prod_{i=1}^{N} a_i$ product

$\sum_{i=1}^{N} a_i$ sum

(a, b) scalar product

$\det A$ determinant of matrix A

A^{-1} inverse of matrix A

∇f gradient vector of function f

$\#A$ cardinality of set A

a^{+} positive part of a

$:=$ defined as or denoted by

$\equiv$ identically equal to

$\approx$ approximately equal to

Universitext

Aksoy, A.; Khamsi, M. A.: Methods in Fixed Point Theory

Alevras, D.; Padberg M. W.: Linear Optimization and Extensions

Andersson, M.: Topics in Complex Analysis

Aoki, M.: State Space Modeling of Time Series

Audin, M.: Geometry

Aupetit, B.: A Primer on Spectral Theory

Bachem, A.; Kern, W.: Linear Programming Duality

Bachmann, G.; Narici, L.; Beckenstein, E.: Fourier and Wavelet Analysis

Badescu, L.: Algebraic Surfaces

Balakrishnan, R.; Ranganathan, K.: A Textbook of Graph Theory

Balser, W.: Formal Power Series and Linear Systems of Meromorphic Ordinary Differential Equations

Bapat, R.B.: Linear Algebra and Linear Models

Benedetti, R.; Petronio, C.: Lectures on Hyperbolic Geometry

Berberian, S. K.: Fundamentals of Real Analysis

Berger, M.: Geometry I, and II

Bliedtner, J.; Hansen, W.: Potential Theory

Blowey, J. F.; Coleman, J. P.; Craig, A. W. (Eds.): Theory and Numerics of Differential Equations

Börger, E.; Grädel, E.; Gurevich, Y.: The Classical Decision Problem

Böttcher, A; Silbermann, B.: Introduction to Large Truncated Toeplitz Matrices

Boltyanski, V.; Martini, H.; Soltan, P. S.: Excursions into Combinatorial Geometry

Boltyanskii, V. G.; Efremovich, V. A.: Intuitive Combinatorial Topology

Booss, B.; Bleecker, D. D.: Topology and Analysis

Borkar, V. S.: Probability Theory

Carleson, L.; Gamelin, T. W.: Complex Dynamics

Cecil, T. E.: Lie Sphere Geometry: With Applications of Submanifolds

Chae, S. B.: Lebesgue Integration

Chandrasekharan, K.: Classical Fourier Transform

Charlap, L. S.: Bieberbach Groups and Flat Manifolds

Chern, S.: Complex Manifolds without Potential Theory

Chorin, A. J.; Marsden, J. E.: Mathematical Introduction to Fluid Mechanics

Cohn, H.: A Classical Invitation to Algebraic Numbers and Class Fields

Curtis, M. L.: Abstract Linear Algebra

Curtis, M. L.: Matrix Groups

Cyganowski, S.; Kloeden, P.; Ombach, J.: From Elementary Probability to Stochastic Differential Equations with MAPLE

Dalen, D. van: Logic and Structure

Das, A.: The Special Theory of Relativity: A Mathematical Exposition

Debarre, O.: Higher-Dimensional Algebraic Geometry

Deitmar, A.: A First Course in Harmonic Analysis

Demazure, M.: Bifurcations and Catastrophes

Devlin, K. J.: Fundamentals of Contemporary Set Theory

DiBenedetto, E.: Degenerate Parabolic Equations

Diener, F.; Diener, M. (Eds.): Nonstandard Analysis in Practice

Dimca, A.: Singularities and Topology of Hypersurfaces

DoCarmo, M. P.: Differential Forms and Applications

Duistermaat, J. J.; Kolk, J. A. C.: Lie Groups

Edwards, R. E.: A Formal Background to Higher Mathematics Ia, and Ib

Edwards, R. E.: A Formal Background to Higher Mathematics IIa, and IIb

Emery, M.: Stochastic Calculus in Manifolds

Endler, O.: Valuation Theory

Erez, B.: Galois Modules in Arithmetic

Everest, G.; Ward, T.: Heights of Polynomials and Entropy in Algebraic Dynamics

Farenick, D. R.: Algebras of Linear Transformations

Foulds, L. R.: Graph Theory Applications

Frauenthal, J. C.: Mathematical Modeling in Epidemiology

Friedman, R.: Algebraic Surfaces and Holomorphic Vector Bundles

Fuks, D. B.; Rokhlin, V. A.: Beginner's Course in Topology

Fuhrmann, P. A.: A Polynomial Approach to Linear Algebra

Gallot, S.; Hulin, D.; Lafontaine, J.: Riemannian Geometry

Gardiner, C. F.: A First Course in Group Theory

Gårding, L.; Tambour, T.: Algebra for Computer Science

Godbillon, C.: Dynamical Systems on Surfaces

Goldblatt, R.: Orthogonality and Spacetime Geometry

Gouvêa, F. Q.: p-Adic Numbers

Gustafson, K. E.; Rao, D. K. M.: Numerical Range. The Field of Values of Linear Operators and Matrices

Hahn, A. J.: Quadratic Algebras, Clifford Algebras, and Arithmetic Witt Groups

Hájek, P.; Havránek, T.: Mechanizing Hypothesis Formation

Heinonen, J.: Lectures on Analysis on Metric Spaces

Hlawka, E.; Schoißengeier, J.; Taschner, R.: Geometric and Analytic Number Theory

Holmgren, R. A.: A First Course in Discrete Dynamical Systems

Howe, R., Tan, E. Ch.: Non-Abelian Harmonic Analysis

Howes, N. R.: Modern Analysis and Topology

Hsieh, P.-F.; Sibuya, Y. (Eds.): Basic Theory of Ordinary Differential Equations

Humi, M., Miller, W.: Second Course in Ordinary Differential Equations for Scientists and Engineers

Hurwitz, A.; Kritikos, N.: Lectures on Number Theory

Iversen, B.: Cohomology of Sheaves

Jacod, J.; Protter, P.: Probability Essentials

Jennings, G. A.: Modern Geometry with Applications

Jones, A.; Morris, S. A.; Pearson, K. R.: Abstract Algebra and Famous Inpossibilities

Jost, J.: Compact Riemann Surfaces

Jost, J.: Postmodern Analysis

Jost, J.: Riemannian Geometry and Geometric Analysis

Kac, V.; Cheung, P.: Quantum Calculus

Kannan, R.; Krueger, C. K.: Advanced Analysis on the Real Line

Kelly, P.; Matthews, G.: The Non-Euclidean Hyperbolic Plane

Kempf, G.: Complex Abelian Varieties and Theta Functions

Kitchens, B. P.: Symbolic Dynamics

Kloeden, P.; Ombach, J.; Cyganowski, S.: From Elementary Probability to Stochastic Differential Equations with MAPLE

Kloeden, P. E.; Platen; E.; Schurz, H.: Numerical Solution of SDE Through Computer Experiments

Kostrikin, A. I.: Introduction to Algebra

Krasnoselskii, M. A.; Pokrovskii, A. V.: Systems with Hysteresis

Luecking, D. H., Rubel, L. A.: Complex Analysis. A Functional Analysis Approach

Ma, Zhi-Ming; Roeckner, M.: Introduction to the Theory of (non-symmetric) Dirichlet Forms

Mac Lane, S.; Moerdijk, I.: Sheaves in Geometry and Logic

Druck und Bindung: Strauss Offsetdruck GmbH